Communications and Control Engineering Series

Vadim I. Utkin

Sliding Modes in Control and Optimization

With 24 Figures

Springer-Verlag

Berlin Heidelberg New York
London Paris Tokyo
Hong Kong Barcelona Budapest

Prof. VADIM I. UTKIN
Institute of Control Sciences
65 Profsyuznaya
Moscow GSP-312
USSR

Original Russian language edition, entitled:
В. И. Уткин «Скользящие режимы и их применение в задачах оптимизации и управления» НАУКА, Москва, 1981 г.

ISBN 3-540-53516-0 Springer-Verlag Berlin Heidelberg New York
ISBN 0-387-53516-0 Springer-Verlag New York Berlin Heidelberg

Printed in the United States of America

Typesetting: Thomson India Ltd., New Delhi, India.

2161/3020 543210 – Printed on acid-free paper

Preface

The book is devoted to systems with discontinuous control. The study of discontinuous dynamic systems is a multifacet problem which embraces mathematical, control theoretic and application aspects. Times and again, this problem has been approached by mathematicians, physicists and engineers, each profession treating it from its own positions. Interestingly, the results obtained by specialists in different disciplines have almost always had a significant effect upon the development of the control theory. It suffices to mention works on the theory of oscillations of discontinuous nonlinear systems, mathematical studies in ordinary differential equations with discontinuous righthand parts or variational problems in nonclassic statements.

The unremitting interest to discontinuous control systems enhanced by their effective application to solution of problems most diverse in their physical nature and functional purpose is, in the author's opinion, a cogent argument in favour of the importance of this area of studies. It seems a useful effort to consider, from a control theoretic viewpoint, the mathematical and application aspects of the theory of discontinuous dynamic systems and determine their place within the scope of the present-day control theory. The first attempt was made by the author in 1975–1976 in his course on "The Theory of Discontinuous Dynamic Systems" and "The Theory of Variable Structure Systems" read to post-graduates at the University of Illinois, USA, and then presented in 1978–1979 at the seminars held in the Laboratory of Systems with Discontinous Control at the Institute of Control Sciences in Moscow.

First of all, the scope of problems to be dealt with should be outlined. The object of our attention will be a system in which the discontinuity of control in each control component is "prescribed" prior to the stages of selection of the criterion and design unlike, for instance, optimal systems where the need for jumpwise control variations at a finite or infinite frequency appears as a result of the solution of the variational problem.

Let us present major arguments showing why the class of discontinuous control systems provides such an effective tool for solving the entire family of control problems for complex dynamic plants. The discontinuity of control results in a discontinuity of the righthand part of the differential equations describing the system motions. If such discontinuities are deliberately introduced on certain surfaces in the system state space, then motions in a sliding mode may occur in the system. This type of motion features some attractive properties and has long since become applied in relay systems. Sliding modes are the basic motions in variable structure systems.

Consider the properties of sliding modes in some greater detail. First, the trajectories of the state vector belong to manifolds of lower dimension than that of the whole state space, therefore the order of differential equations describing sliding motions is also reduced. Second, in most of practical systems the sliding motion is control-independent and is determined merely by the properties of the control plant and the position (or equations) of the discontinuity surfaces. This allows the initial problem to be decoupled into independent lower dimension subproblems wherein the control is "spent" only on creating a sliding mode while the required character of motion over the intersection of discontinuity surfaces is provided by an appropriate choice of their equations. These properties turn to be quite essential for solving many application problems characterized by high order differential equations which prohibits the use of efficient analytic techniques and computer technology. A third specific feature of a sliding mode is that under certain conditions it may become invariant to variations of dynamic characteristics of the control plant which poses a central problem dealt with in the theory of automatic control. It is essential that, unlike continuous systems with nonmeasurable disturbances in which, the conditions of invariancy require the use of infinitely high gains, the same effect in discontinuous systems is attained by using finite control actions.

Finally, a purely technological aspect of using discontinuous control systems should be mentioned. To improve performance, electric inertialess actuators are increasingly employed now, built around power electronic elements which may operate in a switching mode only. Therefore even if we employ continuous control algorithms the control itself is shaped as a high frequency discontinuous signal whose mean value is equal to the desired continuous control. A more natural way then will be to employ such algorithms which are deliberately oriented toward the use of discontinuous controls.

Let us describe the problem faced in an attempt to employ the properties of sliding modes for the design of automatic control systems. Consider first the mathematical problems treated in Part 1. Discontinuous dynamic systems are outside the scope of the classical theory of differential equations and require the development of ad-hoc techniques to study their behaviour. Various publications on the matter have shown a diversity of viewpoints of their authors as to how the motion on a discontinuity boundary should be described, thus leading to diverse sliding mode equations. An approach followed by this author implies regularization through an introduction of a boundary layer which, on the one hand, allows the reason for ambiguity to be revealed and, on the other hand, outlines the class of systems whose sliding equations can be written quite unambiguously. Various methods of designing automatic control systems treated in this book are applied to exactly such class of systems.

As was already noted, the approach used in the book is oriented toward a deliberate introduction of sliding modes over the intersection of surfaces on which the control vector components undergo discontinuity. Realization of such approach obviously implies the knowledge of the conditions of the occurrence of sliding modes, rarely discussed in the literature. From the point of view of mathematics, the problem may be reduced to that of finding the area of attraction to the manifold of the discontinuity surfaces intersection. The solutions suggested are formulated in terms of the stability theory and are obtained via generalization of the classical Lyapunov theorems to discontinuous systems featuring not merely individual motions but, rather, a whole set of motion trajectories from a certain domain on the discontinuity surface.

Part 1 of the book is concluded with problems of robustness of discontinuous systems with respect to small dynamic imperfections disregarded in an idealized model yet always present in a real-life system due to small inertialities of measuring instruments, actuators, data processing devices, etc. In the theory of singularly perturbed differential equations, the control in a continuous system may be regarded as a continuous function of small time constants, provided the fast decaying motions are neglected; consequently, the same property is featured by solutions of the equations describing the motion in the system. This conclusion provides grounds for applying simplified models to solve various control problems in the class of continuous systems. In a discontinuous control system operating in a sliding mode small additional time constants neglected in the idealized model may cause a shift in the switching times thus making the controls in the idealized and real systems essentially different. As a result, the solution obtained may prove to be nonrobust to some insignificant changes in the model of our process which, in its turn, makes very doubtful the entire applicability of the control algorithms to the considered class of discontinuous systems. In this connection, a study of singularly perturbed discontinuous systems is carried out to show that despite the lack of any continuous correlation between the control and the small time constants, the systems with unambiguous equations of their sliding mode motions are robust with respect to small dynamic imperfections.

The major focus in the book is on control systems design methods, discussed in Part II. Despite the diversity of the control goals in the problems considered in this part and in their solution techniques, all the design procedures are built around a common principle: the initial system is decoupled into independent subsystems of lower dimension and sliding motions with required properties are designed at the final stage of a control process. The methods of the analysis of discontinuous dynamic systems suggested in the first part of the book serve as a mathematical background for the realization of this principle.

Part II of the book is aimed not only at stating and developing the results obtained in the sphere of discontinuous control system design, but also at presenting these results in close correlation with the basic concepts, problems and methods of the present-day control theory. This refers both to newly obtained results and to those which have appeared earlier in the literature.

Some problems traditionally posed in the linear control theory such as eigenvalue allocation and quadratic and parametric optimization have been treated with the use of the fundamental control theory concepts of controllability, stabilizability, observability and detectability. The analogs of the well-known theorems on the uniqueness and existence of solutions to such problems are given, and procedures of obtaining the solutions for systems with sliding modes are suggested.

The problem of invariance as applied to linear systems with vector-valued controls and disturbances is treated under the assumption that a dynamic model of disturbances with unknown initial conditions is available. However, in contrast to the well-known methods, the approach suggested does not require the knowledge of accurate values of the model parameters, for it is quite sufficient to know just the range of their possible variations.

Since all design methods are based on the idea of decoupling the system, it seems interesting to compare discontinuous systems with continuous ones realizing the same idea through the use of infinitely high gains. As is well known, the increase of the gain in an open-loop system decouples its overall motion into fast and slow components which may be synthesized independently. The comparison shows that systems with infinitely high gains are particular cases of singularly perturbed systems and their slow motions may be described by the same equations as used for sliding motions in discontinuous systems. The only difference is that in continuous systems these motions may be attained only asymptotically in tending the gains to infinity, while in discontinuous systems the same is attainable in a finite time and with a finite control.

Besides finite-dimensional control problems, a design method for a system with a mobile control in a distributed parameter control plant is discussed in the book. Most probably, this may be regarded as the first attempt to apply sliding modes to control plants of this class. The problem is in steering the controlled parameter to the required spatial distribution in the system with a mobile control subjected to uncontrollable disturbances. A deliberate introduction of sliding modes in the loops responsible for the control of the system motion and the intensity of the source permits such a problem to be solved using the information on the current state of the system, provided the magnitudes the spatially distributed disturbances and their velocities are bounded. The same statement is used to solve the problem in the case of a distributed or lumped control both for the control plants with a single spatial variable and for a set of interconnected distributed plants.

The problems of control and optimization under incomplete information on the operator and the system state as well as computational problems associated with the search for the optimal set of system parameters are considered in the final chapters of Part II. In solving these problems, continuous or discontinuous asymptotic observers and filters, nominal or adjustable models, search procedures, etc. are used wherein the dynamic processes of adaptation, identification or tending to an extremum are conducted in a sliding mode.

A sufficiently great deal of experience has been acquired up to the present in using sliding modes in various application problems. It seemed expedient to devote a separate section of the book (Part III) to applications and provide some examples, most convincing in the author's opinion. The first example dealing with the control of a robot arm is given to illustrate the whole idea. The use of sliding modes for this multivariable high order nonlinear design problem has made it possible to realize motions invariant to load torques and mechanical parameters which could be described by independent linear uniform differential equations with respect to each of the controlled parameters.

Perhaps the most suitable for the application of sliding modes proved to be one of the basic engineering problems, that of control of electrical machines and, in particular, of electrical motors. Increasingly dominating nowadays are AC motors built around power electronic elements operating in a switching mode, which puts the problem of the algorithmic supply in the forefront. The technological aspect of this problem has already been discussed and the considerations speaking in favour of the control algorithms oriented toward introduction of sliding modes. The principles of designing multivariable discontinuous control systems suggested in Part II may be easily interpreted in terms of electrical engineering: serving as the control vector components in this case are discontinuous voltages at the output of power semiconductor converters which are then fed to the motor phases and to the excitation winding (if any); the motion differential equations for all types of motors which are generally nonlinear turn to be linear with respect to the control (in which case the sliding equations are written unambiguously); only some components of the state vector are to be controlled (for instance, angular position of the motor shaft, angular velocity and torque, magnetic flux, power coefficient, currents, etc.) Wide-scale experimental tests have testified to the efficiency of sliding mode control for any types of electrical motors, including induction motors, which are known to be the most reliable and economic yet the hardest to control.

Considered in Chapter 17 of Part III is a set of problems associated with control algorithms for DC, induction and synchronous motors and with methods of acquiring information on the controlled process state.

The closing chapter of this part presents the results obtained in implementing sliding mode control for electrical drives of metal-cutting machine tools and transport vehicles utilizing both independent and external power sources. The potential of sliding mode systems is demonstrated for physical processes most diverse in their nature such as chemical fibre production, metal melting in electric arc furnaces, processes in petroleum refining and petrochemical industries, automation in fishery, stabilization of the resonant frequency of an accelerator intended for physical experiments.

It is obvious that all feasible applications of systems with sliding modes could not be possibly covered in one book. The examples given in Part III have made a stress upon the algorithms not touching the problems of their technical implementation. Consideration of such problems would somehow fall out of the general theoretical line of the book. Nonetheless, the author believes that it is some specific examples that makes the "Mathematical Tools" and "Design" parts of the book appropriately completed.

The author takes a chance to express his deep gratitude to all of his colleagues who have contributed, directly or indirectly, to the book. Most fruitful have been his discussions of the mathematical and technical aspects of the theory of singularly perturbed systems with Prof. A. Vasilyeva and Prof. P. Kokotovic. The results of these discussions have formed a framework of Chapters 5 and 11 where a comparison is made between singularly perturbed systems, high gain systems and discontinuous control systems. The results of Section 3 in Chapter 4 are a "direct corollary" of a discussion with Prof. A. Filippov of the problem of existence of multidimensional sliding modes. Prof. D. Siljak has drawn the author's attention to a possibility of constructing an upper bound for the Lyapunov function using the comparison principle (Section 6, Chapter 4). The statement of the problem of mobile control for a distributed system has been suggested by Prof. A. Butkovsky with whom the author have subsequently solved this problem in a joint effort.

The book has gained from the results published in recent years by Dr. K. Young (Section 1, Chapter 13 and Chapter 15) who had started enthusiastic work in this field while a post-graduate student.

In the works conducted at the Institute of Control Sciences in the design of control algorithms for electrical machines, the basic contribution was made by Dr. D. Izosimov who helped the author to write Chapter 17. The control principles developed in these studies have been extended by post-graduate student S. Ryvkin to synchronous electrical motors (Section 4, Chapter 17).

Unquestionably, it was a good luck for the author that the Editorial Board of the Publishing House made a decision to send the book for review to Prof. A. Filippov, who studied it with a mathematician's rigour and made a number of useful remarks and advices to improve the manuscript.

The appearance of this book could be hardly possible without a permanent help from N. Kostyleva throughout all of its stages starting with the idea to write it and finishing with the preparation of the final version of the manuscript. The seminars held in the Laboratory of Discontinuous Control Systems of the Institute of Control Sciences and discussions with his co-workers have significantly helped the author shape his ideas on the methodology of presenting this material and its correlation with various parts of the control theory.

The author is indebted to L. Govorova who has faultlessly accomplished a hard task of typing the manuscript.

In preparing the English version of the book, the author has gratefully used a chance given by Springer-Verlag to revise the book, and to include some results obtained in the theory of sliding modes and its applications after the publication of the book in Russian in 1981. These refer to stochastic regularization of discontinuous dynamic systems (Section 5, Chapter 2), control of distributed systems (Sections 2 through 4, Chapter 12), the method of system parameter identification employing discontinuous dynamic models (Section 2, Chapter 13), robustness of sliding modes to dynamic discrepancies between the system and its model (Sections 3 and 4, Chapter 14). Part III was enriched by a new Chapter 18 describing direct application of the sliding mode control to a large class of technological systems.

The project of publishing the book in the English translation could hardly ever be realized without constant friendly support and help from Prof. M. Thoma, to whom the author is sincerely indebted.

Moscow, May 1991 Vadim I. Utkin

Contents

Part I. Mathematical Tools

Chapter 1

Scope of the Theory of Sliding Modes

1 Shaping the Problem

A number of processes in mechanics, electrical engineering, and other areas, are characterized by the fact that the righthand sides of the differential equations describing their dynamics feature discontinuities with respect to the current process state. A typical example of such a system is a dry (Coulomb) friction mechanical system whose force of resistance may take up either of the two sign-opposite values depending on the direction of the motion. This situation is often the case in automatic control systems where the wish to improve the system performance, minimize power consumed for the control purposes, restrict the range of possible variations of control parameters, etc. leads to controls in the form of discontinuous functions of the system state vector and the system input actions.

Formally, such discontinuous dynamic systems may be described by the equation

$$\dot{x} = f(x, t), \tag{1.1}$$

where the system state vector is $x \in \mathbb{R}^n$, t is time, and $f(x, t)$ has discontinuities at a certain set within the $(n + 1)$-dimensional space (x, t). Let us confine ourselves to consideration of only those cases when the set of discontinuity points is a set of, possibly time-varying, discontinuity surfaces[1] of n-dimensional state space

[1] Hereinafter we shall also refer these surfaces to as discontinuity boundaries.

defined by the equations

$$s_i(x,t)=0, \quad s_i(x,t)\in\mathbb{R}^1, \quad i=1,\dots,m. \tag{1.2}$$

Discontinuity in the righthand parts of the motion equations is the reason of certain peculiarities in the system behaviour. These peculiarities are observed even in the simplest case of the above Coulomb friction system (Fig. 1) described by equations of the type (1.1) and (1.2):

$$m\ddot{x}+P(\dot{x})+kx=0, \tag{1.3}$$

where x is displacement, m is mass, k is the spring rigidity,

$$P(\dot{x})=\begin{cases} P_0 & \text{with} \quad \dot{x}>0 \\ -P_0 & \text{with} \quad \dot{x}<0 \end{cases} \tag{1.4}$$

and P_0 is a positive constant. It is quite obvious that the discontinuity surface (1.2) in this case is the x-axis

$$s=\dot{x}=0. \tag{1.5}$$

Notice the fact that the Coulomb friction $P(\dot{x})$ is not defined in points where velocity $\dot{x}$ equals zero.

Consider the behaviour of the mechanical system (1.3)–(1.5) on the plane with coordinates x and $\dot{x}$ (Fig. 2). As evidenced from the figure, the description of this behaviour may be obtained quite easily if $|x|>P_0/k$, i.e. if the discontinuity points are isolated. In this case one may apply, for instance, the point-to-point transformation technique. If the state vector appears to stay within the segment. $|x|\leqq P_0/k$ of the discontinuity straight line (1.5) (stagnation zone, as termed by A.A. Andronov in [8]), it will not leave this segment. Since in this case the function $P(\dot{x})$ is not defined on the discontinuity straight line, an immediate problem of an appropriate description of this motion arises. A fixed value of function $P(\dot{x})$ with $\dot{x}=0$ is unsatisfactory for the purpose because the set $|x|\leqq P_0/k$ is a set of equilibrium points feasible only when $P(\dot{x})+kx=0$. Therefore an additional assumption is needed to enable the description of the behaviour of the system at the discontinuity boundary: $P(\dot{x})$ should be a multivalued function of the system state.

Another way is yet possible. The motion equation may be changed by using a more accurate model (Fig. 3) recognizing viscous friction (or deformability of links) which eliminates ambiguity. Then for $|x|\leqq P_0/k$ and $|\dot{x}|<\varepsilon$

$$m\ddot{x}+\frac{P_0}{\varepsilon}\dot{x}+kx=0 \tag{1.6}$$

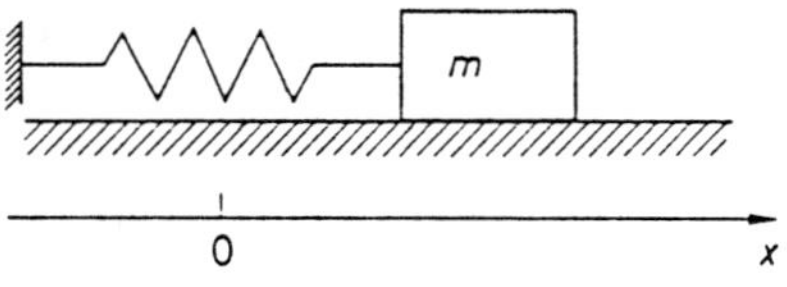

Fig. 1

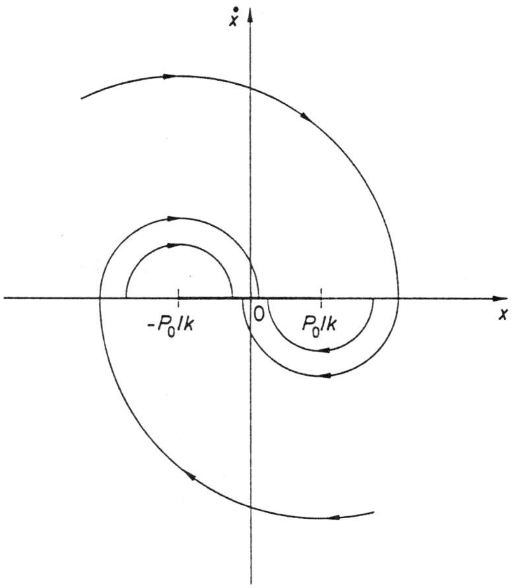

Fig. 2

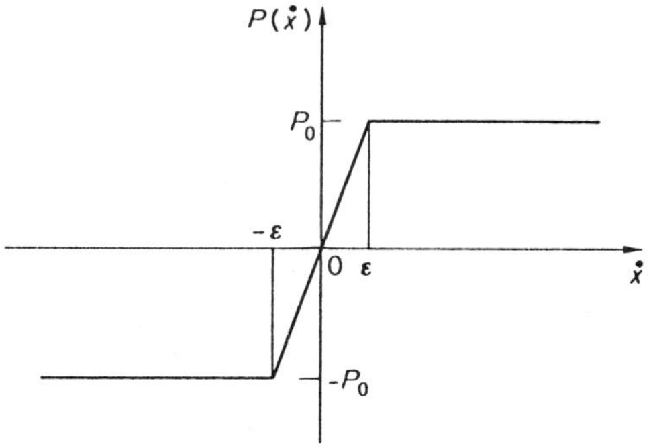

Fig. 3

and the solution to (1.6) tends to zero the slower the less is the linearity zone in Fig. 3.

The impossibility to describe the behaviour of a system without introduction of an additional assumption or design of a more accurate model was, most probably, the reason for an increased interest to Coulomb friction systems, which has manifested itself at the beginning of this century and has kept on strong for several decades[1]. Subsequently the problem of describing

[1] See, for instance, Academician L.I. Mandel'shtam. Collection of Works. Nauka, Moscow, 1979, pp. 121–125 (in Russian).

discontinuous dynamic systems became vital for automatic control people as well.

A distinguished feature of differential equations describing any control system is known to be the presence of a scalar or vector parameter u referred to as *control*:

$$\dot{x} = f(x, t, u), \quad u \in \mathbb{R}^m. \tag{1.7}$$

In early regulators, the controls have mostly been of relay type which was dictated by the need to make their implementation as simple as possible. As a result, the righthand part of the differential equation of the system motion proved to be a discontinuous function of the system state vector. This has forced the control theory specialists think of an adequate description of the behaviour of systems at discontinuity boundaries. For systems with isolated discontinuity points, some analysis and synthesis methods have been designed based on the classical theory of differential equations with the use of point-to-point transformations and averaging at the occurrence of high frequency switching (see, for instance, [42, 109]).

However, in attempts to mathematically describe certain application problems the same case as in the Coulomb friction mechanical system were often faced when the totality of discontinuity points proved to be a nonzero measure set in time. This fact is easily revealed for a sufficiently general class of discontinuous controls defined by the relationships

$$u_i(x, t) = \begin{cases} u_i^+(x, t) & \text{with} \quad s_i(x) > 0 \\ u_i^-(x, t) & \text{with} \quad s_i(x) < 0, \quad i = 1, \ldots, m, \end{cases} \tag{1.8}$$

where $u^T = (u_1, \ldots, u_m)$ and all functions $u_i^+(x, t)$ and $u_i^-(x, t)$ are continuous. The state vector of such systems may stay on one of the discontinuity surfaces or their intersection within a finite time. For example, the system state vector trajectories belong to some discontinuity surface $s_i(x) = 0$ if in the vicinity of this surface the velocity vectors $f(x, t, u)$ are directed toward each other (Fig. 4).

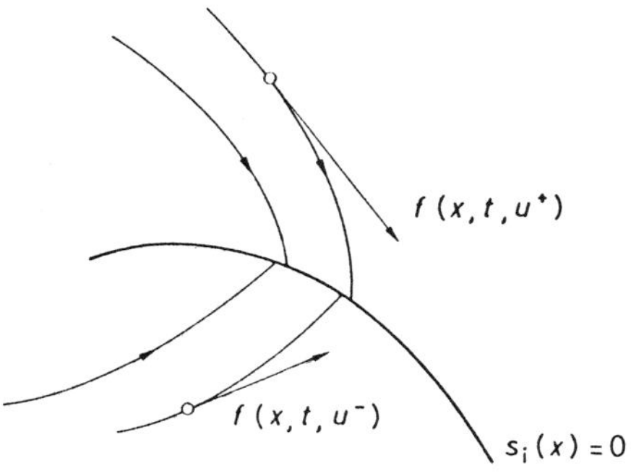

Fig. 4

Another example which refers to a case with two discontinuity surfaces (Fig. 5) illustrates the motion along both a single discontinuity surface (arcs ab and cb) and the intersection of the two surfaces (arc bd).

As evidenced by these examples, the motion trajectories which belong to the set of discontinuity points are singular since in any combination of continuous controls $u_i^+(x,t)$ and $u_i^-(x,t)$ they differ from the system trajectories. An accepted term for the motion on discontinuity surfaces is *sliding mode.* Incidentally, the motion along the segment $|x| \leqq P_0/k$ in the Coulomb friction mechanical system is also a sliding mode motion.

The problem of the sliding mode existence will be treated in full later in the book; here it will be apt to note that a sliding mode does exist on a discontinuity surface whenever the distances to this surface and the velocity of its change $\dot{s}$ are of opposite signs [9], i.e. when

$$\lim_{s \to -0} \dot{s} > 0 \quad \text{and} \quad \lim_{s \to +0} \dot{s} < 0. \tag{1.9}$$

The mathematical description of sliding modes is quite a challenge. It requires the design of special techniques. The solution of Eq. (1.1) is known to exist and be unique if a Lipshitz constant L may be found such that for any two vectors x_1 and x_2

$$\| f(x_1, t) - f(x_2, t) \| \leqq L \| x_1 - x_2 \| . \tag{1.10}$$

It is evident that in the dynamic system (1.7) with the discontinuous control (1.8), condition (1.10) is violated in the vicinity of discontinuity surfaces. Indeed, if points x_1 and x_2 are on different sides of the discontinuity surface and $\| x_1 - x_2 \| \to 0$, inequality (1.10) is not true for any fixed value of L. Therefore, at least formally, some additional effort is needed to find a solution to system (1.7) and (1.8) at an occurrence of a sliding mode. Moreover, let a function $x(t)$ pretend to be a solution lying on the set of discontinuity points. Even in this

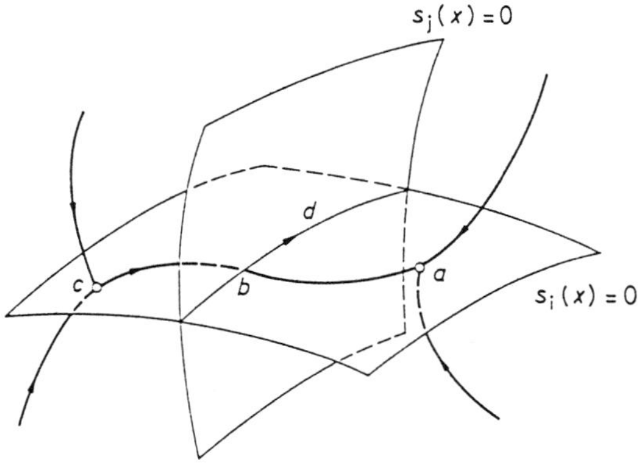

Fig. 5

case the way this function may turn Eq. (1.7) into identity is not clear since control (1.8) is not defined on the surface $s_i(x) = 0$.

Uncertainty of the behaviour of discontinuous systems on the switching surfaces gives freedom in choosing an adequate mathematical model. As a result, the development of the theory of sliding modes gave birth to a number of lively discussions. For example, the description of the sliding mode in a system with a nonlinear discontinuous element using a system of lower-order linear differential equations, a common practice today, seemed at first rather unusual; it was originally suggested in [78, 117] for second-order systems and further extended in [36] for systems of an arbitrary order.

The discussion that took place at the First IFAC Congress in 1960 [49, 107] proved most fruitful in stating new problems in the theory of sliding modes. The discussion has led to a conclusion that the sliding mode motion equations may be different for two dynamic systems with identical equations of motion outside the discontinuity surface if these systems have several discontinuous controls. Most probably, it was during that discussion that the problem of unambiguous description of the control system behaviour on the discontinuity boundary has been first brought to light.

The development of the theory of optimal systems suggested ways to improve efficiency of control in systems with sliding modes. It was found that, generally speaking, sliding modes expand the set of feasible motions in systems with continuous or piecewise-continuous controls. Consequently, it became possible to isolate a class of systems in which the introduction of the sliding mode minimized the optimality criterion [50, 53, 86].

The design of appropriate mathematical tools to describe sliding modes was perhaps most needed for the so-called *variable structure systems* whose intensive theoretical studies have been carried out at the Institute of Control Sciences of the USSR Academy of Sciences since the beginning of 1960's [130]. This special need was dictated by the fact that the sliding mode is the major mode of operation in the variable structure systems. The feasibility of occurrence of such mode stems from the design principle of such systems: a variable structure system is built around a set of continuous subsystems called *structures* which may switch over in the course of the control making the control discontinuous.

A whole set of problems has been originated from the consideration of multivariable systems with variable structure in which every control vector component is subjected to discontinuities on its "own" surface while the sliding-modes occur at the intersection of such surfaces. These problems, for intance, include those of existence and uniquencess of the solution at the discontinuity boundaries, deriving the motion equations and conditions of occurence of the sliding modes on manifolds of various dimension.

It may be interesting to note that in high-gain control systems the so-called *slow* motions occur on manifolds of various dimension. Some efficient methods for the analysis of these modes have been suggested. These include the asymptotic root-loci methods [91, 104, 123, 126, 127] and singular perturbation techniques [67, 77, 156, 157, 169]. The analysis of sliding modes could have been much simplier, had we succeeded in proving the equivalency of their equations to the equations of slow motion in high-gain systems. This point, however true for linear systems [103, 136, 145] cannot be extended toward systems of a general form: discontinuous control systems may feature motions unfeasible with any continuous controls [142].

In the opinion of this writer, the diversity in the sliding modes analysis techniques available in the literature stems from different viewpoints of the authors to the problem. On the one hand, the problem may be approached axiomatically: if the equations on the discontinuity surfaces are undefined and the velocity vector is known to lie in the tangential plane, any motion equations continuation methods may be used. On the other hand, the problem may be made "solvable" if a more accurate model is employed (recall that this technique has been employed before when viscous friction was substituted for the Coulomb friction). However, even this method allows different ways to be used in designing such models.

Both of these approaches, call them axiomatic and physical, respectively, left certain ambiguity and the equation of which sliding equations were indeed true often remained unanswered in discussions.

Further in the book some ideas will be highlighted which have made it possible to write sliding equations for various discontinuous dynamic systems.

2 Formalization of Sliding Mode Description

Let us give a closer look to the physical approach to obtaining the sliding mode motion equations. Uncertainty in the system behaviour on a discontinuity surface has appeared as a result of the imperfection of a model of the type (1.7), (1.8) which was supposed to idealize the real life system. This model fails to recognize such factors as imperfections of the switching device (time delay, dead zones, hysteresis loops, inertiality of elements, etc.). Besides, the equations of an actual control plant may be of an order higher than those of a model. And, finally, the instruments used to obtain the information on the state vector which is necessary to realize controls (1.8) may also prove inertial. Recognition of all these factors makes all discontinuity points isolated thus removing mathematical (but not analytical) difficulties in describing the system behaviour. The physical approach implies introduction of such imperfections which subsequently tend to zero. A result obtained in such limit transitions was chosen as an appropriate mathematical description of a sliding mode.

One of the first examples of applying the physical approach to describe sliding modes in a control system was given in [7]. The system considered was a second order relay control system whose discontinuity surface was actually a straight line on the plane of coordinates of error x and its derivatives $\dot{x}$:

$$cx + \dot{x} = 0, \quad c - \text{const}. \tag{1.11}$$

The behaviour of the system was studied under the assumption that a time delay was inherent in the switching device and, consequently, the discontinuity points were isolated. It was found that irrespective of the control plant parameters and disturbances affecting it, the solution of the second order equation in sliding mode always tends to the solution of the first order linear differential Eq. (1.11) which depends only on the angle factor c of the switching straight line.

Similar limiting procedures were then obtained for linear time-varying systems of an arbitrary order whose relay element features some delay or hysteresis [108]. It was shown that sliding mode motion equations may be obtained in the following way: take the convolution integral correlating the control with the relay element input value, equate this integral to zero, find the solution to the integral equation with respect to the control and, finally, substitute this solution into the initial system.

A pioneering paper in the studies of discontinuous nonlinear systems of the general form featuring a single discontinuity surface was, most probably, Ref. [5] which offered thoroughly performed limiting procedures.

Like some of its predecessors, for instance [108], it considered cases when switching from one continuous system to another goes with a delay or when the switch has a hysteresis loop.

In all of the above cases, the sliding equations are not postulated, their validity being proved with the help of limiting procedures. However, the scope of applications of these results is restricted by a special form of the switching device model. Therefore the question of their applicability to models of other types remains unanswered (for intance, piecewise linear approximation of the discontinuous characteristic). Besides, the limiting procedures treated in the above papers have been designed exclusively around the point-to-point transformation technique, too analytically difficult to be applied to the study of any systems but those with a single discontinuity surface.

A more general concept requires the introduction of a boundary layer. Essentially, this concept implies that (a) the ideal control (1.8) in some vicinity of the discontinuity surfaces is substituted with another control, the only thing we know of which is that a solution of the differential Eq. (1.7) with this control does exist, and (b) the equation obtained in tending the boundary layer to zero is taken as the sliding equation [142]. The use of the boundary layer permits the sliding modes on the discontinuity boundary to be described without specifying the nature of the imperfections, be that in systems with a single discontinuity surface or with a number of intersecting discontinuity surfaces.

Consider now methods of describing sliding modes in systems with discontinuous righthand parts which rest upon the use of the axiomatic approach. It should be noted right away that even though the sliding equations in such methods were postulated, the authors often substantiated the expediency of their continuation techniques by some physical considerations. Some papers have already been mentioned [36, 78, 117] which suggested linear sliding equations for a special type of linear bang-band control systems. The suggestions to substitute the infinite gain element for the discontinuous element realizing control (1.8) [103, 136] or to employ slow motion equations as sliding equations also refer to such continuation techniques. From the physical viewpoint, such a substitution is most natural, for the input value $s_i(x)$ of the switching device in sliding mode is close to zero while the output value $u_i(x, t)$ is generally different from zero.

Instead of trying to obtain a unique sliding equation, the authors of a large number of papers prefer to write down the equations in contigencies, i.e. to specify a set to which the velocity vector $\dot{x}$ should belong when moving along the discontinuity boundary. Thus for instance, the study of a nonlinear system in [4] is restricted by the requirement to have the Lyapunov function be negative definite for any values of the control between its two extreme values.

Nonlinear dynamic systems are presently often described with the use of A.F. Filippov's continuation method [48], which in the general case leads to equations in contigencies. It must be pointed out that the dynamic systems treated in [48] lacked the control and vector $f(x,t)$ in (1.1) was given as a system state discontinuous function. According to Filippov, in each point of the discontinuity surface the velocity vector $f^0(x,t)$ belongs to a minimal convex closed set containing all values of $f(x,t)$, when x covers the entire δ-neighbourhood of the point under consideration (with δ tending to zero) except possibly a zero measure set. The possibility of rejecting a zero measure set permits determining the velocity vector for sliding modes, too. Indeed, although in the points on the discontinuity surface vector $f(x,t)$ in system (1.1) is uncertain, these points in an n-dimensional vicinity make a zero measure set and may be disregarded. Thus, in systems with a single discontinuity surface, the Filippov method yields the following result: (a) the minimal convex set of all vectors $f(x,t)$ in the vicinity of some point (x,t) on the discontinuity surface (Fig. 6) is actually a straight line connecting the ends of vectors f^+ and f^- (to which vector $f(x,t)$ tends with x tending to the point under consideration from either side of the discontinuity surface), and (b) since vector f^0 in sliding mode lies

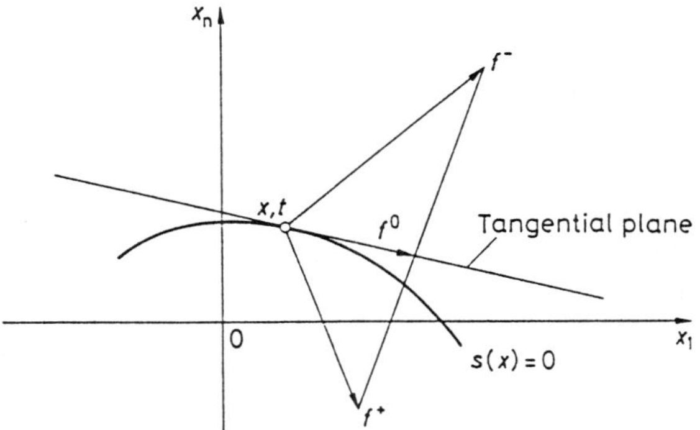

Fig. 6

on a plane tangential to the discontinuity surface, the end of this vector may be found as a point of intersection of this plane and the straight line connecting the ends of vectors f^+ and f^-.

At first sight, the Filippov continuation method seems quite applicable to treat controls of the type (1.7), (1.8) since relationships (1.8) provide unambiguous definition of vector $f(x,t,u)$ outside the discontinuity surface and allow a minimal convex set to be constructed for any point on such surfaces. Discontinuous systems, however, could realize some types of motions which were unfeasible, should Filippov's continuation have been applied (recall the discussion at the First IFAC Congress mentioned in Sect. 1). The reason for this phenomenon is as follows. The sliding mode in a real life system actually occurs not on its discontinuity surface, but within some boundary layer where the control components may take up values differing from u_i^+ and u_i^- (if, for instance, the discontinuous characteristic is approximated by a continuous one with a narrow zone of linearity and saturation). Vector $f(x,t,u)$ may therefore take up values which differ from those obtained with $u_i = u_i^+$ and $u_i = u_i^-$. This results in a wider convex set in the Filippov continuation method and, consequently, in a richer set of motions in the sliding mode.

Sometimes the use of the above techniques for the description of dynamic systems behaviour at the discontinuity boundaries has resulted in different motion equations. Most probably, there is just no need in trying to find a single "true" solution: the nature of the phenomenon we deal with is such that an unambiguous solution simply does not exist. It is important however that it became possible to outline the scope of applicability of each of possible solutions and make practical use of sliding modes in various control applications.

3 Sliding Modes in Control Systems

Mathematical tools developed to study sliding modes have been initially oriented towards analysing the behaviour of systems with discontinuous controls. At the same time, it was already at the early stage of this development that attention was drawn to a number of useful properties of this type of motion, namely, system linearization, reduction of the system differential equation order, and applicability of sliding modes to designing high accuracy follow-up and stabilization systems. Therefore, along with continuous linear controllers, relay-type controllers came into practice which were called *vibrational* if operated in a high-frequency switching mode, or sliding mode, in the present-day terminology. Such controllers were widely used in various applications, for instance, in designing control systems for aircraft DC generators [88]. The theory of vibrational controllers was treated in a great number of papers which analyzed their behaviour mostly with the use of time representation of motion. An overview and a bibliography on the subject were offered in [118].

A promising way of applying sliding modes in the theory of optimal systems was already mentioned in Sect. 2. It is due to be added here that the use of sliding modes in the so-called *singular optimal problems* which will be considered further in Chapt. 9, also minimizes the performance criterion in the system motion on some manifold in the state space [28, 52, 58].

Let us consider now a class of control systems designed with a deliberate introduction of sliding modes. These cover variable structure systems including, by virtue of the definition given in Sect. 2, systems of the type (1.7), (1.8). An example of such a system given above was a second order system whose sliding mode motion was described by the first order Eq. (1.11) and depended on the switching straight line position only. This property of the sliding modes invariancy in the so-called *canonical space* consisting of some coordinate and its derivatives was actively exploited at the first stage of the development of the variable structure systems devoted, for the most part, to consideration of single input–single output problems [130].

Certain limitations in applicability of the above approach have significantly boosted the development of the general type variable systems theory. Implementation of control actions for systems described in the canonical form is rather complicated technically since it requires time derivatives of different orders. Since all available differential devices are dynamic elements, attempt to describe the behaviour of a system in the space of any of its coordinate and its derivatives may prove to be an unfeasible idealization. Besides, present-day control problems are featured by the need to simultaneously control several variables rather than just one, hence by a vector-valued rather than scalar control. At the same time, measurable in most cases are not only the controlled parameters, but also some other physical parameters which characterize the control process. These considerations necessitated the design of vector-valued discontinuous control systems whose sliding modes are realized in the arbitrary state coordinates space. The advent of the mathematical apparatus for the description of sliding modes in discontinuous systems has made it possible to work out design techniques for variable structure systems of a general form [141, 143, 144].

Considered below are the control systems and optimization procedures whose design rests upon a deliberate introduction of sliding modes.

Part 1 outlines mathematical tools intended for studying multivariable discontinuous systems. Thus, it suggests the procedures of obtaining sliding equations and gives proof of their validity, treats the uniqueness of the solution on the discontinuity boundary and the sliding motion robustness and stability.

The principle of decoupling discontinuous control systems is considered in Part 2 in application to a wide circle of control problems. According to this principle, the required motions may be independently designed in lower dimensional spaces, i.e. outside the discontinuity surfaces and along their intersection.

Part 3 is devoted to application problems most diverse in their physical nature and control goals which lend themselves to a successful solution with the help of a deliberately introduced sliding mode.

Chapter 2
Mathematical Description of Motions on Discontinuity Boundaries

1 Regularization Problem

As already noted, the behaviour of system (1.7), (1.8) on discontinuity boundaries cannot be adequately described in terms of the classical theory of differential equations. To solve this problem, various special ways are usually suggested, in order to reduce the original problem to a form which yields a solution close, in a sense, to that of the original problem, and which allows the use of classical analysis techniques. Such a substitution of the problem is usually called *regularization*. The physical ways described in Chap. 2, Part 1 are, in essence, the ways of regularization of the problem of discontinuous dynamic system behaviour.

The theoretical foundation of the regularization techniques which gave solution to many nonclassical problems has been laid by A.N. Tikhonov and further developed by his school. Of particular importance for the control theory were such problems as, for instance, the study of dynamic systems with small parameters multiplying derivatives in differential equations [131, 132, 156–158] and some ill-posed mathematical programming problems [133, 134]. An example of a regularization technique is the introduction of a small parameter into the functional in solving singular problems of optimal control [100].

All regularization techniques considered earlier in the book have only one thing in common: whenever a sliding mode appears, the velocity vector in the state space lies on a manifold tangential to one or several discontinuity surfaces.

In such a case, the use of different techniques may result in different sliding equations. Furthermore, any regularization technique may yield a continuum of sliding equations rather than just a single equation. The problem of uniqueness will be considered later, right now it is due to note that each regularization technique normally corresponds to some real-life system, therefore it hardly makes any sense to speak of generally "correct" or "incorrect" techniques. Most probably, it makes more sense to define the scope of applicability of such techniques and develop those which permit a sufficiently wide class of discontinuous systems to be treated. The words "sufficiently wide" in this case mean that the sliding equations may be found without the use of an accurate model of the system behaviour in the vicinity of the discontinuity boundaries, even if these sliding equations are obtained in a somewhat ambiguous form. At the same time, the set of feasible sliding equations may be significantly narrowed (perhaps down to a single equation) if we are capable of obtaining a more accurate model of the system.

Let us give consideration to regularization scheme which makes use of a boundary layer. This scheme will be employed throughout the book for obtaining the sliding mode equations on the discontinuity boundaries in system (1.7), (1.8). Assume a sliding motion occurs on the intersections of all discontinuity surfaces

$$s(x) = 0, \quad s^{\mathrm{T}}(x) = [s_1(x), \ldots, s_m(x)]. \tag{2.1}$$

Substitute the ideal model (1.7), (1.8) with a more accurate one

$$\dot{x} = f(x, t, \tilde{u}), \tag{2.2}$$

which takes into account of all possible imperfections in the new control, $\tilde{u}$, e.g. delay, hysteresis, switching device inertiality, incomplete correspondence between the model and the actual control plant, imprecision and inertiality of measuring instruments, etc. As a result of the account of such imperfections, the solution to Eq. (2.2) exists in the usual meaning (for example, the righthand part discontinuity points prove to be isolated, or the Lipshitz constant exists for function $f(x, t, \tilde{u})$). However, the "cost" of such regularization is the motion occurring not strictly on the manifold $s(x) = 0$ (2.1), but in some neighbourhood of this manifold

$$\| s(x) \| \leqq \Delta, \quad \| s(x) \| = (s^{\mathrm{T}} s)^{1/2}, \tag{2.3}$$

where Δ is a small number depending on the introduced imperfections. In contrast to the *ideal* motion, this one will be referred to as *real* sliding motion. Whichever imperfections may exist in a real life system, their account will not violate the assumptions made on the $\tilde{u}(x, t)$ function.

We do not specify the types of imperfections and consider the entire class of functions $\tilde{u}(x, t)$ leading to a motion in the vicinity of (2.3). If the value of Δ tends to zero the real sliding mode will tend to the ideal one. The motion equations obtained in such a limiting procedure will be regarded as the ideal equation of the sliding motion along the intersections of all discontinuity surfaces.

The solution on the discontinuity boundary does exist and is unique if the above limiting procedure yields an unambiguous result regardless of the type of function $\tilde{u}(x,t)$ and of the way the limiting procedure is performed. Otherwise, the equations describing the system motion outside the discontinuity surfaces (1.7), (1.8) give an ambiguous description of this motion along the intersections.

Note. The above regularization technique may certainly be used to obtain the sliding equations describing motion along the intersection of just a part of discontinuity surfaces, for instance, along the intersection of k surfaces

$$s^k(x)=0, \quad [s^k(x)]^{\mathrm{T}}=[s_1(x),\ldots,s_k(x)].$$

In this case the limiting procedure should be obtained for the following system

$$\dot{x}=f[x,t,\tilde{u}^k(x,t),u^{m-k}(x,t)],\ \|s^k(x)\|\leqq \Delta, \tag{2.4}$$

where $[\tilde{u}^k]^{\mathrm{T}}=(\tilde{u}_1,\ldots,\tilde{u}_k)$, $[u^{m-k}]^{\mathrm{T}}=[u_{k+1},\ldots,u_m]$. The control components $u_{k+1},\ldots,u_m$ in (2.4) are continuous since functions $s_{k+1}(x),\ldots,s_m(x)$ on manifold $s^k(x)=0$ differ from zero.

2 Equivalent Control Method

A formal procedure will be suggested below to obtain sliding equations along the intersection of a set of discontinuity surfaces for system (1.7), (1.8).

Assume that a sliding mode exists on manifold (2.1). Let us find a continuous control such that under the initial position of the state vector on this manifold, it yields identical equality to zero of the time derivative of vector $s(x)$ along system (1.7) trajectories:

$$\dot{s}=Gf(x,t,u)=0, \tag{2.5}$$

where the rows of the $(m\times n)$ matrix $G=\{\partial s/\partial x\}$ are the gradients of the functions $s_i(x)$.

Assume that a solution (or a number of solutions) of the system of algebraic Eqs. (2.5) with respect to m-dimensional control does (or do) exist. Use this solution, hereinafter referred to as *equivalent control* $u_{\mathrm{eq}}(x,t)$, in system (1.7) in place in u:

$$\dot{x}=f[x,t,u_{\mathrm{eq}}(x,t)]. \tag{2.6}$$

It is quite obvious that, by virtue of condition (2.5), a motion starting in $s[x(t_0)]=0$ will proceed along the trajectories which lie on the manifold $s(x)=0$.

The above procedure will be called the *equivalent control method* and Eq. (2.6) obtained as a result of applying this method will be regarded as the sliding mode equation describing the motion on the intersection of discontinuity surfaces $s_i(x)=0$, $i=1,\ldots,m$.

From the geometric viewpoint, the equivalent control method implies a replacement of the undefined discontinued control on the discontinuity boundary with a continuous control which directs the velocity vector in the system state space along the discontinuity surfaces intersection.

For example, in order to find this vector in a system with a single discontinuity surface $s(x)=0$ at some point (x,t) (Fig. 7) one should vary the scalar control from u^- to u^+, plot the locus of $f(x,t,u)$ and find the point where it intersects the tangential plane. The point of intersection determines the equivalent control $u_{eq}(x,t)$ and the righthand part $f(x,t,u_{eq})$ of the sliding mode differential Eq. (2.6).

Consider now the equivalent control procedure for an important particular case of a nonlinear discontinuous system, the righthand part of whose differential equation is a linear function of the control

$$\dot{x}=f(x,t)+B(x,t)u, \tag{2.7}$$

where $f(x,t)$ and $B(x,t)$ are all argument continuous vector and matrix of dimension $(n\times 1)$ and $(n\times m)$, respectively, and the discontinuous control u changes in compliance with (1.8). The equivalent control Eq. (2.5) for the case (2.7) may be written as

$$\dot{s}=Gf+GBu=0. \tag{2.8}$$

Assuming that matrix GB is nonsingular for all x and t, we find the following equivalent control from (2.8):

$$u_{eq}(x,t)=-[G(x)B(x,t)]^{-1}G(x)f(x,t). \tag{2.9}$$

Substitution of this control into (2.7) yields the equation

$$\dot{x}=f-B(GB)^{-1}Gf, \tag{2.10}$$

which describes the sliding mode motion on the manifold $s=0$.

The above procedures for obtaining the sliding equations are just postulated here. Further in the book we shall discuss independently the rightfulness of

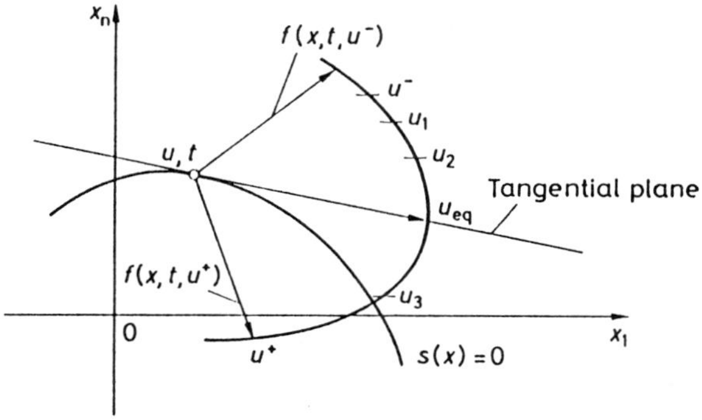

Fig. 7

their usage for system linear and nonlinear with respect to control, applying the regularization technique built around the analysis of the system behaviour in the boundary layer.

3 Regularization of Systems Linear with Respect to Control

The suggested regularization technique allows unambiguous description of the sliding modes for dynamic systems whose differential equation linearly depends on control (2.7). For such systems, the sliding Eq. (2.10) may be obtained using the equivalent control method. To verify its validity, in compliance with the reasonings given in Sect. 1, one should introduce the boundary layer (2.3) where the motion is described by an equation similar to (2.2):

$$\dot{x} = f(x, t) + B(x, t)\tilde{u}, \tag{2.11}$$

where $\tilde{u}$ is a new control which allows regularization of the problem. Next, one should tend the boundary layer width Δ to zero and make certain the solution of Eq. (2.11) tends to that of the equivalent control Eq. (2.10). To distinguish the solution of (2.10) from (2.11), let us mark the solution of the equivalent control Eq. (2.10) as x^*. Assume that the distance $s(x)$ from any point in the vicinity of (2.3) to the manifold $s(x) = 0$ is estimated by the inequality

$$r(s, x) \leqq P\Delta, \tag{2.12}$$

where P is a positive number. Such a P always exists if, for instance, all gradients of functions $s_i(x)$ are linearly independent (it follows from the above assumption that det GB $\neq 0$) and are lower bounded in the norm by some positive number.

Let us present the above assumption in the form of the following theorem.

Theorem. *If*

(1) *there is a solution* $x(t)$ *of system (2.11) which, on the interval* $[0, T]$, *lies in the* Δ*-neighbourhood of the manifold* $s(x) = 0$, *or the inequality (2.3) is valid;*

(2) *for the righthand part of the differential Eq. (2.10) obtained with respect to* x^* *using the equivalent control method (under the assumption* det $GB \neq 0$)

$$\dot{x}^* = f(x^*, t) - B(x^*, t)[G(x^*)B(x^*, t)]^{-1}G(x^*)f(x^*, t) \tag{2.13}$$

a Lipschitz constant L exists;

(3) *partial derivatives of the function* $B(x, t)[G(x)B(x, t)]^{-1}$ *with respect to all arguments exist and are bounded in any bounded domain, and*

(4) *for the righthand part of the real sliding Eq. (2.11)* $f(x, t) + B(x, t)\tilde{u}$ *such positive numbers M and N exist that*

$$\| f(x, t) + B(x, t)\tilde{u}(x, t) \| \leqq M + N \| x \|, \tag{2.14}$$

then for any pair of solutions to Eqs. (2.11) and (2.13) under the initial conditions

$$\| x(0) - x^*(0) \| \leqq P\Delta \tag{2.15}$$

there exists a positive number H such that

$$\| x(t) - x^*(t) \| \leqq H\Delta \quad \text{for} \quad t \in [0, T]. \tag{2.16}$$

To prove the theorem, the norm of the difference of the solutions of equations describing motion in the ideal and prelimit systems should be estimated. In a system with a boundary layer, $\dot{s}$ is not zero any more as was assumed in the equivalent control method. Therefore, the control $\tilde{u}$ defined by virture of system (2.11) will differ from u_{eq}:

$$\tilde{u} = -(GB)^{-1}Gf + (GB)^{-1}\dot{s}$$

and, consequently, the equation for a real sliding mode obtained by substituting $\tilde{u}$ into (2.11)

$$\dot{x} = f(x,t) - B(x,t)[G(x)B(x,t)]^{-1}G(x)f(x,t) + B(x,t)[G(x)B(x,t)]^{-1}\dot{s} \tag{2.17}$$

is also different from Eq. (2.10) or (2.13) in the additional term $B(x,t)[G(x)B(x,t)]^{-1}\dot{s}$. Equation (2.17) certainly represents all imperfections recognized in the control since the value of $\dot{s}$ depends on them. Equation (2.17) describes motions with due regard for all nonspecified imperfections and with the boundary layer width tending to zero its solution describes an ideal sliding. (It should be remembered that, generally speaking, with Δ tending to zero $\dot{s}$ does not necessarily follow this tendency).

Before comparing the solutions of Eqs. (2.10) and (2.17), let us write their equivalent integral equations

$$x^*(t) = x^*(0) + \int_0^t \{f(x^*,\gamma) - B(x^*,\gamma)[G(x^*)B(x^*,\gamma)]^{-1}G(x^*)f(x^*,\gamma)\}\,d\gamma, \tag{2.18}$$

$$\begin{aligned} x(t) = x(0) &+ \int_0^t \{f(x,\gamma) - B(x,\gamma)[G(x)B(x,\gamma)]^{-1}G(x)f(x,\gamma)\}\,d\gamma \\ &+ \int_0^t B(x,\gamma)[G(x)B(x,\gamma)]^{-1}\dot{s}\,d\gamma. \end{aligned} \tag{2.19}$$

Integrating the last term in (2.19) by parts, estimate the norm of the difference between the solutions to (2.18) and (2.19) with due regard for the Theorem conditions:

$$\begin{aligned} \| x(t) - x^*(t) \| \leqq P\Delta &+ \int_0^t L \| x - x^* \|\,d\gamma + \| B(x,\gamma)[G(x)B(x,\gamma)]^{-1}s \| \Big|_0^t \\ &+ \int_0^t \left\| \frac{d}{d\gamma} B(x,\gamma)[G(x)B(x,\gamma)]^{-1} \right\| \cdot \| s \|\,d\gamma. \end{aligned} \tag{2.20}$$

The solution of Eq. (2.19) or its equivalent (2.11) is bounded on a finite time interval $[0, T]$. This follows from the fact that, with condition (2.14) true, the solution to Eq. (2.11) satisfies the inequality

$$\| x(t) \| \leqq \| x(0) \| + MT + \int_0^t N \| x \| dt.$$

According to the Bellman-Gronwall lemma [9] this very inequality makes the solution to (2.11) or (2.19) bounded:

$$\| x(t) \leqq (\| x(0) \| + MT) e^{NT} (t \in [0, T]).$$

Then by virtue of the fact that the solution $x(t)$ is bounded, and using items (1)–(3) of the Theorem, the inequality (2.20) may be represented as

$$\| x(t) - x^*(t) \| \leqq S\Delta + L \int_0^t \| x - x^* \| d\gamma, \tag{2.21}$$

where S is a positive value depending on the functions in the righthand parts of Eqs. (2.11) and (2.17), on the initial conditions, time T and the constant P. Applying the Bellman-Gronwall lemma to inequality (2.21) yields

$$\| x(t) - x^*(t) \| \leqq H\Delta,$$

where $H = Se^{LT}$, which proves the theorem.

The theorem implies that with the initial conditions of systems (2.11) and (2.13) sufficiently close, their solutions are also close. Since Eq. (2.13) describes the ideal system motion along the intersection of discontinuity surfaces, i.e. with $s[x^*(0)] = 0$, and Eq. (2.11) describes the real system motion in the boundary layer of that intersection, the value of $\| x(0) - x^*(0) \|$ is essentially the distance from the point $x(0)$ to some point on the manifold $s = 0$. By virtue of the assumption (2.12) on the functions $s_i(x)$ for any point $x(0)$ belonging to the boundary layer one can always find a point $x^*(0)$ on the manifold $s = 0$ such that condition (2.15) holds. Consequently, by virtue of (2.16), for any solution to Eq. (2.11) describing the motion in the boundary layer, there is a solution to Eq. (2.13) describing the motion along the intersection of discontinuity surfaces which differs from it within the range of Δ, and

$$\lim_{\Delta \to 0} x(t) = x^*(t) \tag{2.22}$$

for any finite time interval.

This result means that regardless of the nature of the imperfections that have generated the real sliding mode in the boundary layer of the intersection of discontinuity surfaces and regardless of the way we make the width of this layer rend to zero, the solution of the nonideal sliding Eq. (2.11) tends to the solution of Eq. (2.13) resulting from formal application of the equivalent control method. In compliance with the reasonings of Sect. 1, this fact substantiates the validity of the method for obtaining ideal sliding equations.

Physically, these considerations may be given the following interpretation. Real sliding modes have a finite frequency at which the components of the control vector are switched. The state velocity vectors of each of the continuous systems are not generally found on the manifold of discontinuity surface intersection. Therefore in order to keep the state vector in the vicinity of (2.3) with Δ decreasing, the switching frequency should increase. Equation (2.17) treated in the theorem describes a real filter whose response to external disturbances reduces with the growth of frequency, starting with a certain frequency value (this disturbance is actually the addend $B(GB)^{-1}\dot{s}$ in (2.17) which makes this equation different from the ideal sliding Eq. (2.13)). Therefore the effect of the additional terms caused by imperfections is reduced with Δ tending to zero, and they do not affect ideal sliding.

The theorem has been proved for a finite time interval. The complexities of the limiting process for an infinite time interval are caused by general properties of solutions of ordinary differential equations rather than by the specifics of our problem. If the solution of Eq. (2.13) is unstable $\left(\lim_{t\to\infty} x = \infty\right)$, then the trajectories of both systems under consideration, (2.11) and (2.13), taken as close to each other as possible at the beginning will stay close within a finite time interval, but may nevertheless diverge at the end. Since the "proximity interval" is infinitely expanding with Δ tending to zero, and since in any real life process Δ is not zero (although it may be close to it), we have to either take the Eqs. (2.10) as "true" or give up the very idea of trying to obtain sliding equations because there are no such equations true for an infinite time interval (like, generally, these are no equations which give precise description of any unstable physical process).

In studying automatic control systems, a researcher is usually concerned with their behaviour at an infinite time interval, but this problem is certainly meaningful only when the system motion is stable.

Let us now consider applicability of the equivalent control method at an infinite time interval to control systems whose design methods will be discussed in detail in Part 2 of this book. Let us start with the sliding mode of a constant parameter linear system:

$$\dot{x} = Ax + Bu + g(t),$$

where A and B are constant $(n \times n)$ and $(n \times m)$ matrices, $g(t)$ is a column vector depending only on time and characterizing the external disturbances, and u is the m-dimensional control that varies according to (1.8). Let discontinuity surfaces $s_i = 0$ be planes, in other words, let the m-dimensional vector s be represented as $s = Cx$, where C is a constant $(m \times n)$ matrix and $\det CB \neq 0$. In compliance with the equivalent control method, to obtain the equations for a sliding mode along the intersection of the surfaces $s_i = 0$ $(i = 1, \ldots, m)$ one should solve the equation $\dot{s} = CAx + CBu + Cg(t) = 0$ with respect to control

$$u_{eq} = -(CB)^{-1}(CAx + Cg(t))$$

and substitute this into the original equation

$$\dot{x}^* = [I_n - B(CB)^{-1}C]Ax^* + [I_n - B(CB)^{-1}C]g(t), \tag{2.23}$$

where I_n is an identity matrix. Under the initial conditions $Cx(0)=0$ this is the ideal sliding equation[1]. As to the motion inside the boundary layer (2.3) the control $\tilde{u}$ introduced for the purpose of regularization of the problem may be expressed through $\dot{s}$, which is no longer taken equal to zero like it was formally done in the equivalent control method:

$$\tilde{u} = -(CB)^{-1}C(Ax+g)+(CB)^{-1}\dot{s}.$$

Consequently, the equations of motion in the vicinity of (2.3) are of the form

$$\dot{x} = [I_n - B(CB)^{-1}C]Ax + [I_n - B(CB)^{-1}C]g + B(CB)^{-1}\dot{s}. \tag{2.24}$$

Let us now demonstrate that if any solution of the ideal sliding Eq. (2.23) is asymptotically stable it will be close to the solution of the real sliding Eq. (2.24) within the range of Δ in an infinite time interval, starting with the time instant when the sliding mode begins. If $\varphi(t)$ is the transition matrix of the solutions to ideal sliding Eq. (2.33) with $g(t)=0$ (e.g. $\varphi(0)=I_n$), then the general solution to this equation is of the form

$$x^*(t) = \varphi(t)x^*(0) + \int_0^t \varphi(t-\tau)[I_n - B(CB)^{-1}C]g(\tau)d\tau$$

Write down the solution of Eq. (2.24):

$$x(t) = \varphi(t)x(0) + \int_0^t \varphi(t-\tau)[I_n - B(CB)^{-1}C]g(\tau)d\tau + \int_0^t \varphi(t-\tau)B(CB)^{-1}\frac{ds}{d\tau}d\tau.$$

Integrating the last term by parts, estimate the norm of the difference in these two solutions:

$$\begin{aligned}\|x(t)-x^*(t)\| \leqq \|\varphi(t)\|\cdot\|x(0)-x^*(0)\| &+ \|B\|\cdot\|(CB)^{-1}\|\cdot\|s(t)\| \\ &+ \|B\|\cdot\|(CB)^{-1}\|\cdot\|\varphi(t)\|\cdot\|s(0)\| \\ &+ \int_0^t \left\|\frac{d}{d\tau}\varphi(t-\tau)\right\|\cdot\|B\|\cdot\|(CB)^{-1}\|\cdot\|s\|d\tau.\end{aligned}$$

Since for any solution of (2.24) the condition $\|s\| \leqq \Delta$ holds and for any solution of (2.23) $s=0$, then in accordance with (2.12) for any vector $x(0)$ a vector $x^*(0)$ exists such that $s[x^*(0)]=0$, $\|x(0)-x^*(0)\| \leqq P\Delta$. If functions

$$\|\varphi(t)\| \quad \text{and} \quad \int_0^t \left\|\frac{d}{d\tau}\varphi(t-\tau)\right\| d\tau$$

prove to be bounded for any $0 \leqq t < \infty$, then a positive value N can be found such that

$$\|x(t)-x^*(t)\| \leqq N\Delta$$

[1] Like in (2.13), the state vector is denoted here as x^* to tell the solution of the ideal sliding equation from that of the real sliding equation.

and

$$\lim_{\Delta\to 0} x(t) = x^*(t) \quad \text{for} \quad 0 \leqq t < \infty.$$

Since $\dot{s}=0$ on the trajectories of system (2.33) the system has m distinct zero eigenvalues and its $(n-m)$ eigenvalues correspond to the sliding mode in the manifold $s=0$. If the sliding motion is asymptotically stable then matrix $\varphi(t)$ consists of constant values and decaying exponential functions, which results in the boundness of $\|\varphi(t)\|$ and $\int_0^t \left\| \frac{d}{d\tau}\varphi(t-\tau) \right\| d\tau$ for any $0 \leqq t < \infty$. As it has been established, this condition means that $x(t)$ tends to $x^*(t)$ with $\Delta \to 0$ for $0 \leqq t < \infty$.

The above result proves applicability of the equivalent control method on an infinite time interval to time-invariant linear systems with an asymptotically stable sliding motion. A similar conclusion will evidently be true for time-varying linear systems where, first, all time-varying matrices and their time derivatives are bounded and, second, the transition matrix $\varphi(t,\tau)$ of the solutions for the sliding equation is such that the functions

$$\|\varphi(t,\tau)\| \quad \text{and} \quad \int_0^t \left\| \frac{d}{d\tau}\varphi(t-\tau) \right\| d\tau$$

are bounded on an infinite time interval.

Thus we have found that the sliding equation for discontinuous control-linear systems (2.7) is unambiguously obtained, for the case $\det GB \neq 0$, by the equations which describe the behaviour of the system outside discontinuity surfaces. It may be interesting to note that all the techniques outlines in Chap. 1 for describing sliding modes in systems linear with respect to control yield the same equation.

Let us show, for example, the way Filippov's technique can be applied to the problem and prove that in the case of a control-linear system the result of using this technique coinsides with Eq. (2.10) obtained by the equivalent control method an consequently, with the result of the limiting procedure described above. Following Filippov's technique, let us make an equation for minimal convex hull for all velocity vectors that describe the motion in the vicinity of any point on the intersection of discontinuity surfaces:

$$f^0 = f(x,t) + B(x,t) \sum_{i=1}^{2m} \mu_i u^i$$

where $\mu_i \geqq O\left(\sum_{i=1}^{2m} \mu_i = 1\right)$ are the parameters defining the points on the convex hull and u^i are all possible control vectors in different regions of the space that are adjacent to the point under study. The sliding motion velocity vector belongs to this hull and lies on the intersection of planes tangential to the discontinuity surfaces in that point. Therefore the direction of the vector is defined by the condition $Gf^0 = 0$, or

$$G\left(f + B\sum_{i=1}^{2m} \mu_i u^i\right) = 0.$$

Solving this equality with respect to the vector

$$z = \sum_{i=1}^{2m} \mu_i u^i$$

(and assuming, like before, that $\det GB \neq 0$) and substituting z into the expression defining vector f^0 yield a sliding motion equation which coincides with Eq. (2.10).

Thus for systems linear with respect to control the equivalent control method provides a standard way of obtaining the sliding equations resulting from Filippov's approach.

The above regularization technique for control-linear systems based on the introduction of a boundary layer verifies the validity of sliding equations obtained at the intersection of discontinuity surfaces. Still remaining to be considered are the so-called *singular* cases of systems (2.7) when $\det GB = 0$ and the equivalent control cannot be unambiguously found from Eq. (2.8).

Before we proceed to examination of ambiguous behaviour at discontinuity boundaries (treated in detail in Chap. 3), let us find the actual meaning of the continuous function $u_{eq}(x, t)$ which has formally appeared in the equivalent control method.

4 Physical Meaning of the Equivalent Control

A sliding motion can be regarded as a kind of idealization, since the velocity vector in the system state space is directed right along the intersection of discontinuity surfaces. In a real life system, however, the motion occurs in a neighbourhood of this intersection or in the boundary layer. If, for instance, the switching device features small inertialities, delay or hysteresis, then in moving within the boundary layer the real control components will switch over at a high rate alternatively taking up the extreme values u_i^+ and u_i^-. The state vector, accordingly, will oscillate at a high rate about the manifold $s = 0$. We cannot disregard one more possibility: if the real control components take up some intermediate values besides extreme ones, u_i^+ and u_i^-, then in moving inside the boundary layer these extreme values may not be attained.

Regardless of the nature of its imperfections, a real control always includes a slow component to which a high rate component may be added. Since a control plant is a dynamic object, its behaviour is largely determined by the slow component while its response to the high rate component is negligible. On the other hand, the equivalent control method demands a substitution of the real control in the motion Eq. (2.7) with a continuous function $u_{\text{eq}}(x, t)$ (2.9) which does not contain any high rate component.

This suggests the following assumption: the equivalent control coincides with the slow component of the real control. To find out if this assumption is true we have to filter out the high rate component of the real control, i.e. make the averaging and isolate the slowly changing component denoted as u_{av}. This component may be measured, for instance, by applying the real control to the input of a filter whose time constant is small enough as compared with the slow component, yet large enough to filter out the high rate component[1]. Let us show that the u_{eq} used in the equivalent control method coincides with the average value of the control u_{av} physically realizable at the output of such a filter if we regard the ideal sliding as a result of the limiting procedure when the boundary layer width tends to zero.

Further reasoning will refer to a system linear with respect to control for which the method of equivalent control is substantiated. According to the regularization technique used, the real control $\tilde{u}$ provides motion in the boundary layer (2.3) of the intersection of discontinuity surfaces. Assume that the conditions of the theorem stated in Sect. 3 hold for function f, B, s_i and $\tilde{u}$ and, besides, the first derivatives of u_{eq} and $(GB)^{-1}$ in all arguments exist and are bounded in any bounded domain.

Let the average control u_{av} be the output of the linear filter

$$\tau \dot{u}_{\text{av}} + u_{\text{av}} = \tilde{u}, \tag{2.25}$$

where τ is a time constant. Equation (2.25) describes m first order filters with the components of vector $\tilde{u}$ fed to their inputs the components of vector u_{av} are the output values of these elements. Finding $\dot{s}$ from Eq. (2.7) and substituting $\tilde{u}$ for u in this equation

$$\dot{s} = Gf + GB\tilde{u}$$

we obtain the control function $\tilde{u}$ as

$$\tilde{u} = -(GB)^{-1}Gf + (GB)^{-1}\dot{s}.$$

Like before, it was assumed here that det $GB \neq 0$. Recall that u_{eq} is formally a solution of the equation $\dot{s} = 0$ for the control, therefore the value of $\tilde{u}$ may be represented in the form

$$\tilde{u} = u_{\text{eq}} + (GB)^{-1}\dot{s}. \tag{2.26}$$

The properties of Eq. (2.25) with (2.26) as its righthand part may be established through the following lemma.

Lemma. *If in the differential equation*

$$\tau \dot{z} + z = h(t) + H(t)\dot{s}$$

---

[1] A filter equation and a more detailed definition of what is meant by "small enough" and "large enough" will be clarified below.

τ is a constant and z, h and s are m-dimensional vectors,

(1) the functions $h(t)$ and $H(t)$, and their first order derivatives are bounded in magnitude by a certain number M and

(2) $\| s(t) \| \leqq \Delta$, Δ being a constant positive value, then – for any pair of positive numbers Δt and ε there exist such a number $\delta(\varepsilon, \Delta t, z(0))$ that $\| z(t) - h(t) \| \leqq \varepsilon$ with $0 < \tau \leqq \delta, \Delta/\tau \leqq \delta$ and $t \geqq \Delta t$.

For proof, write the solution of this equation:

$$z(t) = z(0)e^{-t/\tau} + \frac{1}{\tau} e^{-t/\tau} \int_0^t e^{\gamma/\tau}[h(\gamma) + H(\gamma)\dot{s}(\gamma)]\, d\gamma.$$

Integration by parts yields

$$\begin{aligned} z(t) &= z(0)e^{-t/\tau} + h(t) - h(0)e^{-t/\tau} \\ &\quad - e^{-t/\tau} \int_0^t \dot{h}(\gamma)e^{\gamma/\tau}d\gamma + H(t)\frac{s}{\tau} - H(0)e^{-t/\tau}\frac{s(0)}{\tau} \\ &\quad - \frac{1}{\tau} e^{-t/\tau} \int_0^t \left[\dot{H}(\gamma)e^{\gamma/\tau} + \frac{1}{\tau} H(\gamma)e^{\gamma/\tau} \right] s\, d\gamma, \end{aligned}$$

$$\| z(t) - h(t) \| \leqq \| z(0) - h(0) \| e^{-t/\tau} + M\tau + \frac{2M\Delta}{\tau} + M\Delta + \frac{M\Delta}{\tau},$$

or

$$\| z(t) - h(t) \| \leqq \| z(0) - h(0) \| e^{-t/\tau} + M(\tau + \Delta) + 3M\frac{\Delta}{\tau}.$$

Evidently, for any $\Delta t > 0$

$$\lim_{\substack{\tau \to 0 \\ \Delta/\tau \to 0}} z(t) = h(t)$$

with $t \geqq \Delta t$. Let us apply this result to Eq. (2.25) bearing in mind that the above assumptions concerning the functions $f, B, s_i, \tilde{u}, u_{\text{eq}}, (GB)^{-1}$ hold and that $\tilde{u}$ is in the form (2.26). According to the theorem in Sect. 3, the solution $x(t)$ at any finite time interval $[0, T]$ differs from $x^*(t)$ – the solution of the ideal sliding Eq. (2.13) – by a value of magnitude Δ and a Lipschitz constant exists for the righthand part of this equation. Consequently, the function $x(t)$ is bounded with $t \in [0, T]$. This means that the functions u_{eq} and $(GB)^{-1}$ are norm-bounded along with their first time derivatives and that the conditions of the lemma hold for the righthand part of Eq. (2.25) with $t \in [0, T]$. As a result we find that for a finite time interval $[0, T]$ the value of u_{av} tends to u_{eq} in the sense that for any pair of positive numbers $\Delta t < T$ and ε there is a $\delta > 0$ such that with $0 < \tau \leqq \delta$ and $\Delta/\tau \leqq \delta$ the inequality $\| u_{\text{av}} - u_{\text{eq}} \| \leqq \varepsilon$ holds for $\Delta t \leqq t \leqq T$.

Such a definition of u_{av} tending, in the limit, to u_{eq} is not something involved and follows quite naturally from the system's physical properties. Indeed, in order to bring a real sliding mode nearer to the ideal one, we have to reduce width Δ of the discontinuity surface neighbourhood in which the state vector oscillates. At a decrease of Δ, the control switching frequency, should increase, otherwise the amplitude of the high frequency components of the state vector will be certainly more than Δ. To enable the filtering of high frequency term determined by switching in the sliding mode, the value inverse to the switching frequency should be taken essentially smaller than the time constant τ. However, as has just been established above, this value is proportional to Δ, hence the ratio of Δ/τ should tend to zero. And, finally, the time constant τ should also be tended to zero to prohibit the linear filter from distorting the slow control component which equals exactly u_{av}.

Thus the notion of an "equivalent control" actually has a quite definite physical meaning. This function equals the average control value and may be measured by a first order linear filter provided its time constant is appropriately matched with the boundary layer width.

The result obtained is not valid for systems nonlinear with respect to control. In such systems, in contrast to the case just considered, the dynamic plant response to the high frequency control term generally cannot be neglected.

5 Stochastic Regularization

The regularization method discussed in this chapter implies introduction of a boundary layer in the vicinity of the intersection of discontinuity surfaces and substitution of an ideal discontinuous control, u, with a real control recognizing all possible imperfections. However this regularization method proves inapplicable to a wide class of practical problems due to random disturbances affecting the control plant or measurement noise; even when the intensity of noise is quite low its magnitude, generally speaking, is not limited and therefore the assumption on the finiteness of the boundary layer is not true. Besides, the influence of noise upon the behaviour of the system may be quite essential because of discontinuity of control function.

A new regularization method is now under discussion which does not imply an introduction of a boundary layer and takes account of the class of imperfections resulting from the stochastic nature of the system behaviour. The mathematics underlying in this regularization method is given below permitting an asymptotically accurate solution to be obtained under general assumptions on a sufficiently wide noise spectrum. It is shown first that the behaviour of a discontinuous system disturbed by this kind of noise may be described by a deterministic equation with a continuous righthand part whose solution does exist in the usual sense. Making this equation undergo a limiting procedure with the noise variance tending to zero yields the unknown sliding equation.

This regularization technique which demands no introduction of a boundary layer and is called stochastic is applicable to control-nonlinear systems of the form (1.7), (1.8) in which the discontinuity points make the set

$$\mathscr{U} = \bigcup_{i=1}^{m} \{(x,t): s_i(x,t) = 0\} \tag{2.27}$$

The channels measuring and forming the s_i functions are assumed to be disturbed by additive noise:

$$u_i = \begin{cases} u_i^+(x,t) & \text{if} \quad s_i(x,t) + \eta_i(t) > 0, \\ u_i^-(x,t) & \text{if} \quad s_i(x,t) + \eta_i(t) < 0, \quad i = 1, \dots, m. \end{cases} \tag{2.28}$$

Correspondingly, the discontinuity points of the righthand part of the system (1.7), (2.28) make the set

$$\mathscr{U}_1 = \bigcup_{i=1}^{m} \{(x,t): s_i(x,t) + \eta_i(t) = 0\},$$

where $\eta(t) = (\eta_1(t), \eta_2(t), \dots, \eta_m(t))^{\mathrm{T}}$ is an m-dimensional random process.

From the application viewpoint, it will be interesting to consider the case when process $\eta(t)$ is stationary and features a sufficiently wide spectrum. Possible formalization in this case consists in studying the asymptotic properties of the solution of system (1.7), (2.28) with $\eta(t)$ being of the form

$$\eta(t) = \xi(t\varepsilon^{-1}),$$

if $\varepsilon \to 0$ and $\xi(t)$ is a stationary measurable random process satisfying the condition of a strong uniform mixing [75, 160] which is satisfied for a Gaussian process if, for instance, its covariance is

$$|K(\tau)| \leqq De^{-\lambda|\tau|}, \tag{2.29}$$

where D and λ are positive constants.

In application to the class of systems with a continuous righthand part such a choice of the family of processes $\eta = \eta(t)$ allows the use of asymptotic methods of the averaging theory [75, 160]. The same idea of averaging is also employed in the below approach to the analysis of discontinuous systems of the form (1.7), (2.28). The need to design an ad-hoc technique was dictated by the fact that the righthand part in (1.7), (2.28) is subjected to discontinuities while the available averaging-based techniques were designed under the assumption of the existence of the Lipshitz constant.

Generally speaking, all solution continuation problems hold for system (1.7) with control (2.28) when its solution hits the discontinuity set of the righthand part of the system. However the continuation method employed has no effect upon the behaviour of the solution with $\varepsilon \to 0$ provided the condition

$$\|\dot{x}(t)\| < K \tag{2.30}$$

is satisfied at any finite time interval and the constant K in this inequality is independent of characteristics of process $\eta(t)$. All available continuation techniques including ambiguous methods using equations in contingencies satisfy this condition.

Independence of the solution behaviour with $\varepsilon \to 0$ from the actually applied continuation method belonging to class (2.30) results from the assertion that the Lebegue measure tends to zero with probability 1 for those points of the time interval $[t_0, t_1]$ where $(x(t),t) \in \mathscr{U}_1$ and the system (1.7), (2.28) is not defined (the proof of this assertion is based on the fact that with $\varepsilon \to 0$, the norm of the variation rate of $\eta(t) = \xi(t\varepsilon^{-1})$ exceeds $\dot{s} = (\operatorname{grad} s)\dot{x}$ almost everywhere and the functions $s_i(x,t) + \eta_i(t)$ cannot be equal to zero on a nonzero measure set).

Since outside the set $\mathscr{U}_1$ the righthand parts of the system (1.7), (2.28) are identical to one of the functions $f(x,t,v)$ (each component v_i of vector $v^{\mathrm{T}} = (v_1, \dots, v_m)$ being equal to either u_j^+ or u_j^-), i.e. it is independent of the continuation method, the limit behaviour of the system should obviously be invariant to the continuation method when condition (2.30) is satisfied.

Let $U_k, k = 1, 2, \dots, 2^m$ be a complete set of 2^m quadrants of space $R^m \ni s$, and let $f_k(x,t) = f(x,t,v)$ if $s(x,t) \in U_k$ (i.e. $v_j = u_j^+$ with $s_i(x,t) > 0$ and $v_j = u_j^-$ with $s(x,t) < 0$). For Gaussian processes of the type (2.29), the following assertion is true.

Theorem. *Let $\xi(t)$ be a stationary measurable random process. Then for any $\delta > 0$ the following is true*:

$$\lim_{\varepsilon \to 0} P\left\{ \sup_{t \in [t_0, t_1]} |z(t) - x^{\varepsilon}(t)| \geqq \delta \right\} = 0, \tag{2.31}$$

where $x^{\varepsilon}(t), z(t)$ are the solutions to the system (1.7), (2.28) and system

$$\dot{z} = \sum_{k=1}^{2m} f_k(z,t)\varphi_k(s(z)), \quad z(t_0) = x_0, \tag{2.32}$$

and the functions

$$\varphi_k(y) = P\{y + \xi(t) \in U_k\}, \quad y \in \mathbb{R}^m, \quad k = 1, 2, \dots, 2^m \tag{2.33}$$

satisfy the Lipschitz condition.

Let us give a qualitative explanation of the meaning of this assertion. The function $\varphi_k(y)$ is the probability of the fact that the m-dimensional vector $y + \xi(t)$ proves to be in the k-th quadrant of the m-dimensional space. Since the random process $s_i(x,t) + \xi_i(t\varepsilon^{-1})$ with small values of ε changes very rapidly, then averaging of the velocity vectors $f_k(z,t)$ is possible with respect to all quardants taking the weights equal to the probabilities of appearance in the respective quadrant, i.e. equal to $\varphi_k(s(z))$. As a result of such averaging with $\varepsilon \to 0$, we obtain deterministic Eq. (2.33) with a continuous righthand part which describes the behaviour of the discontinuous system disturbed by high frequency noise in the channels for measuring and shaping the switching function. An accurate statement of the theorem and its rigorous proof are given in [151].

Note that functions φ_k depend upon the characteristics of the random process $\xi(t)$. When the variance of $\xi(t)$ tends to zero, $\varphi_k(t)$ tends to the indicator function of the quadrant U_k and, consequently, the righthand part of (2.33) coinsides with the righthand part of the system (1.7), (1.8) in the limit outside the set $\mathscr{U}$. Therefore, if in this case the solution of system (2.32) exists, then the equation resulting from tending the noise variance to zero may be naturally regarded as the solution of (1.7), (1.8) under small values of noise. If $\|s\| \to 0$ for all t's belonging to some time interval in vanishing the variance of $\xi(t)$, then, in the limit, the motion is a sliding mode. The righthand part of system (2.32) belongs to a convex hull of all vectors $f(x, t, v)$ where the components of vector v are either u_j^+ or u_j^-, since by virture of (2.33) $\varphi_k(y) \geqq 0$ and

$$\sum_{k=1}^{2m} \varphi_k = 1.$$

Consequently, the stochastic regularization method appears to be in good agreement with Filippov's continuation method (see Chap. 1, Sect. 2). In essence, it substantiates the validity of a sliding equation chosen from the set of such equations depending on the actual way of tending the variances of the components of vector ξ to zero. Since application of Filippov's continuation and the equivalent control methods to control-linear systems yields identical sliding equations (see Sect. 3 of this Chapter), stochastic regularization gives substantiation of validity of the equivalent control method for this class of systems as well.

As it will be established in the next chapter, regularization based on the use of the boundary layer allows one to obtain a class of sliding equations wider than that obtained by means of Filippov's continuation. However, if the major imperfection is known to be caused by a noise, then Filippov's continuation method guarantees all feasible sliding equations to be found.

Chapter 3

The Uniqueness Problems

1 Examples of Discontinuous Systems with Ambiguous Sliding Equations

A system nonlinear with respect to control (1.7), (1.8) also permits formal application of the equivalent control method procedure, yielding some differential Eq. (2.6) which is true along the intersection of its discontinuity surfaces and may be regarded as its sliding mode equation.

Before we try to substantiate the validity of the equation obtained in the described manner – via introduction of a boundary layer and a subsequent limiting procedure – let us compare this equation in a scalar control system against the equation resulting from Filippov's procedure. Shown in Fig. 8 for point (x, t) are velocity vectors f^0 and $f(x, t, u_{eq})$ obtained following these two procedures; the way they could be obtained have been discussed in Chap. 1, Sect. 2 (Fig. 6) and in Chap. 2, Sect. 2 (Fig. 7). Since the locus of $f(x, t, u)$ is a spatial curve, the two vectors – vector f^0 obtained by Filippov's technique and vector $f(x, t, u_{eq})$ found with the use of the equivalent control technique – are, generally speaking, not identical and moreover, not even colinear. These vectors are known to coincide only when the locus is a straight line (which is the case in a control-linear system).

Of low probability is the case when the locus is not a straight line but, accidentally, crosses the tangential plane in the same point as does the straight line connecting the ends of vectors f^+ and f^- (a dotted line in Fig. 8). Therefore it follows from the above reasoning that, disregarding such unlikely cases, a

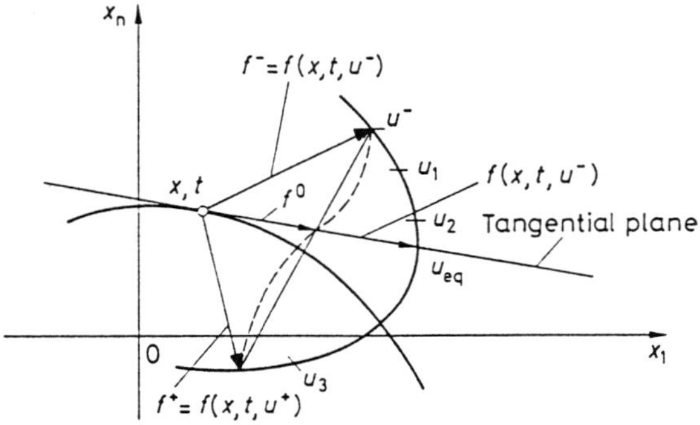

Fig. 8

system linear with respect to control u is actually the only situation when the equivalent control technique and Filippov's procedure yield the same result.

Trying to use the regularization method to find out which equation is the best for the description of the motion along discontinuity boundaries does not yield an unambiguous answer to this question either, for the answer depends on the concrete type of control $\tilde{u}$ used for regularization of the problem. Let us consider several examples illustrating the cases when such ambiguity is possible, and later use them to find the totality of motions feasible along discontinuity boundaries.

1.1 Systems with Scalar Control

Consider system (1.7), with the scalar control u subject to discontinuities on some surface $s(x)=0$. Assume that the switch performing a stepwise change of the control has a hysteresis. Then the function u in Eq. (1.7) will be of the form

$$u=\begin{cases} u^+ & \text{with} \quad s(x)>\Delta \\ u^- & \text{with} \quad s(x)<-\Delta, \quad u^+, u^-, \Delta - \text{const} \end{cases}$$

and in the region $|s(x)| \leqq \Delta$ the function u will maintain the value it had when $|s|$ was for the last time equal to Δ. We still assume that in an ideal system (with $\Delta=0$), the sliding mode condition (1.9) holds which implies that in the vicinity of the discontinuity surface $s(x)=0$ the trajectories are directed toward each other. The presence of a hysteresis is responsible for the fact that once it hits the discontinuity surface, the state vector does not move exactly along this surface but, rather oscillates in its vicinity of width 2Δ (Fig. 9).

To determine the average velocity associated with that motion, find first the shifts of the state, Δx, over two neighboring intervals, one with $u=u^-$ and

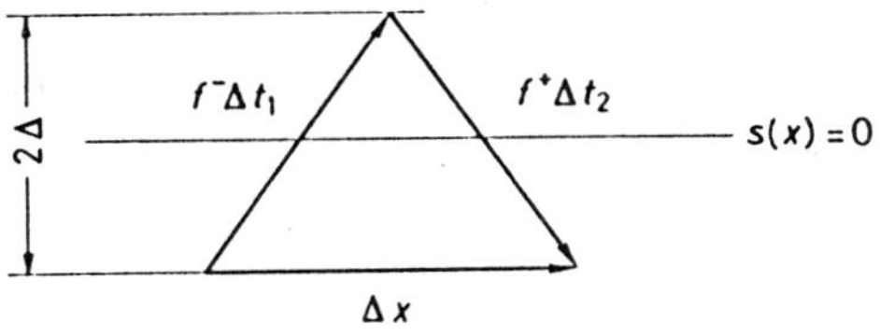

Fig. 9

the other with $u = u^+$:

$$\Delta x = f^- \Delta t_1 + f^+ \Delta t_2, \tag{3.1}$$

where Δt_1 and Δt_2 are the durations of the intervals found from the relations

$$\Delta t_1 = \frac{2\Delta}{\operatorname{grad} s \cdot f^-}, \quad \Delta t_2 = -\frac{2\Delta}{\operatorname{grad} s \cdot f^+}. \tag{3.2}$$

Equations (3.1) and (3.2) determine the average velocity[1]

$$f^0 = \frac{\operatorname{grad} s \cdot f^-}{\operatorname{grad} s \cdot (f^- - f^+)} f^+ - \frac{\operatorname{grad} s \cdot f^+}{\operatorname{grad} s \cdot (f^- - f^+)} f^- \tag{3.3}$$

for the motion equation $\dot{x} = f^0(x, t)$ in the sliding mode. Let us show now that Filippov's procdure yields the same result. The velocity vector, f^0 (Fig. 6), belongs to a straight line which connects the ends of vectors f^+ and f^-, i.e.

$$\dot{x} = f^0(x, t),$$
$$f^0 = \mu f^+ + (1 - \mu) f^-, \quad (0 \leqq \mu \leqq 1)$$

where μ is a parameter depending on the mutual position and magnitudes of the column vectors f^+ and f^- and the row vector grad s, the gradient of function $s(x)$. Finding μ from the condition $\operatorname{grad} s \cdot f^0 = 0$, we obtain the equation

$$\dot{x} = \frac{\operatorname{grad} s \cdot f^-}{\operatorname{grad} s \cdot (f^- - f^+)} f^+ + \frac{\operatorname{grad} s \cdot f^+}{\operatorname{grad} s \cdot (f^- - f^+)} f^-, \tag{3.4}$$

describing the sliding motion, the righthand part of the equation being identical to (3.3).

Consider now a different limiting procedure and show that it leads to the equations of the equivalent control method. This way of regularization implies substitution of the discontinuous control function with a continuous three-zone characteristic with a linear medium zone. Outside this zone the discontinuous and continuous functions are identical. In the limit, the width of the medium zone is zero and the slope factor is infinite.

[1] This discussion is of a qualitative nature. Strictly speaking, in order to obtain this result we should assume that the functions f, u^+ and u^- are continuous and differentiable and apply a standard technique introducing infinitesimals and tending Δ to zero.

By our assumption, a sliding mode exists on the discontinuity surface $s = 0$, i.e. in compliance with (1.9) the trajectories of the state vector x for $u = u^+$ and $u = u^-$ are directed toward this surface. Consequently, after attaining the boundary layer determined by the width of the linear zone in the above discontinuous characteristic approximation, the state vector will remain in that zone. The control u may then be replaced with a linear function ks subsequently tending the coefficient k to infinity. In essence, this limiting process was suggested in [103, 136] as a way to obtain the sliding equation by replacing the discontinuous control with a linear one featuring an unlimited gain.

The authors of monographs [103, 136] have considered equations linear with respect to all x, rather than just with respect to u. In this case it is easy to obtain the necessary equations substituting ks for u and tending k to infinity. This technique is hardly useful in obtaining an equation of the general form. However if the derivatives of the function u_{eq} with respect to all of its arguments are bounded the result of such a limiting procedure may be obtained as follows. Like before, let us present $\dot{s}$ as

$$\dot{s} = \operatorname{grad} s \cdot f(x, t, u) \tag{3.5}$$

and recall that $\dot{s} = 0$ with $u = u_{eq}$, i.e.

$$\operatorname{grad} s \cdot f(x, t, u_{eq}) = 0.$$

Rewrite Eq. (3.5) as

$$\dot{s} = -\theta(u - u_{eq}). \tag{3.6}$$

Regarding the function

$$\theta(x, t, u) = \frac{\operatorname{grad} s \cdot f(x, t, u)}{u - u_{eq}}$$

it is assumed that $|\theta(x, t, u)| \geqq \theta_0 > 0$, $\theta_0 = \text{const.}$[1] Let us take $u^+ > u^-$. Then the very fact that the sliding mode does exist (according to which the trajectories run toward each other in the vicinity of the surface $s(x) = 0$, i.e. $\dot{s}(u^+) < 0$ and $\dot{s}(u^-) > 0$) leads us to a conclusion that $\dot{s}$ decreases when u changes from u^- to u^+ turning to zero when $u = u_{eq}$. Besides according to (3.6) the function $\theta(x, t, u)$ is positive and $\theta(x, t, u) \geqq \theta_0$.

In compliance with the suggested line of reasoning, the equation of motion in the vicinity of the surface $s(x) = 0$ may be found if in Eq. (3.6) the control u will be replaced with ks, or, which is the same, s will be replaced with u/k:

$$\tau\dot{u} + \theta u = \theta u_{eq}, \quad \text{where} \quad \tau = 1/k. \tag{3.7}$$

When k tends to infinity, or τ to zero, the solution $u(t)$ of Eq. (3.7) for a finite time interval $[0, T]$ tends to u_{eq} in the following sense: for any pair of

---

[1] This condition excludes merely the case when the locus either touches the tangential plane or comes as close to it as desired outside the vicinity of the point of their intersection.

positive numbers $\Delta t < T$ and ε there exists such δ that the inequality $|u - u_{eq}| \leqq \varepsilon$ holds for any $t \in (\Delta t, T)$, if $0 < \tau \leqq \delta$.

To prove this assertion, consider a solution to equation $\tau \dot{u} + \theta u = \theta u_{eq}$ regarding θ as a time function:

$$u = u(0)e^{-1/\tau \int_0^t \theta(\gamma)d\gamma} + \frac{1}{\tau} e^{-1/\tau \int_0^t \theta(\gamma)d\gamma} \int_0^t \theta(\gamma) u_{eq} e^{-1/\tau \int_0^\gamma \theta(\lambda)d\lambda} d\gamma$$

$$= u(0)e^{-1/\tau \int_0^t \theta(\gamma)d\gamma} + e^{-1/\tau \int_0^t \theta(\gamma)d\gamma} \int_0^t u_{eq} de^{-1/\tau \int_0^\gamma \theta(\lambda)d\lambda}$$

After integrating the second term by parts, we obtain

$$u = u(0)e^{-1/\tau \int_0^t \theta(\gamma)d\gamma} + u_{eq} - u_{eq}(0)e^{-1/\tau \int_0^t \theta(\gamma)d\gamma} - e^{-1/\tau \int_0^t \theta(\gamma)d\gamma} \int_0^t \frac{du_{eq}}{d\gamma} e^{-1/\tau \int_0^\gamma \theta(\lambda)d\lambda} d\gamma.$$

By virtue of the assumptions made on all the functions used in the original system, the magnitude $\left|\frac{du_{eq}}{d\gamma}\right|$ is bounded within any finite time interval $[0, T]$ by some number M, and θ is lower bounded by a positive number θ_0. Therefore

$$|u - u_{eq}| \leqq |u(0) - u_{eq}(0)| e^{-1/\tau \int_0^t \theta(\gamma)d\gamma} + \frac{M\tau}{\theta_0}.$$

As a result we obtain that for an arbitrarily small Δt

$$\lim_{\tau \to 0} u = u_{eq} \quad \text{with} \quad \Delta t \leqq t \leqq T.$$

This shows that the substitution of ks for u and tending k to infinity is actually equivalent to replacing control u with u_{eq}, the results being identical to those of the equivalent control method but differing from Filippov's technique.

Let us consider a numerical example in which different sliding equations are obtained. For the discontinuous system

$$\dot{x}_1 = 0,3x_2 + ux_1, \quad \dot{x}_2 = -0,7x_1 + 4u^3 x_1,$$

$$u = \begin{cases} +1 & \text{with} \quad sx_1 < 0 \\ -1 & \text{with} \quad sx_1 > 0, \quad s = x_1 + x_2 \end{cases}$$

the conditions of the sliding mode occurrence (1.9) hold in any point of the discontinuity straight line (i.e. with $x_1 = x_2$) while the equivalent control method Eq. (2.5) has the form

$$\dot{s} = (-1 + u_{eq} + 4u_{eq}^3)x_1 = 0.$$

Find the equivalent control $u_{eq} = 0.5$ and substitute it into the first equation of the system (bearing in mind that $x_2 = -x_1$ in a sliding mode): $\dot{x}_1 = 0.2x_1$. According to (3.4), the application of the Filippov technique to the system under consideration yields the equation $\dot{x}_1 = -0.1x_1$ pertaining to the motion along the straight line $s = 0$.

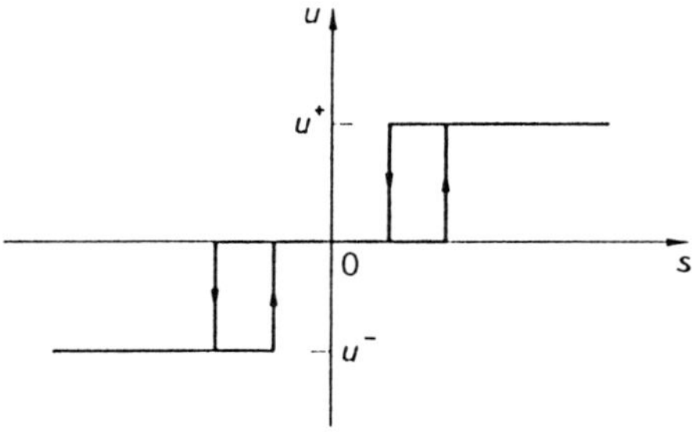

Fig. 10

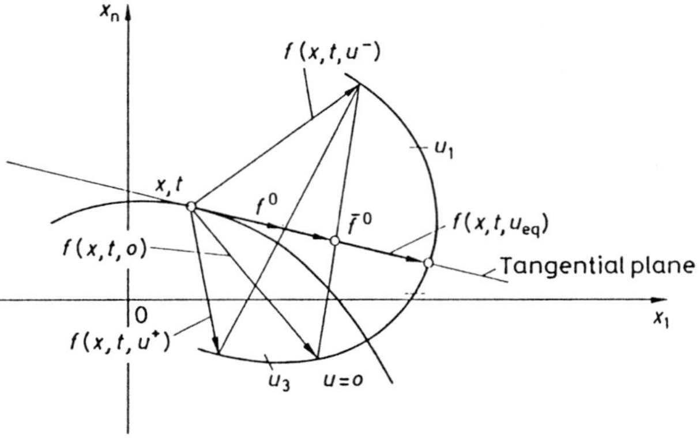

Fig. 11

The comparison of the two last equations shows that a continuous approximation of the discontinuity (in which case the equivalent control method works) yields an unstable sliding motion, while the motion along the same line in case of switching with a hysteresis (when Filippov's continuation method is applicable) turns out to be stable[1].

Let us consider again a system with a scalar discontinuous control and show that, besides the above two types of motion, some other motions may occur on the discontinuity boundary depending on the way the function $\tilde{u}$ is implemented. Let the switch of this system feature a dead zone besides a hysteresis (Fig. 10). Assume for certainty that for $u=0$ (i.e. in the dead zone), the projection of vector $f(x,t,0)$ on the gradient to the discontinuity surface is negative, i.e.

$$\dot{s} = \operatorname{grad} s \cdot f(x,t,0) < 0$$

as shown in Fig. 11. In such a case a sliding mode occurs in which the control alternately takes up the values 0 and u^-. Consequently, the velocity vector $\bar{f}^0$

[1] This effect was observed experimentally on an analog computer for very thin boundary layers.

in the state space may be found with the help of Filippov's procedure yielding the point of intersection of the tangential plane with the straight line which connects the ends of vectors $f(x, t, u^-)$ and $f(x, t, 0)$ (Fig. 11).

If the control in the dead zone differs from zero, one may obtain some other velocity vectors whose ends lie on the intersection of the tangential plane and a straight line connecting any two points of the $f(x, t, u)$ locus. For second order systems, the totality of such velocities lie on the segment of the line running between the ends of vectors f^0 and $f(x, t, u_{eq})$ (which refers to the locus presented in Fig. 11, of course). For a system of an arbitrary order, this totality covers some area on the tangential plane.

Thus, depending on the regularization technique used in treating a system nonlinear with respect to scalar control we may obtain different sliding motion equations. This proves that, generally speaking, the equations describing motion in nonlinear systems outside the discontinuity surface do not allow one to unambiguously obtain the equations of the sliding motion over that surface.

1.2 Systems Nonlinear with Respect to Vector-Valued Control

Like systems with a scalar control, general-type systems (1.7), (1.8) with vector-valued controls and a number of discontinuity surfaces feature certain ambiguity of motion in sliding mode. The specific form of the sliding equation is determined not only by the form of function $\tilde{u}$ (as is true of the scalar case), but also by the way the boundary layer tends to zero.

The feasibility of such ambiguity may be explained using an example of the dynamic system described by the equations[1]

$$\dot{x}_1 = u_1, \quad \dot{x}_2 = u_2, \quad \dot{x}_3 = u_1 u_2,$$

where

$$u_1 = \begin{cases} +1 & \text{with} \quad x_1 < 0 \\ -1 & \text{with} \quad x_1 > 0 \end{cases} \quad \text{and} \quad u_2 = \begin{cases} +1 & \text{with} \quad x_2 < 0 \\ -1 & \text{with} \quad x_2 > 0. \end{cases}$$

Assume that both relay elements implementing controls u_1 and u_2 have identical hysteresis loops. With regard for the hysteresis, in moving in the boundary layer of the discontinuity surface $x_1 = 0$ and $x_2 = 0$ the controls, u_1 and u_2 are actually periodic functions with a zero average value, their product being also a periodic function with the average value possibly differing from zero. This average value may take up any value of A ranging from -1 to $+1$, depending on the phase correlation between u_1 and u_2 at the time the sliding motion occurs. If we now tend both hysterisis loops to zero the switch-off

---

[1] This example was authored by D. Izosimov.

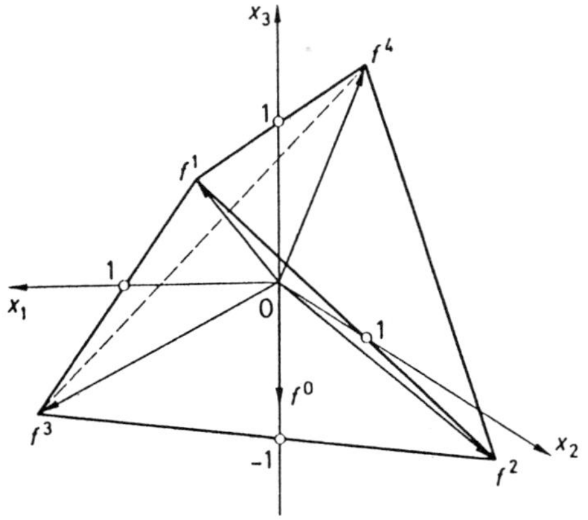

Fig. 12

frequency will tend to infinity and the limit motion will depend only on the average control values, i.e. will be described by the equations

$$\dot{x}_1 = 0, \quad \dot{x}_2 = 0, \quad \dot{x}_3 = A.$$

By virtue of the uncertainty of A, the sliding mode is bound to be uncertain too.

All possible velocity vectors in the vicinity of the origin lying at the intersection of the discontinuity surfaces $x_1 = 0$ and $x_2 = 0$ are shown in Fig. 12:

$$f^1 = \begin{pmatrix} 1 \\ 1 \\ 1 \end{pmatrix}, \quad f^2 = \begin{pmatrix} -1 \\ 1 \\ -1 \end{pmatrix}, \quad f^3 = \begin{pmatrix} 1 \\ -1 \\ -1 \end{pmatrix}, \quad f^4 = \begin{pmatrix} -1 \\ -1 \\ 1 \end{pmatrix}.$$

The sliding mode velocity vector f^0 is directed along the intersection of the discontinuity surfaces (or along axis x_3) and may take up any position between -1 and 1.

The above example of vector-valued discontinuous control shows ambiguity even in the simplest case of a single imperfection in controls $\tilde{u}_1$ and $\tilde{u}_2$ (a hysteresis); a real-life realization of motion in sliding mode depends on the correlation of instants the sliding motion starts over each of the discontinuity surfaces.

1.3 Example of Ambiguity in a System Linear with Respect to Control

Consider a dynamic system with the two-dimensional control

$$\dot{x} = f(x,t) + b^1(x,t)u_1 + b^2(x,t)u_2, \quad f, b^1, b^2 \in \mathbb{R}^n$$

and assume each component of this control is subjected to discontinuities over the same surface $s(x) = 0$. Outside the discontinuity surface, the velocity vectors are defined unambiguously:

$$f^+ = f + b^1 u_1^+ + b^2 u_2^+, \quad f^- = f + b^1 u_1^- + b^2 u_2^-$$

At first sight, the sliding mode equations may be found in this control-linear case with the use of Filippov's procedure. The end of the velocity vector f^0 in the sliding mode lies in the point of intersection of the tangential plane and the straight line connecting the ends of vectors f^+ and f^- (Fig. 13). However, as evidenced by the analysis of the system behaviour in the boundary layer which appears, for example, when elements with hysteresis loop of width Δ_1 and Δ_2 are used to implement controls $\tilde{u}_1$ and $\tilde{u}_2$, sliding equations cannot be obtained unambigously. Next, it is assumed that

$$u_1^+ = u_2^+ = u^+, \quad u_1^- = u_2^- = u^-.$$

If we find that $\Delta_1 < \Delta_2$ and, in the initial instant of time, $\|s\| \leqq \Delta_1$ and $\tilde{u}_2 = u_2^-$, then control $\tilde{u}_2$ will never switch over due to the fact that (see Fig. 13) vectors f^- for $s < 0$ and $f_1^+ = f + b^1 u^+ + b^2 u^-$ for $s > 0$ are directed toward the discontinuity surface and the state vector will oscillate in the Δ_1-neighbourhood of this surface. Consequently, the sliding motion velocity vector will equal f_1^0 (in this sliding mode, only control u_1 is switched over while in a system with the scalar control implemented by a hysteresis element the Filippov procedure is applicable).

If we still have $\Delta_1 < \Delta_2$ and in the initial instant of time, $\|s\| \leqq \Delta_1$ again but $u_2 = u_2^+$, then the velocity vectors will equal, in turn, f^+ and $f_1^- = f + b^1 u_1^- + b^2 u_2^+$. As a result we shall obtain yet another sliding motion with a velocity of f_2^0.

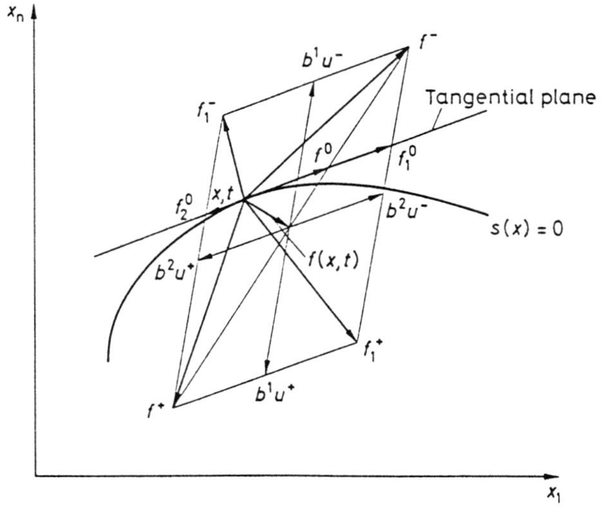

Fig. 13

If, finally, $\Delta_2 < \Delta_1$, then the motion will take place in the boundary layer of width Δ_1 yielding a switch in control $\tilde{u}_2$ with its average value staying between u_2^+ and u_2^-. As a result, the velocity vector in the sliding mode will stay between the ends of vectors f_1^0 and f_2^0 in the tangential plane. This means that the sliding motion equation in this controllinear case depends on the way this control is realized in the boundary layer.

In contrast to Filippov's procedure, the formal application of the equivalent control method to this problem does not unambiguously yield a sliding equation. Both control components in this case are, indeed, discontinuous over the same surface and, consequently, rank $G = 1$ and $\det GB = 0$ (let us recall that G is the matrix whose rows are the gradients of functions $s_i(x)$ and B is the matrix whose columns are b^1 and b^2), hence the equivalent control method Eq. (2.8) has an infinite number of solutions.

In all of the above examples, problem regularization resting upon the introduction of a boundary layer does not allow one to obtain the sliding mode equations without specifying the type of vector $\tilde{u}$ used. This sets forward a problem of finding the set of feasible sliding equations and its subsets if the way control is realized in the boundary layer is known.

2 Minimal Convex Sets

One important regularity should be noticed in the examples of Sect. 1: in a certain point on the intersection of the discontinuity surfaces, the sliding motion velocity vector always belongs to the minimal convex set which contains all velocity vectors of the system motion in the boundary layer in the neighbourhood of that point[1].

If in the first example (Fig. 11) $\tilde{u}$ in the boundary layer is allowed to take up any value between u^+ and u^-, then such a minimal convex set should include a spatial curve, locus of $f(x,t,u)$ for $u^- \leqslant u \leqslant u^+$. The intersection of this set with the plane tanential to the discontinuity surface determines the totality of righthand parts of the sliding mode differential equations. In particular, a minimal convex set for a second order system will be a segment confined between the locus and the straight line that passes through the ends of vectors $f(x,t,u^+)$ and $f(x,t,u^-)$. The section of the straight line connecting the ends of vectors f^0 and $f(x,t,u_{eq})$ determines the totality of velocities in sliding mode. If, however, the control $\tilde{u}$ in the boundary layer may take up either of the two values, u^+ or u^-, then our procedure will evidently be equivalent to Filippov's procedure.

[1] This assertion is qualitative in its nature. A more rigorous statement should read as follows: a minimal convex set includes all velocity vectors of a system if x runs a δ-neighbourhood of the point under consideration, $\delta > P\Delta$ (i.e. in compliance with (2.3) and (2.12), δ-neighbourhood containing points which are outside the boundary layer) with δ and Δ tending to zero.

In the second example, the controls u_1 and u_2 in the boundary layer are allowed to take up only their extreme values $+1$ and -1, making the righthand part of the differential equation be equal to either of the four vectors, f^1, f^2, f^3 or f^4, respectively. The minimal convex set containing these four vectors will be a tetrahedron (Fig. 12) which actually cuts off a segment $[-1, 1]$ on the intersection of the discontinuity surfaces or axis x_3 that includes the entire totality of the sliding motion velocity vectors.

And finally, the totality of the sliding motion velocity vectors in the third example (Fig. 13), a straight line segment between the ends of vectors f_1^0 and f_2^0, does also belong to the minimal convex set (a parallelogram) which contains all feasible velocity vectors f^+, f^-, f_1^+ and f_1^- in the boundary layer of the discontinuity surfaces intersection.

The above displays one of the most interesting properties of systems nonlinear with respect to discontinuous control, the capability to produce motions which never occur with any continuous control. The essence of this property may be explained with the help of the well known theorem stating that the vector

$$f^0 = \frac{1}{b-a}\int_a^b f(t)dt$$

($a, b = \text{const}$, f^0 and $f(t)$ being n-dimensional vectors) belongs to a minimal convex set containing the locus of vector $f(t)$ if t runs through all its values from a to b.

Treating discontinuous systems, let us refer to that part of our reasoning only which substantiates the peculiarity of system behaviour on discontinuity boundaries exposed in the examples of Sect. 1. Assuming that function $f(x(t), t, u(t))$ in (1.7) is integrable in t we may write this equation in the integral form as

$$x(t+\Delta t) = x(t) + \int_t^{t+\Delta t} f(x(\tau), \tau, u(\tau))d\tau = x(t) + \int_t^{t+\Delta t} (f(x(t), t, u(\tau)) + O(\Delta t))d\tau.$$

Then the velocity vector in point x, t will be

$$\dot{x} = \lim_{\Delta t \to 0} \frac{x(t+\Delta t) - x(t)}{\Delta t} = \lim_{\Delta t \to 0} \frac{1}{\Delta t} \int_t^{t+\Delta t} (f(x, t, u(\tau)) + O(\Delta t))d\tau,$$

or, by virtue of the property established for vector f^0,

$$\dot{x} \in F_0(x, t),$$

where $F_0(x, t)$ is the minimal convex set containing the set F which consists of all the $f(x, t, u)$ vectors provided u takes up all values within a given area. Rigorous proofs pertaining to the conditions of existence of the limit and to finding the sequence of solutions to Eq. (1.7) tending to any of solutions to equation $\dot{x} = F_0(x, t)$ are furnished to Refs. [53, 58, 162].

The above reasonings lead to the following conclusion. If, in some boundary layer of the intersection of discontinuity surfaces, the control components take up the values belonging to a certain set, rather than just u_i^+ and u_i^-, then the sliding motion velocity vector should be sought on the convex hull of velocities determined by the totality of the control values. This technique was used in [3] to obtain sliding equations in contingencies under an assumption that in the boundary layer the control components may take up any values between u_i^+ and u_i^-.

Thus, if the problem of mathematical description of a discontinuous dynamic system is regularized by means of introducing a boundary layer within which the totality of the control values is known to be, generally speaking, different from those outside this layer, then one may only assert that the sliding mode velocity vector for some point on the discontinuity surfaces intersection does belong to a certain set. In the neighbourhood of that point the minimal convex set of vectors $f(x,t,u)$ defined by all feasible control values in the boundary layer is actually the set of feasible sliding mode state vector velocities[1]. Such a convex set may be constructed on a set of discrete values of vector $f(x,t,u)$ as was the case in the first example of Sect. 1 when the discontinuous control was implemented by an element with one or two hysteresis loops.

In the cases treated in [3] the control components in the boundary layer could take up any values between u_i^+ and u_i^-; consequently the minimal convex set should be constructed on a totality of vectors $f(x,t,u)$ of all control values belonging to a bounded full-dimensional domain in the m-dimensional space. Sometimes even this set constructed for all the values $u_i^- \leqq u_i \leqq u_i^+$ may turn insufficient. For example, if the ideal model fails to recognize the "fast dynamics" of the device which realizes discontinuous control and if this dynamics is oscillatory by nature, then in the sliding mode some values of the control vector components may lie outside the u_i^- and u_i^+ range. In this case, the convex hull should be constructed for a more extended set of controls which leads to a wider range of feasible sliding mode motions than in the case $u_i^- \leqq u_i \leqq u_i^+$.

After the minimal convex set is constructed, one should find a set of feasible motions in the sliding mode. The sliding mode equation is found unambiguously for a certain point on the intersection of the discontinuity surfaces if just one of the vectors from the above minimal convex set lies on the intersection of the tangential planes. This is the case with control-linear systems (2.7) when Eq.(2.8) solved with respect to the equivalent control (2.9) yields just a single solution. When there is more than one common vector for the minimal convex set and the discontinuity surface intersection, then a more detailed analysis of the system behaviour in the boundary layer is needed to find out which sliding equation

[1] Note again the qualitative nature of our reasonings (see the footnote in the beginning of this section when the first mention was made of the minimal convex set of velocity vectors in the boundary layer).

from the subset obtained is actually realized. This conclusion is verified by the examples given in Sect. 1.[1]

If, finally, the minimal convex set has no velocity vectors directed along the discontinuity surfaces intersection, then the state vector will leave the boundary layer and there will be no sliding motion occurring on the manifold considered.

In conclusion, let us compare the above technique of finding the totality of sliding equations with Filippov's continuation method (see Chap. 1, Sect. 2). The models treated by Filippov had no control vector; instead, velocity vectors were given directly. Such models contained no information pertaining to the dynamic system behaviour in the vicinity of the intersection of discontinuity surfaces, therefore the use of the minimal convex set of the velocity vectors specified outside of the discontinuity surfaces could prove insufficient for finding all feasible sliding mode motions.

The availability of a control in the motion equation and the knowledge of the totality of its values feasible in the boundary layer provide the additional information used in constructing an extended minimal convex set which depends on the regularization method, and in finding realizable sliding equations. It is obvious that the Filippov technique and the method resting upon the introduction of the boundary layer applied to a scalar control system yield identical results if the control in the boundary layer takes up only two values, u_i^- and u_i^+ (for example when the switch features a delay or hysteresis).

3 Ambiguity in Systems Linear with Respect to Control

A case of ambiguity of the sliding equation in control-linear system (2.7) has been considered in Sect. 1 of this chapter. The reason for ambiguity was in the fact that det $GB = 0$; singular cases of this type have no unambiguous solution to the equivalent control method Eq. (2.8).

Using the linear time-invariant system

$$\dot{x} = Ax + Bu, \quad x \in \mathbb{R}^n, \quad u \in \mathbb{R}^m, \quad A, B = \text{const} \tag{3.8}$$

with linear discontinuity surfaces

$$s = 0, \quad s = Cx \in \mathbb{R}^m, \quad C = \text{const} \tag{3.9}$$

[1] It should be noted that for models of the system behaviour in the boundary layer more complex than discussed in Sect. 1, the procedure of finding the sliding equation cannot be reduced to the Filippov or the equivalent control method procedures. If for instance a discontinuous control characteristic is approximated by a continuous one and small inertialities are recognized in the system equation, then complex nonlinear oscillations will occur in the boundary layer and it will be hard to average them. It is only in the case when the domain where the real control differing from the ideal one is tended to zero that we can obtain the equation resulting from Filippov's procedure [56].

as an example, consider some standard situtions which may occur if matrix CB in the equation

$$\dot{s} = CAx + CBu_{eq} = 0 \tag{3.10}$$

is singular.

Let us start with the case when there is no solution to Eq. (3.10) for a point x residing on the manifold (3.9) at the time instant t, i.e.

$$s = Cx(t) = 0. \tag{3.11}$$

$$\operatorname{rank}(CB) < \operatorname{rank}(CB, CAx). \tag{3.12}$$

Let us show that such a boundary layer $\|s\| \leqq \Delta$ may always be found that the state vector trajectory starting in point $x(t)$ will go beyond its limits. Recall that, in compliance with our discontinuity problem regularization technique (Chap. 2), control u in the boundary layer is replaced with function $\tilde{u}$ for which a solution of Eq. (3.8) exists. After integrating the relationship $\dot{s} = CAx + CB\tilde{u}$ pertaining to motion in the boundary layer under the initial condition (3.11), we have

$$s(t + \Delta t) = \int_t^{t\Delta t} CAx(\tau)dt + CB \int_t^{t+\Delta t} \tilde{u}(\tau)dt.$$

Assuming the function $\tilde{u}(\tau)$ is bounded and, consequently, the components of vector $x(\tau)$ within the interval $[t, t + \Delta t]$ are bounded too, we may find the norm of the vector

$$\|s(t + \Delta t)\| = \|CBu_{av} + CAx(t) + O(\Delta t)\| \Delta t, \tag{3.13}$$

where $u_{av} = \int_t^{t+\Delta t} \tilde{u}(\tau)d\tau$ and $O(\Delta t)$ is a first order infinitesimal. The condition (3.12) means that the columns of matrix CB and vector CAx are linearly independent and therefore, with Δt small enough, we have

$$\min_{u_{av}} \|C\,Bu_{av} + CAx(t) + O(\Delta t)\| \Delta t \geqq \varkappa\Delta t, \tag{3.14}$$

where $\varkappa$ is a positive value. Comparing (3.13) with (3.14) we see that a finite increment is gained by $\|s\|$ during time Δt. Consequently, a sliding motion understood as a limit motion in the boundary layer $\|s\| \leqq \Delta$ with Δ tending to zero is impossible if an equivalent control does not exist. Such a case is typical for systems with linearly dependent columns of matrix B. In such systems the dimension of the space where the ends of the velocity vectors $Ax + Bu$ reside for some point x is less than m while the dimension of the manifold of the discontinuity surfaces intersection (3.9) equals $n - m$ (provided matrix C has a maximum rank). Generally speaking, these spaces do not intersect, thus the control cannot "direct" the velocity vector along the intersection of the discontinuity surfaces.

Assume now that rank $C < m$, i.e. that the discontinuity surfaces equations are linearly dependent, rank $B = m$ and rank $(CB) = \text{rank}\, C$. In this case the columns of matrix CB, like column CAx, are actually the linear combinations of the columns of matrix C. Since rank $(CB) = \text{rank}\, C$, the addition of column CAx to matrix CB does not augment its rank which means that rank. $(CB) = \text{rank}\ (CB, CAx)$ leading to an infinite number of solutions to the equivalent control method Eq. (3.9). To obtain the totality of sliding motion equations, (m-rank C) components of the control vector should be arbitrarily chosen so that the remaining components could be unambiguously defined from the equation $\dot{s} = 0$, and thus found solutions should be substituted into the initial system (3.8). Since the columns of matrix B are linearly independent, the above procedure cannot vanish the arbitrarily chosen (m-rank C) components. It is the totality of these components' values that determines the set of feasible sliding motion equations.

Consider, finally, the case when the sliding motion equation is defined unambiguously in spite of singularity of Eq. (3.10). If in this case rank $C =$ rank $B < m$, then Eq. (3.10) appears to have an infinite number of solutions. Assume that if C' and B' are the matrices made of base rows and columns of matrices C and B, respectively, then det $C'B' \neq 0$.

Let matrices Λ_C and Λ_B consist of the expansion coefficients of the remaining rows and columns of matrices C and B in terms of base rows and columns. Under the assumptions made, Eqs. (3.8) and (3.10) have the form

$$\dot{x} = Ax + B'(u' + \Lambda_B u''), \tag{3.15}$$

$$\dot{s} = \begin{pmatrix} C'A \\ \Lambda_C C'A \end{pmatrix} x + \begin{pmatrix} C'B' \\ \Lambda_C C'B' \end{pmatrix} (u' + \Lambda_B u'') = 0, \tag{3.16}$$

where $u^{\mathrm{T}} = [(u')^{\mathrm{T}}, (u'')^{\mathrm{T}}]$. The equivalent control Eq. (3.16) has obviously a single solution with respect to vector $u' + \Lambda_B u''$ (but not a single one with respect to all components of vector u); therefore, a substitution of this solution into (3.15) yields an unambiguously defined sliding mode equation. All the reasonings of Sect. 3, Chap. 2 substantiating the equivalent control method in nonsingular cases are applicable to prove the validity of the above sliding equation as well as the equations for systems characterized with ambiguous sliding mode motions.

Thus, in case of singularity of the equation $\dot{s} = 0$ used to find the equivalent control, the system equation specified outside the discontinuity surfaces will give either ambiguous or unambiguous sliding motion equation along these surfaces. Moreover, singularity may result in the fact that the sliding motion will be impossible at all.

Chapter 4

Stability of Sliding Modes

1 Problem Statement, Definitions, Necessary Conditions for Stability

Major attention in the previous chapters was paid to methods of describing the motion along a discontinuity surface or an intersection of such surfaces. The equations of sliding mode, or the set of such equations in ambiguous cases, found in those chapters only indicate the possibility for this type of motion to exist. Let us now proceed directly to finding the coditions for further motion to be a sliding mode, should the initial state be on the intersection of discontinuity surfaces.

From the viewpoint of geometrical representation of motion, we are interested in the trajectories in the neighbourhood of the intersection of discontinuity surfaces such that in a small deviation from this intersection the state vector would always come back to the intersection. This exactly is the statement of the problem of asymptotic stability of motion in nonlinear dynamic systems. Hence, the considered problem is actually that of the sliding mode stability.

The regularization based upon the introduction of a boundary layer has allowed a mathematical description to be obtained for sliding mode motions. Likewise, the problem of their stability should be considered with due regard for the specifics of motion in the boundary layer rather than in the framework of an idealized model.

The stability of sliding modes will be considered in application to the control-linear system (2.7). Here, again, the sliding mode is assumed to be a limit motion in system (2.11) with the boundary layer (2.3) tending to zero in which the ideal control (1.8) is substituted with a function $\tilde{u}$ allowing regularization of the problem. Two more assumptions should be made for control $\tilde{u}$:

$$\tilde{u}_i = u_i \quad \text{if} \quad |s_i| \geqq \Delta_0 \tag{4.1}$$

Δ_0 being a positive value, and

$$\min(u_i^-, u_i^+) \leqq \tilde{u}_i \leqq \max(u_i^-, u_i^+), \quad \text{if} \quad |s_i| \leqq \Delta_0. \tag{4.2}$$

Both of these assumptions refer to the type of function $\tilde{u}$ and mean that outside the boundary layer $\tilde{u}$ coincides with the ideal control while inside that layer it does not exceed the extreme values taken by the ideal control.

In terms of a system with a boundary layer, the well-known Lyapunov's definition of stability [98] may be reverbalized for our specific problem of the sliding mode stability in system (2.7) with control (1.8).

Definition. In the state space of the dynamic system (2.7), the domain $S(t)$ belonging to the manifold of the intersection of discontinuity surfaces (2.1) is a sliding domain, or a domain comprising stable sliding mode trajectories, if
(1) this domain does not contain the trajectories of any of 2^m continuous systems adjacent to the manifold, and
(2) for any positive ε such positive numbers Δ_0 and δ may be found that any motion of system (2.7) with control (4.1), (4.2) starting in an n-dimensional δ-neighbourhood of domain S may leave the n-dimensional ε-neighbourhood of this domain only in the ε-neighbourhood of its boundaries.

The above definition isolates a generally time-varying domain of the coordinate system values such that for the projection of overall motion on the m-dimensional subspace $s_1, \ldots, s_m$ its origin is a stable equilibrium point "in the small". The first condition of the definition prohibits the state vector trajectories to lie in the manifold of the intersection of discontinuity boundaries, if each control component takes up either of the two possible values, u_i^+ or u_i^-.

Proceeding from this definition let us show the necessary stability conditions of a sliding mode. In order the domain $S(t)$ to be a sliding domain, it is necessary that the following conditions be satisfied: first, the equivalent control equation (2.8) should have at least a single solution and, second, each component of this solution should satisfy the inequality

$$\min(u_i^-, u_i^+) \leqq u_{i\text{eq}} \leqq \max(u_i^-, u_i^+). \tag{4.3}$$

The first of these conditions has been proved in Chap. 3, Sect. 3 where various singular cases with $\det GB = 0$ were treated. The second condition may be substantiated in about the same way. If, however, condition (4.3) is violated, then by virtue of (4.2) the value of $\dot{s}$ in a motion in the boundary layer is always different from zero. Using relationships of the type (3.13) and (3.14), it is easy to show that the state vector is bound to leave some finite neighbourhood of the manifold $s = 0$ which contradicts the sliding mode stability definition.

2 An Analog of Lyapunov's Theorem to Determine the Sliding Mode Domain

We have just found that the problem of finding the sliding domain is reducible to the specific problem of stability of a nonlinear system. As is usually done in the theory of stability, let us try to find conditions sufficient for a certain domain S on the intersection of discontinuity surfaces in the state space of the system (2.7), (1.8) to be a sliding domain in the sense of the definition given in Sect. 1. Such an effort is necessary either due to discontinuity of the righthand part of the idealized system (2.7), (1.8) or due to uncertainty of control $\tilde{u}$ in system (2.11) as it moves inside the boundary layer (4.2). Another specific feature is that stability is actually needed for the projection of the motion on the subspace $s_1, \ldots, s_m$ rather than for the overall motion. Let us now formulate as a theorem the sufficient conditions recognizing these features and allowing the sliding domain to be found.

Theorem. *For the domain $S(t)$ on the intersection of discontinuity surfaces of the system (2.7), (1.8) to be a sliding domain, it is sufficient that for all x belonging to this domain there exists, in a certain region Ω of the subspace $s_1, \ldots, s_m$ containing the origin, a function $v(s, x, t)$ continuously differentiable with respect to all its arguments such that:*

(1) *$v(s, x, t)$ is positive definite with respect to s, i.e. $v(s, x, t) > 0$ with $s \neq 0$ and with arbitrary x and t; $v(0, x, t) \equiv 0$; on the sphere $\| s \| = R$ for all x from the region considered and for any t the following relations hold:*

$$\inf_{\|s\|=R} v = h_R, \quad \sup_{\|s\|=R} v = H_R, \quad R \neq 0, \quad \lim_{R \to 0} H_R = 0, \tag{4.4}$$

where h_R and H_R are some positive quantities depending only on R;

(2) *the total time derivative of the function v, by virtue of the system (2.7), (1.8),*

$$\dot{v} = \operatorname{grad}_s v \cdot (Gf + GBu) + \operatorname{grad}_x v(f + Bu) + \frac{\partial v}{\partial t};$$

$$\operatorname{grad}_s v = \left(\frac{\partial v}{\partial s_1}, \ldots, \frac{\partial v}{\partial s_m}\right), \quad \operatorname{grad}_x v = \left(\frac{\partial v}{\partial x_1}, \ldots, \frac{\partial v}{\partial x_n}\right)$$

is negative everywhere except the discontinuity surfaces where this function is not defined, and on the entire surface of the sphere $\| s \| = R$ except for the points *of discontinuity, the relationship*

$$\sup_{\|s\|=R} \dot{v} = -m_R, \quad m_R > 0, \quad \sup_{R\in[a,b]} m_R < 0 \tag{4.5}$$

is true for any quantities $0 < a < b$ with $s \in \Omega$ if $\|s\| \leqq b$.

The upper bound for the function $\dot{v}$ is sought for all $\dot{x}$ from the domain and for any t while the quantity m_R itself is positive and depends only on R. If condition (4.5) is true for the function $\dot{v}$, then it will be referred to as *negative definite* in further discussion.

To prove the theorem we need a few relationships for finding the value of $\dot{v}$ which, for the idealized system (2.7), (1.8) can be represented as

$$\dot{v} = G_0(s, x, t) + \sum_{i=1}^{m} G_i(s, x, t) u_i, \tag{4.6}$$

where $G_0, G_1, \ldots, G_m$ are certain functions continuous with respect to all arguments and depending on f, B, G, $\text{grad}_s v$, $\text{grad}_x v$ and $\partial v/\partial t$.

Let us take an arbitrary point x_0 on one or several discontinuity surfaces, e.g. $s_i = 0 (i = 1, \ldots, k, k < m)$ and on the sphere $\|s\| = R$. Although the function $\dot{v}$ is not defined in this point, let us find it in a formal way selecting the values of $u_{k+1}, \ldots, u_m$ in compliance with (1.8) and substituting either u_i^+ or u_i^- for u_i $(i = 1, \ldots, k)$:

$$\dot{v}(x_0) = G_0(s, x_0, t) + \sum_{i=1}^{k} G_i(s, x_0, t) u_i^{\pm} + \sum_{i=k+1}^{m} G_i(s, x_0, t) u_i \tag{4.7}$$

where $u_i^{\pm}$ denotes either u_i^+ or u_i^-. Note that in (4.7) the components $s_1, \ldots, s_k$ of vector s are zero. Since all the functions G_i, u_i^+ and u_i^- are continuous and condition (4.5) holds for the function $\dot{v}(x_0)$, the estimate

$$\dot{v}(x_0) \leqq -m_R \tag{4.8}$$

is valid in any combination of the values $u_i (i = 1, \ldots, k)$.

Let us now consider the function $\dot{v}$ along the trajectories of system (2.11) which was used in definition of the sliding mode. If the trajectories of this system go outside the Δ_0-neighbourhood of discontinuity surfaces, then by virtue of (4.1) the values of $\dot{v}$ for an idealized system and a system with a boundary layer are the same. It is evident that for all points of the sphere $\|s\| = R$, except the Δ_0-neighbourhoods of discontinuity surfaces, the function $\dot{v}$ for a system with nonidealities does not exceed the values $-m_R$ by virtue of condition 2 of the Theorem. Inside the Δ_0-neighbourhood of the intersection of surfaces $s_i = 0$ $(i = 1, \ldots, k)$, the function $\dot{v}$ for the system (2.11), (4.1), (4.2) can be presented in the form

$$\dot{v}(x) = \dot{v}_0 + O(\Delta_0),$$

$$\dot{v}_0 = G_0(s, x_0, t) + \sum_{i=1}^{k} G_i(s, x_0, t) \tilde{u}_i + \sum_{i=k+1}^{m} G_i(s, x_0, t) u_i, \tag{4.9}$$

where x_0 is the point on the intersection of surfaces $s_i = 0$ $(i = 1, \dots k)$ nearest to the one in question (the distance between the points x and x_0 is obviously a value of about Δ_0), $O(\Delta_0)$ is a value infinitesimal relative to Δ_0.

The value $\dot{v}(x_0)$ is estimated by the inequality (4.8) and its maximal value in the point x_0 is attained if, depending on the signs of functions G_i, the values of u_i^+ or u_i^- for the components of control $\tilde{u}_i(i = 1, \dots, k)$ are properly chosen. The value of function $\dot{v}_0$ maximal in $\tilde{u}_i(i = 1, \dots, k)$ coincides with the value of function $\dot{v}(x_0)$ maximal in u_i $(i = 1, \dots, k)$ because $\tilde{u}_i$ is somewhere between u_i^+ and u_i^- and is linearly represented in function $\dot{v}(x_0)$. Consequently, $\dot{v}_0$ is also less than $-m_R$ and, by virtue of (4.9) for a fixed R one can always select a Δ_0 such that on a sphere of radius R the function, $\dot{v}(x)$, will be nonpositive.

Let us use this property of a nonideal system to show that when the conditions of the theorem are observed the domain S on the intersection of discontinuity surfaces is a sliding domain. By the definition of this domain, we have to verify that for any ε-neighbourhood of the origin of the subspace $s_1, \dots, s_m$ there are Δ_0- and ε-neighbourhoods such that any motion of the system (2.11), (4.1), (4.2) starting in the δ-neighbourhood will stay inside the ε-neighbourhood.

Without loss of generality, a part of the ε-neighbourhood can be played by a sphere $\| s \| \leqq \varepsilon$ in the domain Ω where the conditions of the theorem hold. From the first condition it follows that the minimal value of function v on the sphere $\| \rho \| = \varepsilon$ is h_ε. Select now a sphere of radius δ such that the maximal value of function v inside the sphere is h_ε (this can be done because, by the conditions of the theorem, function v is zero with $s = 0$ and its upper estimate H_R is continuous in the point $s = 0$ and upper-bounded on any sphere). As has been shown above, $\dot{v}$ may be made nonpositive on the surface of any sphere of a finite radius, by appropriate choice of Δ_0 for the system with nonidealities. Evidently, Δ_0 may be selected so that $\dot{v}$ will be nonpositive between the sphere $\| s \| = \delta$ and $\| s \| = \varepsilon$.

Let us now consider an arbitrary motion with this value of Δ_0 starting in the domain $\| s \| \leqq \delta$. As soon as the state leaves this domain the function v will turn to be nonincreasing or it will not exceed h_ε. But the value of the function v on the sphere $\| s \| = \varepsilon$ is not less than h_ε. Consequently, for all x the state trajectories starting in the δ-neighbourhood of the origin of the subspace $s_1, \dots, s_m$ cannot leave the ε-neighbourhood. The theorem is thus proved.

Using the equations of system (2.7) with ideal control (1.8), these sufficient conditions make it possible to find such a domain on the intersection of discontinuity surfaces where the sliding modes are stable. An important point is that the system (2.11), (4.1), (4.2) taking into account the motion in the boundary layer is not needed here. This was a system in terms of which a definition was given to the sliding motion domain and which could not be unambiguously defined in the boundary layer.

Now a few brief remarks on the conditions of the theorem and their verification. When the function v is time invariant and the domain S is bounded and closed, the condition (4.4) holds everywhere. This follows from the fact that any continuous function on a closed set always reaches its upper and lower bounds and, since the function v vanishes only with $s=0$, the lower bound is positive on a sphere of nonzero radius. The condition (4.4) holds everywhere if the function v depends only on s. As for the condition (4.5) for the function $\dot{v}$, then even if this function were negative and dependent only on s, the value of m_R on the sphere $\|s\|=R$ may still be equal to zero. The reason for this is that $\dot{v}$ is estimated in an open domain since the points on the sphere $\|s\|=R$ which belong to the discontinuity surfaces were excluded from consideration. We should note here that if the function $\dot{v}$ is negative definite, then $\lim \dot{v} \neq 0$ with any combination of s_i tending to zero, however not all at a time.

The theorem given above refers to the conditions of the sliding mode stability 'in the small". Let us verbalize and prove similar conditions referring to the sliding mode stability "in the large" assuming the sliding domain $S(t)$ to be identical to the entire manifold $s=0$. Stability "in the large" implies in this case that the conditions of stability "in the small" hold for the manifold $s=0$ and $\lim_{t\to\infty} s=0$ for an arbitrary initial system state. The following theorem allows the conditions of the Barbashin-Krasovsky theorem [9] to be applied to studying the discontinuous systems stability "in the large".

Theorem. *To obtain a stable sliding mode motion along the intersection of discontinuity surfaces $s=0$, it is sufficient that for all x and t such a continuously differentiable function $v(s,x,t)$ exists that conditions 1 and 2 of the theorem of stability "in the small" hold and, moreover,*

$$\lim_{R\to\infty} h_R=\infty. \tag{4.10}$$

It is necessary to prove that, following a certain time instant, a trajectory starting in the initial point $x(t_0)$, $v(s(t_0),x(t_0),t_0)$ will fully lie in an ε-vicinity of the origin of space s for any ε. We shall be interested only in those cases when $h_\varepsilon < v_0$, otherwise $\|s\| \leqq \varepsilon$ in the initial time instant and by virtue of the preceding theorem, the trajectory will stay within this vicinity.

The surface $v(s,x,t)=v_0$ is bounded in space s with all x and t. (If this was not so, we could find a sequence of points belonging to this surface for which $\|s\|$ tends to infinity, which contradicts the conditions (4.10)). Consequently, such an R_0 exists that the surface $v=v_0$ in space s lies fully inside the sphere $\|s\|=R_0$.

As it was shown in proving the theorem on the sliding domain existence conditions, such Δ_0 may be found for which the function $\dot{v}$ will be nonpositive between the spheres $\|s\| \leqq \delta$ and $\|s\| \leqq \varepsilon$. In a similar way one may find a Δ_0 such that the function $\dot{v}$ will be strictly negative within the spherical layer $\delta \leqq \|s\| \leqq R_0$ and

$$\sup_{\delta \leqq \|s\| \leqq R_0} \dot{v} = -m_0 \tag{4.11}$$

within this bounded closed domain, where m_0 is a positive value. Function v

decreases in the spherical layer $\delta \leqq \| s \| \leqq R_0$ at a finite rate, but at the some time $v \geqq 0$, therefore the trajectory of the system should leave this domain. Since $\dot{v} < 0$ and, consequently, $v < v_0$ with $t > t_0$, the state vector cannot leave the sphere $\| s \| \leqq R_0$. As a result, at some time the state will reach the sphere $\| s \| \leqq \delta$. According to the theorem on the sliding domain existence conditions, the value of δ may be chosen such that further motion of the system will take place in the ε-vicinity of the origin of space s. The theorem is proved.

We have just considered general conditions for functions, similar to Lyapunov's function used in finding the sliding mode stability conditions. An essential point is that an idealized system was used to formulate these conditions, although the very fact of stability was defined with the use of a regularized system. Note that the same applies to the mathematical description of sliding modes: an idealized system permits the sliding equations to be obtained using the equivalent control method, while verification of the method itself requires introduction of a boundary layer and analysis of the system motion within this layer.

It is most obvious that the above method, like the Lyapunov functions method, gives no concrete recipe for obtaining function v for a discontinuous system. Therefore the next sections will be devoted to those cases when such a function satisfying the theorem conditions may be found.

3 Piecewise Smooth Lyapunov Functions

Pay attention to the fact that both theorems on the sliding mode stability conditions have required the Lyapunov functions to be continuously differentiable. Consider two examples showing that this requirement is essential in treating the existence of a sliding mode. In both examples the function v is piecewise smooth and positive definite with respect to s. The complete time derivative of this function is negative everywhere except the discontinuity surfaces, but, at the same time, a sliding mode occurs on the intersection of discontinuity boundaries in the first case and fails to occur in the second one.

In the first example a system with a two-dimensional control is considered whose motion projection on the surface s_1, s_2 is described by the equations

$$\dot{s}_1 = -\operatorname{sign} s_1 + 2 \operatorname{sign} s_2,$$

$$\dot{s}_2 = -2 \operatorname{sign} s_1 - \operatorname{sign} s_2,$$

$$\operatorname{sign} s_i = \begin{cases} +1 & \text{if} \quad s_i > 0 \\ -1 & \text{if} \quad s_i < 0, \quad i = 1, 2. \end{cases}$$

Let us take a positive definite function v in the following form: $v = |s_1| + |s_2|$. Its derivative with respect to both arguments becomes discontinuous when any

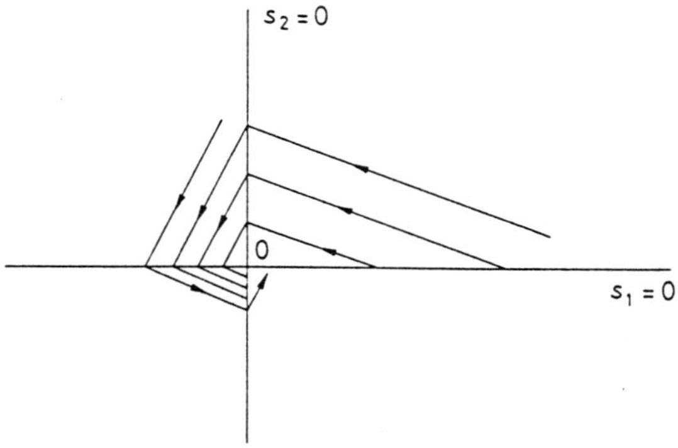

Fig. 14

of these arguments vanishes. The value $\dot{v}$ obtained from the motion equations

$$\dot{v}=\frac{\partial v}{\partial s_1}\dot{s}_1+\frac{\partial v}{\partial s_2}\dot{s}_2=\operatorname{sign} s_1(-\operatorname{sign} s_1+2\operatorname{sign} s_2)$$

$$+\operatorname{sign} s_2(-2\operatorname{sign} s_1-\operatorname{sign} s_2)=-2.$$

is negative everywhere except for the discontinuity surfaces where this function is not defined. Hence, the functions v and $\dot{v}$ have different signs and, as it follows from consideration of the s_1, s_2 plane (Fig. 14), the system is stable indeed, i.e. a sliding mode does occur on the intersection of the discontinuity surfaces.

To be able to correlate these two facts, let us give consideration to the second example dealing with a system with two-dimensional discontinuous control whose projection on the s_1, s_2 plane may be described by the equations

$$\dot{s}_1=-2\operatorname{sign} s_1-\operatorname{sign} s_2,$$

$$\dot{s}_2=-2\operatorname{sign} s_1+\operatorname{sign} s_2.$$

Let us take the positive definite function $v=4|s_1|+|s_2|$. Its derivative

$$\dot{v}=\frac{\partial v}{\partial s_1}\dot{s}_1+\frac{\partial v}{\partial s_2}\dot{s}_2=-7-6\operatorname{sign} s_1 s_2$$

is negative everywhere except for the discontinuity surfaces. But it is evident from direct study of the s_1, s_2 plane of the system (Fig. 15) that despite the difference in signs of functions v and $\dot{v}$ the state vector coming from any initial position reaches the plane $s_1=0$ on which a stable sliding mode occurs. The velocity vector of a sliding motion along the surface $s_1=0$ may be found in the following way. According to the equivalent control method, sign s_1 is found from the equation $\dot{s}_1=0$ and substituted into the second equation. This results in a sliding equation $\dot{s}_2=2\operatorname{sign} s_2$ which shows that the magnitude of s_2 always increases. This implies that no sliding mode occurs at the intersection of

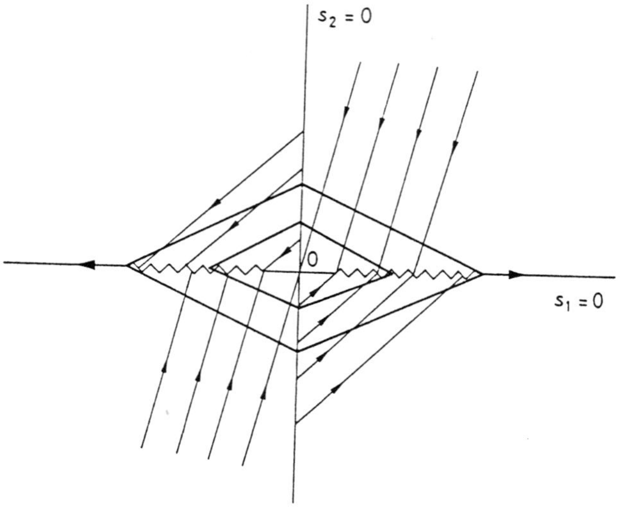

Fig. 15

discontinuity surfaces $s_1 = 0$ and $s_2 = 0$. In our example the trajectories intersect the equivalued surfaces (where $v = \text{const}$) from the outside everywhere except the "angular points" while the trajectories of a sliding motion diverge "through angular points". In the first example there was no sliding motion along one of the discontinuity surfaces; in other words, the totality of points on the trajectories which were on the "angles" of the equivalued surfaces $v = \text{const}$ were of zero measure. It is for this reason that the difference in signs between functions v and $\dot{v}$ led to the conclusion that the origin of coordinates on the s_1, s_2 plane was stable. However, as evidenced by the second example, the knowledge of signs of the piecewise smooth function and its derivative, generally speaking, is not sufficient to ascertain that a sliding mode does exist.

To be able to use the Lyapunov function of the "sum of absolute values" type whenever a sliding mode occurs on some part of the discontinuity surfaces one should replace discontinuous control with equivalent ones and only then find the sign of the function $\dot{v}$. With this peculiarity in mind, let us first give consideration to the problem of finding the sliding domain with the help of the piecewise smooth Lyapunov function for a particular case of the equation describring the motion projection on subspace s

$$\dot{s} = -D(x, t)\operatorname{sign} s, \tag{4.12}$$

where $s \in \mathbb{R}^m$, vector $(\operatorname{sign} s)^{\mathrm{T}} = (\operatorname{sign} s_1, \ldots, \operatorname{sign} s_m)$ and $D(x, t)$ is, generally speaking, a time-verying $(m \times m)$ state vector depending matrix.

Let us find out the class of matrices $D(x, t)$ for which the function

$$v = (\operatorname{sign} s)^{\mathrm{T}} s \tag{4.13}$$

is the Lyapunov function. Assume $s^{\mathrm{T}} = ((s^k)^{\mathrm{T}}, (s^{m-k})^{\mathrm{T}})$ and the sliding motion occurs on the intersection of discontinuity surfaces $s^k = 0$; assume also that the components of vector s^{m-k} are not zero or that the set of points where they vanish is a set of a zero measure and thus may be disregarded.

According to the equivalent control method, vector $\operatorname{sign} s^k$ in the motion Eq. (4.12) must be substituted with such function $(\operatorname{sign} s^k)_{\mathrm{eq}}$ that $\dot{s}^k = 0$. Since in the sliding mode $s^k = 0$, the time derivative of function v (4.13) consists of $(m-k)$ terms:

$$\dot{v} = \frac{d}{dt}((\operatorname{sign} s^{m-k})^{\mathrm{T}} s^{m-k}) = (\operatorname{sign} s^{m-k})^{\mathrm{T}} \dot{s}^{m-k}. \tag{4.14}$$

Add the righthand part of the relation (4.14) with the value $(\operatorname{sign} s^k)^{\mathrm{T}}_{\mathrm{eq}} \dot{s}^k$ which in a sliding mode equals zero:

$$\dot{v} = (\operatorname{sign} s^k)^{\mathrm{T}}_{\mathrm{eq}} \dot{s}^k_k + (\operatorname{sign} s^{m-k})^{\mathrm{T}} \dot{s}^{m-k}.$$

Replacing vector $\dot{s}$ with its value obtained from the motion Eq. (4.12) where vector $\operatorname{sign} s^k$ is substituted by $(\operatorname{sign} s^k)_{\mathrm{eq}}$, we have

$$\dot{v} = -z^{\mathrm{T}} D z, \tag{4.15}$$

where

$$z^{\mathrm{T}} = ((\operatorname{sign} s^k)^{\mathrm{T}}_{\mathrm{eq}}, (\operatorname{sign} s^{m-k})^{\mathrm{T}}).$$

If the symmetric matrix

$$D + D^{\mathrm{T}} > 0 \tag{4.16}$$

then the quadratic form (4.15) is negative definite on any sphere $\| s \| = R$, including the discontinuity points. As it was done in the first theorem of Sect. 2, we may show that the domain $S(t)$ on the intersection of discontinuity boundaries $s = 0$ is a sliding domain, if condition (4.16) holds and

$$\inf_{S(t)} \det(D(x,t) + D^{\mathrm{T}}(x,t)) > 0. \tag{4.17}$$

Finally, if inequality (4.17) is true for any x and t the entire manifold $s = 0$ becomes a sliding domain by virtue of the second theorem and the sliding modes in this domain are stable "in the large".

For a system of the general form (2.7) the equation describing the projection of motion on the subspace s

$$\dot{s} = Gf + GBu \tag{4.18}$$

may be rewritten as

$$\dot{s} = GB(u - u_{\mathrm{eq}}), \tag{4.19}$$

where the equivalent control u_{eq} is found from (2.9).

Let us shown that if, with an arbitrary x and t, the matrix $D = -GB$ satisfies conditions (4.16) and (4.17) and if $\| GB \| \leqq M$ (M being a positive number), a control of the type (1.8) may always be found for which stable "in the large" sliding modes exist throughout the entire manifold $s=0$. Sufficient for finding this control is only the knowledge of the upper-estimate function for the equivalent control components

$$F(x,t) > |u_{ieq}(x,t)|.$$

The time derivative of the Lyapunov function (4.13) on the trajectories of Eq. (4.19) may be written for the control

$$u = \alpha F(x,t)\,\mathrm{sign}\, s \quad (\alpha\text{-positive number}) \tag{4.20}$$

in the form similar to (4.15):

$$\dot{v} = \alpha F(x,t) z^{\mathrm{T}} GB\left(z - \frac{1}{\alpha}\mu(x,t)\right), \tag{4.21}$$

where z is defined in a way identical to (4.15) and vector $\mu(x,t) = u_{\mathrm{eq}}(x,t)/F(x,t)$,

$$\|\mu(x,t)\| \leqq \sqrt{m}. \tag{4.22}$$

Denote as λ_0 the value

$$\lambda_0 = \inf_{x,t} \lambda(x,t) > 0, \tag{4.23}$$

where $\lambda(x,t)$ is the minimal eigenvalue of the matrix $-(GB+(GB)^{\mathrm{T}})$. Since we are interested to know the sign of function $\dot{v}$ outside the manifold $s=0$ or outside the origin of coordinates of the subspace $s_1,\dots,s_m$, then at least one of the components of vecot s will differ from zero and, consequently, one of the components of vector z will equal $+1$ or -1, therefore

$$\|z\| \geqq 1. \tag{4.24}$$

With due regard for the relations (4.22) to (4.24), the following estimate is true of function $\dot{v}$ in (4.21):

$$\dot{v} \leqq -\alpha F(x,t)\left(\lambda_0 \|z\|^2 - \frac{1}{\alpha}\|z\|\cdot\|GB\|\cdot\|\mu\|\right)$$

$$\leqq -\alpha F(x,t)\|z\|\left(\lambda_0 - \frac{1}{\alpha} M\sqrt{m}\right). \tag{4.25}$$

For $\alpha > M\sqrt{m}/\lambda_0$, the derivative of the Lyapunov function is evidently negative definite and control (4.20) provides the sliding modes stable "in the large" throughout the entire manifold of the coscontinuity boundaries intersection.

The above shows that the use of piecewise smooth Lyapunov functions in conjunction with the equivalent control method permits one to obtain the sufficient conditions of stability of sliding modes in the general form systems as well, provided the sum of the matrix preceding the control in the equation written with respect to vector s and the transposed matrix is negative definite.

4 Quadratic Forms Method

Consider a way of using positive definite quadratic forms with constant coefficients

$$v = \tfrac{1}{2} s^{\mathrm{T}} W s, \quad W = W^{\mathrm{T}}, \quad W > 0, \quad W = \text{const} \tag{4.26}$$

to find the sliding domains on the intersection of discontinuity boundaries in system (2.7) featuring control (1.8). As was shown in Sect. 2, this problem is equivalent to that of stability of the equilibrium position $s = 0$ for the solutions of Eq. (4.18) which describes the motion projection on subspace s. Unlike piecewise smooth functions, quadratic forms with x- and t-independent coefficients satisfy the conditions of the first theorem in Sect. 2.

First consider the case when the equation of the motion projection on subspace s has the form (4.12). Find the time derivative of function (4.26) along the solution to system (4.12) in all points except for the discontinuity surfaces:

$$\dot{v} = -s^{\mathrm{T}} L \operatorname{sign} s, \tag{4.27}$$

where $L = WD$. Matrix L is both x- and t-dependent. Should we suceed in finding a range of values for the elements of L for which the function $\dot{v}$ is negative definite in the sense of the second condition of the theorem, then the set of values of vector x associated with this range and, as a result, the sliding domain will be found.

Let us now show that if $D = \text{const}$ in (4.12), the inequalities

$$l_{kk} > \sum_{\substack{i=1 \\ i| \neq k}}^{m} |l_{ki}| \quad (k = 1, \ldots, m) \tag{4.28}$$

represent necessary and sufficient conditions for the function $\dot{v}$ with constant coefficients in L to be negative definite. (In (4.28) the terms l_{ki} indicates elements of the k-th row of matrix L).

The sufficiency of the conditions becomes evident if we present the function $\dot{v}$ as

$$\dot{v} = -\sum_{k=1}^{m} |s_k| \left(l_{kk} + \sum_{\substack{i=1 \\ i \neq k}}^{m} l_{ki} \operatorname{sign} s_k s_i \right).$$

When inequalities (4.28) hold, the expression taken in the parentheses is positive for each term of the sum and, consequently, the function $\dot{v}$ is negative. It upper bound over the entire sphere $\| s \| = R$ except the discontinuity surfaces equals

$$\sup_{\|s\| = R} \dot{v} = -m_R = -\| s \| \Delta l_r$$

where Δl_r is the minimal number of all

$$\Delta l_k = l_{kk} - \sum_{\substack{i=1\\ i\neq k}}^{m} |l_{ki}| \quad (k=1,\ldots,m)$$

and $\lim \dot{v} = -\|s\|\Delta l_r$, if $\lim|s_r| = \|s\|$, $\lim s_i = 0$ $(i=1,\ldots,m, i\neq r)$, $\operatorname{sign} s_i = -\operatorname{sign} s_r l_{ri} (i=1,\ldots,m, i=r)$. (Recall that the upper bound is unattainable in this case since the function $\dot{v}$ is not defined on discontinuity surfaces $s_i=0$ and we therefore deal with an open domain on the sphere).

If we substitute the inequality sign in any of the inequalities (4.28) either with the opposite sign or with equality, then the value of $\Delta l_r'$ will be either negative or zero. The upper bound of the function $\dot{v}$ will consequently be nonnegative which proves the necessary feature of conditions (4.28) for the function to be negative definite. If the inequalities (4.28) do hold, the entire manifold $s=0$ becomes the sliding domain and the sliding modes become stable "in the large".

If matrix D in (4.12) depends both on the state vector and time, matrix L is also a function of both x and t in time derivative of v. Therefore the conditions which make the function $\dot{v}$ negative definite isolate a domain S, generally speaking, timevarying, of such values of x that the inequalities

$$\inf_{x\in S,\, t\in[0,\infty)} \left(l_{kk}(x,W,t) - \sum_{\substack{i=1\\ i\neq k}}^{m} |l_{ki}(x, Wt)| \right) > \Delta l_k > 0 \tag{4.29}$$

are true, where $\Delta l_k = \text{const}$, $k=1,\ldots,m$. For the domain S in the set of points of the sphere $\|s\|=R$ except the discontinuity surfaces, the upper bound of function $\dot{v}$ does not exceed the value

$$\sup_{\|s\|=R} \dot{v} = -\|s\|\Delta l_r, \tag{4.30}$$

where Δl_r is the minimal number of all Δl_k, which means that the second condition of the theorem in Sect. 2 is satisfied in the domain S. Consequently, this domain meeting the requirements of condition (4.29) is a sliding domain at the intersection of discontinuity surfaces.

Consider now the general type systems whose equation of motion with respect to vector s is of the type (4.18). Assume there exists such a symmetric positive definite matrix W for which matrix $L = -WGB$ and condition (4.29) holds inside a certain domain $S(t)$ on the manifold $s=0$ with any x and t. Let us show that in case $\|GB\| \leqq M$ (M being a positive number) a control may be found for which a sliding domain exists on the intersection of discontinuity boundaries. Take vector u in the form (4.20) assuming, like before, that the scalar function $F(x,t)$ is an upper estimate for any component of the equivalent control. Find the time derivative of function (4.26) on trajectories of system (4.18) or that of its equivalent system (4.19):

$$\dot{v} = \alpha F(x,t) s^{\mathrm{T}} L\left(\operatorname{sign} s - \frac{1}{\alpha}\mu(x,t)\right). \tag{4.31}$$

The upper bound of this function may be found from (4.22) and (4.30):

$$\dot{v} \leqq -\alpha F(x,t)\|s\|\left(\Delta l_r - \frac{1}{\alpha}\|W\|M\sqrt{m}\right).$$

It is evident that for $\alpha > \|W\|M\sqrt{m}/\Delta l_r$ the funtion $\dot{v}$ is negative definite and, therefore, $S(t)$ is a sliding domain. If $\Delta l_r > 0$ for any x and t, then the entire manifold $s=0$ presents a sliding domain and sliding modes therein are stable "in the large".

Let us consider an example of using this approach to find a sliding domain in an x- and u-linear system featuring a discontinuous three-dimensional control which is described by the equation $\dot{x} = Ax + Bu$. Here x is an n-dimensional vector, and A and B are constant $(n \times n)$- and $(n \times 3)$-dimensional matrices. The discontinuity surfaces in this case are planes, i.e. $s = Cx$, where C is a constant $(3 \times n)$-dimensional matrix and, consequently, the gradient matrix for this system is C.

Assume that matrix CB which equals GB in (4.19) and symmetrical and positive definite matrix W are of the form

$$CB = -\begin{pmatrix} 1 & 0.8 & 0.5 \\ 0.6 & 1.5 & 1 \\ 0.6 & 1 & 1 \end{pmatrix}, \quad W = \begin{pmatrix} 3 & -1 & -1 \\ -1 & 2 & -1 \\ -1 & -1 & 3 \end{pmatrix}$$

and condition (4.29) is true for the matrix

$$L = W{\cdot}CB = \begin{pmatrix} 1.8 & -0.1 & -0.5 \\ -0.4 & 1.2 & 0.5 \\ 0.2 & 0.7 & 1.5 \end{pmatrix}$$

Find coefficient Δl_r in the estimate (4.30):

$$\Delta l_r = \min(\Delta l_1, \Delta l_2, \Delta l_3) = (1.2, 0.3, 0.6) = 0.3.$$

If the control has the form (4.20), then (4.31) may be presented as

$$\dot{v} = -\alpha F(x,t)\left(s^{\mathrm{T}} L \operatorname{sign} s - \frac{1}{\alpha} s^{\mathrm{T}} L \frac{u_{\mathrm{eq}}}{F(x,t)}\right).$$

Since $u_{\mathrm{eq}} = -(CB)^{-1}CAx$ is a linear function, such a positive number l_0 may be found that the absolute value of each component of vector Lu_{eq} will not exceed the value $l_0|x|$, if $|x| = \sum_{i=1}^{n} |x_i|$.

If the function $F(x,t)$ is equal to $l_0|x|$ and if relationship (4.30) is taken into account, we obtain the following estimate of the derivative of Lyapunov's function

$$\dot{v} = -\alpha l_0 |x| \cdot \|s^{\mathrm{T}}\| \cdot \left(\Delta l_r - \frac{\sqrt{3}}{\alpha}\right).$$

The function $\dot{v}$ is negative definite when $\alpha > \sqrt{3}/0,3$. Consequently, for the piecewise linear control $u = \alpha l_0 |x| \operatorname{sign} s$ the entire manifold of the intersection of discontinuity surfaces may be made a sliding domain featuring sliding modes which are stable "in the large".

In the above example, we have succeeded in finding a symmetric positive definite matrix W such that all diagonal elements of matrix $L = WCB$ exceed the sum of absolute values of the elements contained in the respective rows. In the general case, however, it is still not clear for what class of matrices CB (or matrices GB in nonlinear systems) we can find a matrix W that answers the above property[1]. This actually limits applicability of quadratic forms so widely employed in the analysis of linear systems.

It will be useful to specify the particular cases when the problem of stability of sliding modes is solvable with the help of the Lyapunov function presented in the quadratic form. Let $D(x,t)$ in Eq. (4.12) (or matrix GB in Eq. (4.19)) be a Hadamard matrix or matrix with positive diagonal elements and a dominant diagonal, i.e. such that for its elements $d_{ki}(x,t)$ the inequalities

$$\inf |d_{kk}(x,t)| > \sum_{\substack{i=1 \\ i \neq m}}^{m} |d_{ki}(x,t)| \geqq 0, \quad d_{kk}(x,t) > 0 \tag{4.32}$$

are true with any x and t, or true in some domain $S(t)$. In this obvious case matrix W should be an identity matrix, and condition (4.29) is satisfied for matrix $L = D$ which is a sufficient condition of the sliding mode stability[2].

The problem of sliding mode stability can be solved if only the first of the inequalities (4.32) holds for matrix -GB (whose elements are, likewise, denoted as $d_{ki}(x,t)$) and if the diagonal elements of this matrix are of a constant sign, but not necessarily positive. The control (4.20) for such a system is replaced by $u = D_0 u^*$ where $u^* = \alpha F(x,t) \operatorname{sign} s$ and D_0 is a diagonal matrix with constant elements $d_k = -\operatorname{sign} d_{kk}(x,t)$. In this case Eq. (4.18) is written as $\dot{s} = GBD_0 u^* + Gf$ where the matrix GBD_0 already does satisfy both inequalities of (4.32).

Assume that matrix D in (4.12) is a constant, symmetric and positive definite one. Choose matrix W in the quadratic form to be D^{-1} (that is also symmetric and positive definite). Then $L = WD = I$, and for an identity matrix condition (4.29) is satisfied. Consequently, the entire manifold $s = 0$ in system (4.12) becomes a sliding domain, provided $D = \text{const}$, $D = D^{\mathrm{T}}$ and $D > 0$.

Recall that if we could successfully apply quadratic forms to solve the sliding mode stability problem for system (4.12) with some class of matrices D, then the problem would be solvable for system (4.18) whose matrix $-GB$ belongs to the same class and whose control is of the type (4.20).

---

[1] Moreover, in [143] the author gives an example of matrix CB for which no such matrix W can be found.

[2] Note that there is no need to apply the Lyapunov's function method in this case. Inequalities (4.32) mean that the conditions of existence of sliding modes (1.9) are satisfied for each discontinuity surface in system (4.12).

Notice the difference of the above two cases which stems from the fact that matrix W in quadratic form (4.26) is bound to be constant (this was the requirement permitting the use of inequalities (4.29) for the study of sliding mode stability). In the case of matrices D and GB with a dominant diagonal, matrix W was chosen constant which made it possible to determine the stability conditions both for constant and x- and t-dependent matrices D and GB. In the case of positive definite matrices D and $-GB$ this technique becomes applicable only when these matrices are constant, since $W=D^{-1}$.

Let us show now that, in application to discontinuous systems, the procedure of finding the sliding mode stability conditions "in the small" using quadratic forms does not depend on whether the matrices D or GB are constant or are a function of time and state of the system.

Let matrix $W(x,t)$ be x- and t-dependent and the Lyapunov function (4.26) with such a matrix satisfy the first condition of the theorem in Sect. 2. Assume the complete time derivative of matrix W does exist and is upper bounded in the norm by some number M. Find function $\dot{v}$ on the trajectories of system (4.12) for this case:

$$\dot{v} = -s^{\mathrm{T}}L(x,t)\,\mathrm{sign}\,s + s^{\mathrm{T}}\dot{W}s,$$

where matrix L is still equal to WD. If estimate (4.30) holds for the function $s^{\mathrm{T}}L\,\mathrm{sign}\,s$ with any x and t belonging to the above-specified domain, then

$$\dot{v} \leqq -\|s\|\Delta l_r + \|s\|^2 M.$$

Since the first of these summands is proportional to $\|s\|$ while the second, to $\|s\|^2$, a vicinity around the origin of coordinates $s=0$ may always be found such that the upper bound of function $\dot{v}$ in this vicinity is strictly negative. This means that, regardless of the rate of changing of matrix W, the second condition of the theorem in Sect. 2 holds and, therefore, the domain we consider is a sliding domain.

Thus, if the domain $S(t)$ is a sliding domain for the fixed values of matrix D belonging to a certain region and if this fact may be verified with the help of function v presented in the quadratic form with D-dependent coefficients, then the domain $S(t)$ will remain a sliding domain even under an arbitrary change of elements in D within the region of their values (provided $\|\dot{W}\|$ remains bounded). In essence, this conclusion provides grounds for application of the so-called "frozen coefficients technique" to find the sliding domain in discontinuous systems with the aid of quadratic forms.

5 Systems with a Vector-Valued Control Hierarchy

In this section we will consider systems with discontinuous control (2.7), (1.8) for which the problem of the sliding mode stability may be decoupled into a number of lower dimensional problems. One of such systems was treated in

Sect. 4 where the matrix $D(x, t)$ was assumed to have a dominant diagonal (4.32). Systems featuring this property always have a stable sliding mode that exists on each of their discontinuity surfaces.

Let us discuss a different approach to deriving a sliding domain; this approach is also reducible to a sequential analysis of individual problems of a lower dimension, but in each of these problems a sliding mode does not necessarily arise on the corresponding manifold of the intersection of discontinuity surfaces. Let vector s be presented as a set of subvectors $s^1, \dots, s^k$ $(k \leqq m)$: $s^T = ((s^1)^T, \dots, (s^k)^T)$. Correspondingly, the components of subvectors u^i in the control vector $u^T = ((u^1)^T, \dots, (u^k)^T)$ are subjected to discontinuities on the surfaces which are determined by the components of vectors s^i.

Let the vector-valued hierarchy of controls be

$$s^1 \to s^2 \to \cdots \to s^k \tag{4.33}$$

meaning that the sliding mode occurs on the manifold $s^i = 0$ only after it takes place on the manifold

$$s^1 = 0, \dots, s^{i-1} = 0. \tag{4.34}$$

Assume now that a stable sliding mode occurs in system (2.7) on the manifold (4.34). Then applying the equivalent control method to the equations

$$\dot{s}_1 = 0, \dots, \dot{s}^{i-1} = 0 \tag{4.35}$$

defined on the trajectories of system (2.7), we may find $u^1_{eq}, \dots, u^{i-1}_{eq}$ (let a solution to system (4.35) exist). After substituting these functions into (2.7) we obtain the equation describing the motion of the system on the manifold (4.34):

$$\dot{x} = f^i(x, t) + B_i(x, t)u^i + \bar{B}_i(x, t)\bar{u}^i \tag{4.36}$$

where $(\bar{u}^i)^T = ((u^{i+1})^T, \dots, (U^k)^T)$, and vector f^i and matrices B_i and $\bar{B}_i$ are determined by the initial equation's functions f and B, respectively, and by the matrix which consists of the gradients of functions $s^1, \dots, s^{i-1}$ $(f^1 = f,$ $(B_1, \bar{B}_1) = B$, and $\bar{u}^k = 0)$. Write the equation of the projection of motion in system (4.36) onto subspace s^i:

$$\dot{s}_i = G_i f^i + G_i B_i u^i + G_i \bar{B}_i \bar{u}_i,$$

where $G_i = \{\partial s^i / \partial x\}$. If $\det G_i B_i \neq 0$, then the sliding motion on the manifold $s_i = 0$ will be unambiguously defined by the equivalent control u^i_{eq} which is the solution of the equation $\dot{s}^i = 0$:

$$u^i_{eq} = -(G_i B_i)^{-1} G_i (f^i + \bar{B}_i \bar{u}^i). \tag{4.37}$$

Assume now that we have succeeded in finding the Lyapunov function for matrix $G_i B_i$ in the form of either a piecewise smooth function or a quadratic form. In this case, as was shown in Sects. 3 and 4, the stable modes of sliding on the manifold $s^i = 0$ may be provided by a control of the type

$$u^i = \alpha F_i(x, t) \operatorname{sign} s^i, \tag{4.38}$$

where $F_i(x,t)$ is an upper estimate for the components u_{eq}^i of vector u_{eq}^i:

$$F_i(x,t) > |u_{j\mathrm{eq}}^i|. \tag{4.39}$$

Consequently, if we succeed in finding a Lyapunov function corresponding to matrix G_iB_i for all x and t at every step $i = 1, \ldots, k$, then the existence of sliding modes stable "in the large" will be guaranteed throughout the intersection of discontinuity surfaces $s = 0$.

Notice that vectors u_{eq}^i depend on the remaining vectors $u^{i+1}, \ldots, u^k$ (4.37). Therefore, by virtue of (4.37) through (4.39), the magnitude of each component of vector u^i should exceed some function which additively includes the controls $u^{i+1}, \ldots, u^k$ that are somewhat subordinate in our hierarchy of discontinuity surfaces (4.33). Inequalities (4.39) for $i = 1, \ldots, k-1$ are exactly those conditions that establish a control hierarchy corresponding to (4.33).

The relationships obtained above allow the following sequence of operations to be proposed in order to provide sliding modes throughout the entire manifold $s = 0$. For vector s and matrix GB in Eq. (4.18), find a subvector s^1 and a minor associated with it for which the Lyapunov function may be written. Then write the equation of sliding along the manifold $s^1 = 0$ and the equation of the projection of this motion on the subspace $\bar{s}^1$ ($s^{\mathrm{T}} = ((s^1)^{\mathrm{T}}, (\bar{s}^1)^{\mathrm{T}})$). In this equation, subvector s^2 and the corresponding minor in the matrix that precedes the control are found in a similar manner. As a result, the hierarchy (4.33) is established in k steps. Vectors $u^1, \ldots, u^k$ are found in the reverse order. First we find the control u^k which, according to relationships (4.37) through (4.39), does not depend upon the remaining subvectors u^i; then, in compliance with the same relationships, we find u^{k-1} and so on down to u^1.

This approach suggests that a Lyapunov function may be found at each step. Since methods of obtaining such functions for an arbitrary matrix G_iB_i are not readily available, we shall consider one particular case when a control hierarchy can be established. If all $s^1, \ldots, s^k$ are scalar functions, then $k = m$ and $s^i = s_i$. The conditions of existence of a sliding mode on a single discontinuity surface have the form (1.9). But if we are talking about the sliding modes stability "in the large", then the inequalities

$$\dot{s}_i < 0 \quad \text{with} \quad s_i > 0, \quad \dot{s}_i > 0 \quad \text{with} \quad s_i < 0 \quad \text{and} \quad |\dot{s}_i| > m_i \tag{4.40}$$

must hold for any x and t, the value of m_i depending entirely on $|s_i|$. (It is quite obvious that the value $|s_i|$ decreases when conditions (4.40) are satisfied, and for an arbitrary $\varepsilon > 0$ such a time t_1 may be found that $|s_i(t)| \leqq \varepsilon$ with any $t \geqq t_1$).

Assume again that a stable sliding mode exists on the intersection of surfaces $s_i = 0, \ldots, s_{i-1} = 0$. When $\det GB \neq 0$ for this kind of motion, the coefficient preceding at least one of the remaining $(m - i + 1)$ control components of the first order equation with respect to s_i which describes this motion differs from zero. Then, assigning this component number i and choosing the magnitudes of the function $u_i^+(x,t)$ and $u_i^-(x,t)$ large enough, we can always guarantee, by virtue of (4.40), the occurrence of a sliding mode on the surface $s_i = 0$ and, consequently, the occurrence of such mode on the intersection of the surfaces $s_1, \ldots, s_i = 0$.

As we can see, there is no need to look for the Lyapunov function in case of a scalar control hierarchy; hence the procedure of choosing the controls that guarantee stable sliding motions along the manifold $s = 0$ may always be realized.

Let us refer again to the linear arbitrary order system with a three-dimensional control used as an example in Sect. 4 and use vector hierarchy method to find the conditions for sliding mode that occur on the intersection of the three discontinuity surfaces to be stable. The equation of the system motion projection on subspace s is of the form (see Sect. 4)

$$\dot{s} = CBu + CAx \tag{4.41}$$

where matrix

$$CB = -\begin{pmatrix} 1 & 0.8 & 0.5 \\ 0.6 & 1.5 & 1 \\ 0.6 & 1 & 1 \end{pmatrix}.$$

Corresponding to vector $(s^1)^{\mathrm{T}} = (s_2, s_3)$ is the minor

$$D = \begin{pmatrix} 1.5 & 1 \\ 1 & 1 \end{pmatrix}$$

which is a symmetric positive definite matrix; as was established in Sect. 4, this is exactly the case when the Lyapunov function may be determined. In compliance with the procedure suggested above, let the control hierarchy be as follows:

$$\begin{aligned} &s_1 \to s^2, && u^1 \to u^2, \\ &s^2 = s_1, && u^{\mathrm{T}} = ((u^1)^{\mathrm{T}}, u^2), \\ &(u^1)^{\mathrm{T}} = (u_2, u_3), && u^2 = u_1. \end{aligned}$$

Assume a sliding mode occurs on the manifold $s^1 = 0$. Find u^1_{eq} from the equation $\dot{s}^1 = 0$, and vector $\dot{s}^1$ from the two last equations in (4.41):

$$\dot{s}^1 = -Du^1 - \begin{pmatrix} 0.6 \\ 0.6 \end{pmatrix} u^2 + \begin{pmatrix} r^1 \\ r^2 \end{pmatrix} x, \tag{4.42}$$

$$u^1_{\mathrm{eq}} = -\begin{pmatrix} 0 \\ 0.6 \end{pmatrix} u^2 + \begin{pmatrix} l^1 \\ l^2 \end{pmatrix} x \tag{4.43}$$

Substitution of u^1_{eq} into the first equation of system (4.41) with respect to $s^2 = s_1$ yields

$$\dot{s}^2 = -0.7u^2 + l^3 x.$$

In Eqs. (4.42) through (4.44) r^1, r^2, l^1, l^2 and l^3 are n-dimensional rows with constant coefficients depending on A, B and C.

The conditions under which a sliding mode occurs on the plane $s^2 = 0$, in accordance with (1.9) and (4.44), are satisfied if the scalar control u^2 is chosen as

$$u^2 = \alpha_2 |x| \operatorname{sign} s^2 \quad \text{and} \quad \alpha_2 > |l^3|/0.7 \tag{4.45}$$

The control u^2 has been obtained under the assumption that a sliding mode does exist on the manifold $s^1 = 0$. This condition must now be realized with the help of vector u^1 under the chosen control u^2 (4.45). In compliance with the recommendations given in Sect. 4, the Lyapunov function for system (4.42) is the quadratic form

$$v = \tfrac{1}{2}(s^1)^{\mathrm{T}} D^{-1} s^1$$

its derivative on the system trajectories being

$$\dot{v} = -(s^1)^{\mathrm{T}}\left(u^1 - \begin{pmatrix} 0 \\ 0.6 \end{pmatrix} u^2 + \begin{pmatrix} l^1 \\ l^2 \end{pmatrix} x\right). \tag{4.46}$$

Like control u^2 in (4.45), choose vector u^1 in the form of a piecewise linear function of the state vector: $u^1 = \alpha_1 |x| \operatorname{sign} s^1$. As follows from (4.45) and (4.46), the derivative of the Lyapunov function will be negative definite if

$$\alpha_1 > \max(|l^1|, 0.6\alpha_2 + |l^2|). \tag{4.47}$$

The established control hierarchy shows that the sliding mode stable "in the large" first occurs on the intersection of discontinuity boundaries $s_2 = 0$ and $s_3 = 0$ (or $s^1 = 0$) and only after that it is observed on the intersection of all three discontinuity surfaces $s = 0$.

6 The Finiteness of Lyapunov Functions in Discontinuous Dynamic Systems

When the entire manifold of discontinuity surface intersections presents a domain of sliding modes stable "in the large", this manifold consists of the state vector trajectories. On the other hand, the so-called integral manifolds which also consist of trajectories may occur in the state space of dynamic systems described by differential equations with continuous righthand parts. Let us show, in terms of realizability, the difference of principle between the two sets of motions in continuous and discontinuous systems.

If the conditions of the theorem of existence and uniqueness of continuous systems solutions are met, the integral manifold in such systems cannot be attained in a finite time from an arbitrary initial position outside this manifold. Otherwise two trajectories could run through the same point. Therefore, dealing with continuous systems we can speak of reaching the integral manifold only asymptotically if the state vector is outside this manifold at the initial instant of time.

In this section, the problem of attaining the intersection of discontinuity surfaces $s=0$ or the problem of originating sliding modes in systems described by differential equations with discontinuous righthand parts has actually been solved with the use of special Lyapunov functions. Now we shall consider those cases with discontinuous systems when piecewise smooth Lyapunov functions (4.13) and quadratic-form Lyapunov functions (4.36) may be applied to solve the problem of convergence of the motion trajectories to the sliding manifold. Let us show that for these cases the Lyapunov functions are *finite* i.e. vanishing in a finite time; correspondingly, the sliding motions on the manifold $s=0$ will also occur in a finite time.

For the first case considered in Sect. 3 the decrease rate of function v (4.13) was found from the relationship (4.25). If in the initial instant of time $s \neq 0$, then at least one of the components of this vector s_i differs from zero and, by virtue of (4.15), $\|z\| \geqq 1$. It follows from inequality (4.25) that if function $F(x,t)$ is lower bounded by some number F_0, then

$$\dot{v} < -m_v, \tag{4.48}$$

where $m_v = \alpha F_0\left(\lambda_0 - \frac{1}{\alpha} M\sqrt{m}\right)$ and m_v is a positive value, since coefficient $\alpha > M\sqrt{m}/\lambda_0$ in (4.25). As a result, we obtain an estimate for a positive definite function $v(s)$ in the form $v \leqq v(t_0) - m_v(t-t_0)$. Consequently, the state vector will hit the manifold $s=0$ within a finite time interval $t_s - t_0 \leqq v(t_0)/m_v$. When $t > t_s$, function $v(t)$ becomes identical to zero, that is, the subsequent motion will proceed in the sliding mode along the manifold of the intersection of discontinuity surfaces.

The problem of the sliding mode stability of systems considered in Sect. 4 was treated with the help of the positive definite quadratic form (4.26) whose derivative on the system trajectories satisfied condition (4.31). If λ_M is the maximal eigenvalue of matrix W, then $\|s\| \geqq \sqrt{2v/\lambda_M}$ and the use of (4.31) yields the following estimate for $\dot{v}$:

$$\dot{v} \leqq -m_v \sqrt{\frac{2}{\lambda_M}} \sqrt{v}, \tag{4.49}$$

where $m_v = \alpha F_0\left(\Delta l_r - \frac{1}{\alpha}\|W\| M\sqrt{m}\right)$ and m_v is a positive value, provided $\alpha > \|W\| M\sqrt{m}/\Delta l_r$ as before.

Let us apply the comparison principle [128] to show that the state vector in the considered case also hits the manifold of the intersection of discontinuity surfaces $s=0$ in a finite time. According to this principle, the upper estimate of any solution of the differential inequality (4.49) is found from the equation

$$\dot{\rho} = -m_v \sqrt{\frac{2}{\lambda_M}} \sqrt{\rho}, \quad \rho(t_0) = v(t_0), \tag{4.50}$$

$$\rho(t) \geqq v(t) \quad \text{for} \quad t \geqq t_0. \tag{4.51}$$

The solution of Eq. (4.50) has the form

$$\rho(t)=\left(-\frac{m_v}{\sqrt{2\lambda_M}}(t-t_0)+\sqrt{v(t_0)}\right)^2. \tag{4.52}$$

The function $\rho(t)$ (4.52) is an upper estimate of the positive definite function v (4.51), therefore

$$v(t_s-t_0)=0, \quad t_s \leqq t_0+\frac{\sqrt{2\lambda_M v(t_0)}}{m_v}$$

and $v(t)=0$ with $t \geqq t_s$. Hence, both Lyapunov functions considered here, the piecewise smooth and the quadratic-form ones, are finite and, consequently, a motion along the manifold $s=0$ occurs in discontinuous system in a finite time.

Note that the interval t_s-t_0 following which $v=0$ and a stable sliding mode occurs is bound to decrease with the growth of α. Therefore the increase of the magnitude of control (4.20) permits one to speed up the occurrence of the motion on a certain manifold of the system state space. The greater part of the design methods described in Part 2 further in this book rest upon the use of such motions.

Chapter 5

Singularly Perturbed Discontinuous Systems

1 Separation of Motions in Singularly Perturbed Systems

One of the major obstacles in the use of efficient tools for designing control systems is the high order of equations that describe their behaviour. In many cases they may be reduced to a lower order model by neglecting small time constants or rejecting fast components of the system overall motion. Fast motions may, for instance, be caused by small impedances in equations of electromechanical energy converters [155], time constants of electric motors in systems controlling slow processes [68], nonrigidity of flying vehicles construction [74] and many other reasons. The design of control systems resting upon the use of low-order models may be carried out both by analytical and by various computational techniques. (Application of computational techniques to the design of control systems may be seriously hindered not only by their high dimension, but also by the fact that the computational problems in such systems are generally ill-posed and require ad-hoc methods to be developed).

Verification of rightfulness of rejecting fast motions, use of low-order models for obtaining solutions of different accuracy, estimation of accuracy and independent study of fast and slow motions can be achieved using the mathematical apparatus of singularly perturbed differential Eqs. [27, 99, 131, 155, 156, 157].

A standard problem statement in the singularly perturbed equations theory deals with the analysis of dynamic systems of the form

$$\dot{x} = f(x, y, t, \mu), \tag{5.1}$$

$$\mu\dot{y} = g(x, y, t, \mu), \tag{5.2}$$

where $x \in \mathbb{R}^n$, $y \in \mathbb{R}^l$ and μ is a scalar positive parameter depending on small masses, inductances, capacitances, small time constants of measuring instruments, etc.

Where μ is small enough and $g \neq 0$, vector y changes at a rate essentially higher than that of vector x, which results in different motion rates. When $\mu = 0$, the order of the system (5.1), (5.2) decreases because the differential Eq. (5.2) turns into algebraic:

$$0 = g(\bar{x}, \bar{y}, t, 0) \tag{5.3}$$

with vectors $\bar{x}$ and $\bar{y}$ corresponding to the singular system state. Assuming det $\{\partial g/\partial y\} \neq 0$, we may find the solution to (5.3) with respect to $\bar{y}$:

$$\bar{y} = \varphi(\bar{x}, t). \tag{5.4}$$

Substitution of this solution into Eq. (5.1) yields the equation of the n-th order model of the system

$$\frac{d\bar{x}}{dt} = \bar{f}(\bar{x}, t), \quad \bar{f}(\bar{x}, t) = f(\bar{x}, \varphi(\bar{x}, t), t, 0). \tag{5.5}$$

This model represents slow motion in the initial system.

The above procedure is quite formal in its nature, because when $\mu = 0$ the righthand part of the Eq. (5.2) $\dot{y} = g/\mu$ is not bounded. The above-mentioned papers covering the theory of singularly perturbed systems show that Eq. (5.5) is applicable to describing slow motions along the manifold (5.4) in the $(n + l)$-dimensional space provided the fast motions are asymptotically stable, i.e. a fast change of y results in the convergence of the state vector trajectories to this manifold. Then the solution of system (5.1), (5.2) for a finite time interval $[t_0, t_0 + T]$ may be presented as

$$x(t) = \bar{x}(t) + O(\mu, t), \tag{5.6}$$

$$y(t) = \varphi(\bar{x}, t) + \Pi y(\tau) + O(\mu, t), \quad t \in [t_0, t_0 + T], \tag{5.7}$$

where $\bar{x}(t)$ is the solution to the truncated system (5.5), and

$$\lim_{\mu \to 0} O(\mu, t) = 0, \quad \lim_{\tau \to \infty} \Pi y(\tau) = 0.$$

The function $\Pi y(\tau)$ represents a fast motion whose damping rate grows with a decrease of μ because $t = \mu\tau$.

As follows from (5.6) and (5.7),

$$\lim_{\mu \to 0} x(t) = \bar{x}(t), \quad t_0 \leqq t \leqq T$$

and for any t_1 such that $t_0 < t_1 < t_0 + T$

$$\lim_{\mu \to 0} y(t) = \varphi(\bar{x}, t), \quad t \in [t_1, t_0 + T]$$

Thus, with μ small enough the solution of the initial system of the $(n+l)$-th order with respect to x is quite close to that of the n-th order system (5.5) while vector y tends to $\bar{y}$ (5.4) outside the initial time interval $[t_0, t_1]$ which decreases with μ. This initial time interval corresponds to fast motion which is neglected in the slow motion model (5.5). The conditions sufficient for the fast motions stability are, first, that the equilibrium $z=0$ in the system

$$\frac{dz}{d\tau} = g(\bar{x}, \bar{y} + z(\tau), t, 0), \quad z = y - \bar{y} \tag{5.8}$$

be asymptotically stable in the point $x(t_0)$, $y(t_0)$, t_0; second, that the vector $z(0) = y(t_0) - \bar{y}(t_0)$ belong to the attraction domain and, third, that the eigenvalues of matrix $\{\partial g/\partial y\}$ have negative real parts on the trajectories $\bar{x}(t), \bar{y}(t)$. Equation (5.5) is actually a zero approximation equation. Approximation of higher orders may be obtained via the use of power series expansion of the solution.

2 Problem Statement for Systems with Discontinuous Control

It is essential that all of our conclusions were made under the assumption of continuous differentiability of functions f and g in (5.1), (5.2) with respect to the components of the state vector and time. Most obviously, the systems in which the control is a discontinuous function of state require verification of the methodology described in Sect. 1 above in order to justify the rejection of fast motions in the analysis of their behaviour.

Consider a system having small time constants in the measuring instruments, data processing and switching units, disregarded in ideal model (2.7), (1.8). Its behaviour is described by the equations of a higher order

$$\dot{x} = f(x, y, t) + B(x, y, t)u(x, y, t), \tag{5.9}$$

$$\mu \dot{y} = Ay + A_x x, \tag{5.10}$$

$$u_i(x, y, t) = \begin{cases} u_i^+(x, y, t) & \text{with} \quad s_i(x, y, t) > 0 \\ u_i^-(x, y, t) & \text{with} \quad s_i(x, y, t) < 0, \end{cases} \tag{5.11}$$

where $x \in \mathbb{R}$, $y \in \mathbb{R}^l$, $u \in \mathbb{R}^m$, and A and A_x are constant matrices. Equation (5.10) defining fast motions is assumed linear and the surfaces $s_i = 0 (i = 1, \ldots, m)$ on which the components u_i of the control vector u are subjected to discontinuity are chosen in the $(n+l)$-dimensional space.

Similar to the continuous system (5.1), (5.2), the dimension of system (5.9), (5.10) with $\mu = 0$ decreases since in this case Eq. (5.10) becomes algebraic:

$$Ay + A_x x = 0. \tag{5.12}$$

Assuming det $A \neq 0$ and substituting the solution of (5.12)

$$y = -A^{-1}A_x x \tag{5.13}$$

into (5.9), we obtain a truncated system of the following form:

$$\dot{x} = f(x, -A^{-1}A_x, t) + B(x, -A^{-1}A_x x, t)u(x, -A^{-1}A_x x, t), \tag{5.14}$$

$$u_i(x, -A^{-1}A_x x, t) = \begin{cases} u_i^+(x, -A^{-1}A_x x, t) & \text{with} \quad s_i(x, -A^{-1}A_x x, t) > 0, \\ u_i^-(x, -A^{-1}A_x X, t) & \text{with} \quad s_i(x, -A^{-1}A_x x, t) < 0. \end{cases} \tag{5.15}$$

Before trying to find out if the formally found system (5.14), (5.15) may be regarded as a slow motion model, consider a major difference in substantiating the closeness of the solutions for the complete and truncated systems in both the continuous and discontinuous cases. In the continuous case, the problem was to prove that after the fast motions decay the state vector trejectories will lie in the near vicinity of the manifold (5.3). Then, by virtue of continuity of function $f(x, y, t)$ the righthand parts of the complete system (5.1) will be close to those of the truncated system (5.5) formally obtained with $\mu = 0$, and this implies the closeness of their solutions.

In the discontinuous case, the above way of substantiating the validity of representing slow motions by system (5.14), (5.15) does not work. To show this, let the state vector of the complete system (5.9) through (5.11) stay in the vicinity of the manifold (5.12). Then, regardless of the closeness of functions s_i in the complete and truncated systems, the times of changes of their signs for the opposite or the times of stepwise changes of their controls may not coincide in these systems. Therefore, unlike continuous systems, the righthand part of (5.9) will not be close to that of (5.14) with parameter μ chosen arbitrarily small. This makes a specific peculiarity of problems of decoupling the overall motion into slow and fast components in discontinuous control systems.

In moving outside discontinuity surfaces when the control is a continuous function of the state vector, the methods of the singularly perturbed differential equations discussed in Sect. 1 of this chapter are applicable. Whether or not the truncated system (5.14), (5.15) featuring sliding mode motions may be applied as a mathematical description of slow motions in the initial high-order system (5.9) through (5.11) is a question of special interest. Below the conditions will be shown under which the formal order reduction technique which rests upon zeroing of parameter μ leads to slow motion equations in this case as well.

3 Sliding Modes in Singularly Perturbed Discontinuous Control Systems

The following will be assumed to consider the problem of closeness of solutions for the complete and truncated systems.

1. A Lipshitz constant exists for any of 2^m continuous subsystems comprising the complete system (5.9) through (5.11).
2. The vector-valued function $s(x, y)$ with components $s_1(x, y), \ldots, s_m(x, y)$ is differentiable with respect to all its arguments.
3. The matrix GB is nonsingular for the truncated system (5.14), (5.15).
4. A Lipshitz constant exists for the equation of the sliding motion along the manifold $s = 0$ in the truncated system (5.14), (5.15).
5. The righthand part of Eq. (5.9) satisfies the inequality

$$\| f(x, y, t) + B(x, y, t)u(x, y, t) \| \leqq L_x \| x \| + L_y \| y \| + L_0, \tag{5.16}$$

   where L_x, L_y and L_0 are constant positive values.
6. The Matrix A in the fast motion Eq. (5.10) is Hurwitz.

The first assumption implies that outside its discontinuity surfaces the system has a unique solution; the second, third and fourth assumptions provide the existence and uniquencess of both the sliding equation in the truncated system and its solution; the fifth assumption implies that the righthand part of (5.9) increases no faster than a linear function of the state, and finally, the sixth assumption will be used to prove the convergence of fast motions.

Let domain $S(t)$ on the intersection of discontinuity surfaces $s = 0$ in the truncated system state space be a sliding domain which is verified with the help of function $v(s, x, t)$ satisfying the conditions of the theorem from Sect. 2, Chapt. 4. Then the slow motion of the complete system in subspace x under the same initial condition will be close to the sliding motion of the truncated system. The accurate meaning of closeness of these motions will be cleared up by the theorem below proved on the basis of the following lemma.

Lemma. *If inequality (5.16) holds for the system (5.9) through (5.11), then for any finite time interval $[t_0, t_0 + T]$ and for any $t_0 < \bar{t}_1 < t_0 + T$*

$$y = -A^{-1}A_x + O(\mu, t), \quad \lim_{\mu \to 0} O(\mu, t) = 0 \quad \textit{for} \quad t \in [\bar{t}_1, t_0 + T]. \tag{5.17}$$

Proof. By analogy with (5.8), we introduce variable z to characterize the deviation from the manifold (5.12) consisting of slow motion trajectories:

$$z = y + A^{-1}A_x x. \tag{5.18}$$

The motion equation with respect to z is from (5.9), (5.10) and (5.18):

$$\dot{z} = \frac{1}{\mu} Az + A^{-1}A_x(f + Bu). \tag{5.19}$$

Since matrix A features asymptotic stability, such a positive definite matrix Q may be found that

$$\frac{QA + A^T Q}{2} = -I_l$$

where I_l is an identity matrix. Find the time derivative for the quadratic form $v_1 = \frac{1}{2} z^T Q z$ on the trajectories of system (5.19):

$$\dot{v}_1 = -\frac{1}{\mu} z^T z + z^T Q A^{-1} A_x (f + Bu). \tag{5.20}$$

As follows from (5.16), (5.18) and (5.20), such positive numbers N, L and M exist for which

$$\dot{v}_1 \leqq v_1 \left(-\frac{2}{\mu\lambda_2} + N \right) + \| z \| \cdot (L \| x \| + M), \tag{5.21}$$

where λ_2 is the greatest eigenvalue of matrix Q. Assume that

$$\dot{v}_1(t) < 0, \tag{5.22}$$

then $v_1(t) \leqq v_1(t_0)$ and vector $z(t)$ is norm-bounded

$$\| z(t) \|^2 \leqq \frac{\lambda_2}{\lambda_1} \| z(t_0) \|^2, \tag{5.23}$$

λ_1 being the smallest eigenvalue of matrix Q.

Applying the Bellman-Gronwall lemma, like in the case of system (2.11) with its righthand part (2.14), one may show that if inequalities (5.16) and (5.23) hold, the following estimate of the solution (5.9) is true:

$$\| x(t) \| \leqq L_1 \| x(t_0) \| + M_1 \| z(t_0) \| + N_1, \quad t \in [t_0, t_0 + T], \tag{5.24}$$

where positive values M_1, L_1 and N_1 are determined by the values of L_x, L_y, L_0 and T, as well as matrices A and A_x. Using (5.21), (5.23) and (5.24), we obtain

$$\dot{v}_1 \leqq v_1 \left(-\frac{2}{\mu\lambda_2} + N \right) + \sqrt{\frac{\lambda_2}{\lambda_1}} \| z(t_0) \| \cdot (L L_1 \| x(t_0) \| + L M_1 \| z(t_0) \| + L N_1 + M) \quad \text{for} \quad t \in [t_0, t_0 + T]. \tag{5.25}$$

As follows from (5.25), under the initial conditions $x(t_0)$, $z(t_0)$ and with $t \in [t_0, t_0 + T]$, the value of $\dot{v}$ may be made negative for

$$v_1 \geqq \sqrt{\mu} \tag{5.26}$$

choosing parameter μ small enough. Since in the domain (5.26)

$$\lim_{\mu \to 0} \dot{v}_1(t) = -\infty$$

then such as μ may be found that the relationships

$$v(t) < \sqrt{\mu} \quad \text{with} \quad t_0 < t_1 \leqq t \leqq t_0 + T, \tag{5.27}$$

$$\lim_{\mu \to 0} t_1 = t_0 \tag{5.28}$$

are true for the initial conditions $v(t_0) \geqq \sqrt{\mu}$.

When the motion proceeds over the domain (5.27) inequality (5.23) holds, and for this inequality estimates (5.24) and (5.25) are true. The quadratic form v_1 is positive definite, therefore $\lim_{\mu \to 0} z = 0$ and, by virtue of (5.27),

$$\lim_{\mu \to 0} z = 0 \quad \text{with} \quad t \in [t_1, t_0, t\,]. \tag{5.29}$$

The validity of the assertion of the above lemma follows from the pomparison of (5.18), (5.28) and (5.29).

This result means that, similarly to continuous systems, the righthand part of the fast motion differential Eq. (5.10) in discontinuous systems tends to zero with a decrease of parameter μ for all t's except for the vanishing initial time interval.

Theorem. *Let for the truncated system (5.14) and (5.15) such a function $v(s, x, t)$ exist that the conditions of the first theorem of Sect. 2, Chapt. 4 are satisfied for some domain $S(t)$ on the interesection of discontinuity surfaces $s(x, -A^{-1}A_x x) = 0$, and let a stable sliding motion in this domain occur in the time interval $t \in [t_0, t_0 + T]$. Then*

$$\lim_{\mu \to 0} x(t) = x^*(t) \quad \text{for} \quad t \in [t_0, t_0 + T],$$

where $x^*(t)$ is the solution to the equation describing the sliding motion in the truncated system, $x^*(t) \in S$, $t \in [t_0, t_0 + T]$ and $x(t)$ is the solution to the complete system (5.9) through (5.11) under the initial conditions $\| x^*(t_0) - x(t_0) \| = O(\mu)$ and $\lim_{\mu \to 0} O(\mu) = 0$.[1]

Proof. By virtue of conditions (5.28), (5.16) and the boundness of the function $z(t)$ it may be asserted that

$$\| x(t_1) - x(t_0) \| = O(\mu)$$[2]

for the solution of Eq. (5.14); therefore it suffices to prove the theorem only for $t \in [t_1, t_0 + T]$. Since, according to the lemma, vectors y and x are correlated through (5.17) in this time interval, the complete system Eqs. (5.9) through

[1] Further in the book the same notation $O(\mu)$ or $O(\mu, t)$ may be applied to different scalar and vector-valued quantities infinitesimal with respect to μ.

[2] This assertion also follows from the Bellman-Gronwall lemma.

(5.11) under the above assumptions are presentable in the form

$$\dot{x} = f(x, -A^1 A_x x, t) + B(x, -A^{-1} A_x x, t)u + O(\mu, t), \tag{5.30}$$

$$u_i = \begin{cases} u_i^+(x, -A^{-1}A_x x, t) + O(\mu, t) & \text{with} \quad s_i(x, -A^{-1}A_x x) + O(\mu, t) > 0 \\ u_i^-(x, -A^{-1}A_x x, t) + O(\mu, t) & \text{with} \quad s_i(x, -A^{-1}Ax) + O(\mu, t) < 0 \end{cases} \tag{5.31}$$

In contrast to the truncated system Eqs. (5.14) and (5.15), the above Eqs. (5.30) and (5.31) include infinitesimals $O(\mu, t)$. Recall the theorem of the sliding domain existence that was proved in Sect. 2, Chap. 4 with the help of the boundary layer system analysis substituting $\tilde{u}$ (3.1), (3.2) for the control u. Applying the same line of reasoning to system (5.30), (5.31), find the derivative of the function v

$$\dot{v} = v_0 + O(\Delta_0) + O(\mu, t) \tag{5.32}$$

where the function $\dot{v}_0$ is again found from (4.9). Repeating the proof for the theorem of the stability of sliding "in the small" we can show, that for any ε such numbers Δ_0, δ and μ_0 may be found that any motion of (5.30) and (5.31) starting in $\|s\| \leqq \delta$ will not leave the domain $\|p\| \leqq \varepsilon$ if $\mu \leqq \mu_0$. Reducing the value of μ and the width of the boundary layer, the motion occurring in system (5.30), (5.31) may be brought to the ideal sliding over the manifold $s(x, -A^{-1}A_x x) = 0$ as close as required. Since a decrease of μ makes $O(\mu, t)$ in (5.30) tend to zero, then by virtue of the equivalent control method the righthand parts of the sliding equations in system (5.30) will differ from those in the truncated system (5.14) by the value of $O(\mu, t)$ and, consequently, the same property will be featured by the solutions $x(t)$ and $x^*(t)$ to these two systems. The theorem is proved.

Thus, a general formal procedure for obtaining slow motion equations is applicable to discontinuous systems in the case when sliding modes occur in a truncated system.

Part II. Design

The mathematical tools developed in the first part of the book for the analysis of discontinuous dynamic systems permit the formalization of methods of designing control systems most diverse in types of control plants and criteria of functioning. In this part, theoretical principles of designing multivariable systems will be treated with a unified approach based on deliberate introduction of sliding modes. Along with developing some concrete control design procedures, this part will deal with such problems fundamental for the control theory as the existence and uniqueness of solutions of the considered problems in terms of the initial system.

Chapter 6

Decoupling in Systems with Discontinuous Controls

1 Problem Statement

Since we are going to "unify" the design principles by introducing sliding modes, let us confine ourselves with only those discontinuous systems whose sliding equations may be written quite unambiguously. From the entire variety of nonlinear systems, this limitation isolates a subclass which, generally speaking, may be presented by equations linear in control

$$\dot{x} = f(x,t) + B(x,t)u, \tag{6.1}$$

where $x \in \mathbb{R}^n$ and $u \in \mathbb{R}^m$. Formally, a standard statement of the control theory problems is in choosing such a control u, functionally or operationally dependent upon the system state, time and disturbances, that brings an appropriate transformation of the solution to the initial system (6.1). The idea of the "appropriate" transformation is treated in such a broad sense by the present-day control theory authors that it will be useful to give the most important problem statements in this book.

Most obviously, one should start with the stability problem traditionally posed throughout the entire history of the control theory. The basic apparatus for the mathematical study of stability is based upon the use of Lyapunov's functions for nonlinear systems, and diverse algebraic criteria for linear time-invariant systems.

Closely related to the stability problem is the problem of eigenvalues allocation in linear multivariable systems. Investigation of the solution existence rests upon the use of the concepts of controllability and stabilizability [72] while design is usually based on the reduction of the initial system to its canonical form [165].

Problems of dynamic optimization seem to attract great attention from control specialists today. From the viewpoint of the theory of optimal systems, the design problem is in finding the control which minimizes a certain functional on the system trajectories. Effective tools for obtaining optimal control are provided by special variational methods like the maximum principle or dynamic programming [46].

An important particular case of optimal systems which permits a complete solution of the design problem to be obtained is a linear system with a quadratic functional. The use of the so-called Riccati equation in such systems makes it possible to obtain an optimal control in the form of the state vector linear function [71, 95].

From the application viewpoint, an attractive property of closed-loop control systems is the invariancy of their motion to disturbances and variations of the control plant dynamic characteristics. The design of invariant systems implies either the use of the compensation principle if the information on the system "nonstationarities" is available, or high gain feedback control [90, 103, 113].

Methods of designing self-optimizing and adaptive systems deserve to be mentioned separately. The extremum search problem may be stipulated by the specifics of functioning of the control plant which features an extreme static characteristic [119] and the adaptation process organization requires the controller parameters to be readjusted depending on the dynamic properties of the plant [85]. Design of self-organizing systems generally requires application of various modifications of gradient procedures while design of adaptive systems is based upon the use of identification techniques or reference models.

Quite often, for instance, in treating the programming problems, the realization of a computational algorithm may require the analysis of a dynamic system of the type (6.1) which describes the process of search for the solution. The entire lot of the control theory tools may be effectively used in designing such search procedures [137].

In the sections that follow the above problems will be considered in application to control plants linear and nonlinear, time-varying and time-invariant, those described by ordinary differential equations and by partial differential equations. The idea of principle behind all of the design methods is in decoupling the initial problem into independent problems of lower dimension by means of an introduction of sliding modes and making use of those properties which are not inherent in any motion occurring outside discontinuity surfaces (for instance, invariancy to disturbances, stability, etc.).

2 Invariant Transformations

As suggested above, a multivariable system may be decoupled into lower-dimensional independent subsystems by means of a deliberate introduction of sliding modes and use of their properties. In systems of the type (6.1) featured by vector valued controls each component of which is discontinuous on the corresponding surface, sliding motions occur in some domain belonging to the manifold of the intersection of m discontinuity surfaces $s_i(x) = 0$ $(i = 1, \ldots, m)$ (2.1) having the dimension $(n - m)$.

Below we shall show that the motion occurring on the $(n - m)$-order manifold is always the same, only the ways of organizing the sliding motion on the manifold are different. It turns out that the sliding motion is invariant to the linear transformations of discontinuity surfaces and the initial control vector u. Using invariant transformations, one may choose from the entire set of feasible sliding mode implementation techniques one which provides the easiest realization of the sliding mode. Such transformations will be employed in designing discontinuous dynamic systems.

Consider two such transformations. Introduce a new vector s^* which is obtained as a result of the following linear and, generally, time-varying transformation of the initial vector s:

$$s^* = \Omega(x, t)s(x), \tag{6.2}$$

where s^* and s are, respectively, vectors with components s_i^* and s_i $(i = 1, \ldots, m)$, and $\Omega(x, t)$ is an $(m \times m)$ nonsingular matrix. The intersection of the surfaces $s_i^*(x, t) = 0$ in this case coincides with that of the surfaces $s_i(x) = 0$ $(i = 1, \ldots, m)$.

Let us assume that the surfaces $s^*(x, t) = 0$ are the discontinuity surfaces for m components of the control vector u, and that a sliding mode occurs on the manifold $s^* = 0$ in a certain range of values of x and t. In compliance with the equivalent control method, write the sliding motion equations assuming that $\| \dot{\Omega}$ and $\| \Omega^{-1} \|$ are bounded. First, find u_{eq} from the equation

$$\dot{s}^* = \Omega(x, t)G(f + Bu_{eq}) + \dot{\Omega}s = 0. \tag{6.3}$$

Here G is the $m \times n$ matrix whose rows are gradients of the function $s_i(x)$. Since matrix $\Omega(x, t)$ is nonsingular, (6.3) yields the following:

$$u_{eq} = -(GB)^{-1}Gf - (GB)^{-1}\Omega^{-1}\dot{\Omega}s \tag{6.4}$$

Comparing this u_{eq} with the one obtained from the equivalent control Eq. (2.9) calculated for the system (2.7), (1.8) with the components discontinuous on the surfaces $s_i(x) = 0$, one may see that the two Eqs. (6.4) and (2.9), differ in just one term, $-(GB)^{-1}\Omega^{-1}\dot{\Omega}s$. Since in a real sliding mode vector s^* determined by imperfections is sufficiently small and $\det \Omega \neq 0$, the magnitude of s is also small. Both $\| \Omega \|$ and $\| \Omega^{-1} \|$ are bounded, therefore when all the imperfections tend to zero the term $-(GB)^{-1}\Omega^{-1}\dot{\Omega}s$ will also tend to vanish. Hence this term may be neglected in the expression for u_{eq} in order to have the ideal sliding equations (the same concusion may be made following rigorous reasonings similar to those given in Sect. 3, Chap. 2). This means that a linear transformation of a discontinuity surface has no effect upon the equivalent control value on the manifolds $s=0$ or $s^*=0$.

Thus one can see that the sliding equations in the system with discontinuity surfaces $s_i(x)=0$ and $s_i^*(x,t)=0$ are identical.

Assume now that subjected to discontinuities on the chosen surfaces $s_i=0$ are not the components of the initial control vector, but rather the components of some vector u^* related to u via the following nonsingular transformation:

$$u=\Omega(x,t)u^*, \det\Omega\neq 0. \tag{6.5}$$

To find the equations of sliding over the manifold $s=0$, we should find in this case the equivalent control

$$u^*_{\mathrm{eq}}=-\Omega^{-1}(GB)^{-1}Gf. \tag{6.6}$$

Substitution of this control into the initial system yields the same sliding equations as in the system with the "old" control, u. We have considered two transformation, s into s^* and u into u^*. The sliding mode equations appear to be invariant with respect to both of them. At the same time, the motion in subspace s depends on matrix Ω. In the first case, this motion is described by the equation

$$\dot{s}=Gf+GBu \tag{6.7}$$

with u undergoing discontinuities on the surfaces $s_i^*=0$ depending on $\Omega(s^*=\Omega s)$. In the second case, the motion in subspace s explicitly depends on Ω:

$$\dot{s}=Gf+GB\Omega u^* \tag{6.8}$$

with $s_i=0$ as discontinuity surfaces. In the second case (6.8) the matrix Ω may always be chosen so that the design procedures described in Sects. 3 and 4, Chap. 4, are applicable to the matrix $GB\Omega$.

The design in the first case requiring the discontinuity surfaces transformation may also be carried out with the use of the quadratic-form Lyapunov function [21]

$$v=s^{\mathrm{T}}Ws, \quad W>0.$$

Find the derivative of function v on the trajectories of system (6.7):

$$\dot{v}=(s^*)^{\mathrm{T}}(\Omega^{-1})^{\mathrm{T}}WGf+(s^*)^{\mathrm{T}}(\Omega^{-1})^{\mathrm{T}}WGBu.$$

It is evident that in this case we can choose matrix Ω so that condition (4.29) hold for the matrix $L=(\Omega^{-1})^{\mathrm{T}}WGB$, and then carry on the design in compliance with the procedure described in Sect. 4, Chap. 4.

3 Design Procedure

The basic idea behind the design of systems with discontinuous controls based on the deliberate introduction of a sliding mode, regardless of the criterion of functioning, is in the following: first, a sliding motion desired in a certain sense is obtained by an appropriate choice of discontinuity surfaces, and then a control is chosen so that the sliding modes on the intersection of those discontinuity surfaces are stable, i.e. the trajectories which start on this manifold never leave it.

To realize the first step of this procedure, let us use the equivalent control method to write the equation of sliding over some manifold $s(x) = 0$:

$$\dot{x} = f - B(GB)^{-1}FG, \tag{6.9}$$

$$s(x) = 0. \tag{6.10}$$

Matrix G in (6.9) consists of the gradient vectors of functions $s_i(x)$ which make vector s. Matrix GB of dimension $m \times m$ is assumed to be nonsingular for all x and t. In this case the rank of matrix G is m and, following the implicit function theorem Eq. (6.10) enables on to define m, (for instance, last) components of the state vector, x, as functions of the remaining $n - m$ components. Substituting these functions into system (6.9) and dropping the last m equations in this system we obtain sliding equations in the form of the following $(n - m)$-th order system:

$$\dot{x}_1 = F(x_1, t, G), \tag{6.11}$$

where $x_1, F \in \mathbb{R}^{n-m}$. This poses an $(n - m)$-th order design problem. By varying the equations of the discontinuity surfaces $s(x) = 0$ and, consequently, by changing matrix G, one may "control" the motion in system (6.11).

The second step of the suggested design procedure is in finding such an m-dimensional control that the origin in the m-dimensional subspace $s_1, \ldots, s_m$ is a stable equilibrium point. In order to see if this nonlinear problem is solvable for all discontinuity surfaces chosen at the first step of the design procedure, consider the equation of the motion projection on the subspace s:

$$\dot{s} = Gf + GBu. \tag{6.12}$$

As it was found in Chap. 4 in the analysis of solutions to the nonlinear system (6.12), the point of stability primarily depends upon the type of matrix GB. Since matrix B is determined by the control plant Eq. (6.1) and matrix G has already been chosen at the first step of the design, the motion occurring in the subsapce s may either turn to be unstable, or the stability problem for this motion may not lend itself to a solution. This conclusion, besides, makes decomposability of the design problem into two independent problems of lower dimension somewhat doubtful. Indeed, matrix G must be equally acceptable for both problems, the problem of obtaining the required sliding motion and the problem of its stability. However, application of the invariant transformations (6.2) and (6.3) considered in the previous section makes decoupling quite possible.

With any gradient matrix G we can transform the discontinuity surfaces or the control vector in such a way that $(\Omega^{-1})^{\mathrm{T}}WGB$ will be a Hadamard matrix in the first case and matrix $GB\Omega$ preceding the control in (6.8) in the second case will be, for instance, diagonal or symmetric and positive definite. It was shown in Chap. 4, Sects. 3 and 4 that stability in subspace s, as applied to these cases, may always be attained if the absolute values of u_i^+ and u_i^- and their difference $u_i^+ - u_i^-$ are large enough. It is essential for this m-dimensional problem with the m-dimensional control that the state vector reaches the manifold $s=0$ within a finite time rather than asymptotically, and the time needed to reach this manifold may be made as short as required by increasing the magnitudes of the control components (Chap. 4, Sect. 6). This means that if the control resources of a discontinuous system are sufficient, then its properties are largely defined by its motion in the sliding mode.

Note that, in contrast to continuous systems when an m-dimensional control was used to solve the problem of design in an n-dimensional space, here it suffies to just solve the problem of stability in an m-dimensional space.

Thus a deliberate introduction of sliding modes allows the control problem to be decoupled into two problems of a lower dimension, $(n-m)$ and m, which may be treated independently.

4 Reduction of the Control System Equations to a Regular Form

In compliance with the design procedure described in Sect. 3, the sliding Eq. (6.11) depends not only on the vector function $s(x)$, but also on the gradients of its components. Hence, the elements of matrix G cannot be chosen arbitrarily, which constitutes a specific feature of the design of required motions in a sliding mode. This consideration, generally speaking, does not permit direct use of the conditional design procedures, for instance, the variational methods. It may therefore be useful to consider the class of systems whose sliding equations are not explicitly dependent upon the gradient matrix G.

Let us give a more detailed statement of this problem. Since, by our assumption, $\det GB \neq 0$, then rank $G = m$ and, by virtue of the implicit function theorem, m components of the state vector x may be represented through the remaining $(n-m)$ components under the condition $s=0$. Assume that such m components make vector x^2 while the remaining $(n-m)$ components make vector x^1. Then it is possible to specify the manifold $s=0$ as

$$s(x) = x^2 - s_0(x^1) = 0, \tag{6.13}$$

where $s_0(x^1)$ is the solution of equation $s=0$ with respect to x^2. In compliance with such a decoupling pattern, rewrite the equation of system (6.1) in the following form:

$$\begin{aligned}\dot{x}^1 &= f_1(x^1, x^2) + B_1(x^1, x^2)u,\\ \dot{x}^2 &= f_2(x^1, x^2) + B_2(x^1, x^2)u,\\ x^{\mathrm{T}} &= (x^{1\mathrm{T}}, x^{2\mathrm{T}}).\end{aligned} \tag{6.14}$$

All of our subsequent manipulations will also be valid for time-varying systems which are reducible to the form of (6.14) by expending their state variables space. If it turns out that $B_1(x^1, x^2) = 0$ in the system thus obtained, then the equation of motion along the manifold $s(x) = 0$ (6.13) or $x^2 = s_0(x^1)$ will be of the form

$$\dot{x}^1 = f_1(x^1, s_0(x^1)), \tag{6.15}$$

i.e. the sliding equation does not depend upon the gradients of vector $s(x)$ and the function $s_0(x^1)$ may be therefore chosen arbitrarily. As a result, we face a regular design problem for an $(n-m)$-dimensional system (6.15) with an m-dimensional control $s_0(x^1)$. The system presented as

$$\begin{aligned}\dot{x}^1 &= f_1(x^1, x^2),\\ \dot{x}^2 &= f_2(x^1, x^2) + B_2(x^1, x^2)u\end{aligned} \tag{6.16}$$

will hereinafter be referred to as a *regular form*.

The above reasonings lead us to the following problem. It is necessary, first, to find the classes of systems for which such a generally nonlinear coordinate transformation exists that it brings the initial system to its regular form; second, to find the subclasses of systems for which the transformation functions may be found in an explicit form.

To solve this problem, we introduce a vector of new state variables $y^{\mathrm{T}} = (y_1, \dots, y_n)$, related to the initial variables by the nonlinear transformation

$$y^1 = \varphi(x), \quad y^2 = x^2, \tag{6.17}$$

where $y^{1\mathrm{T}} = (y_1, \dots, y_{n-m}), y^{2\mathrm{T}} = (y_{n-m+1}, \dots, y_n)$ and $\varphi^{\mathrm{T}} = (\varphi_1, \dots, y_{n-m})$, $\varphi_i(x)$ being continuous and continuously differentiable with respect to all their arguments. The equations with respect to y^1

$$\dot{y}^1 = \frac{\partial\varphi(x)}{\partial x} f(x) + \frac{\partial\varphi(x)}{\partial x} B(x)u \tag{6.18}$$

will be independent of the controls if the vector function $\varphi(x)$ is a solution to the matrix partial differential equation

$$\frac{\partial\varphi(x)}{\partial x} B(x) = 0. \tag{6.19}$$

In order to solve Eq. (6.19), one should find an m-dimensional integral manifold of the corresponding Pfaff system. This requirement, as well as all subsequent manipulations, is based on the theory of Pfaff's forms, which constitutes a well developed branch of mathematical analysis [120].

A Pfaff system has the form

$$W(x)\,dx=0, \tag{6.20}$$

where $dx^{\mathrm{T}}=(dx_1,\ldots,dx_n)$ and $W(x)$ is a full rank $(n-m)\times n$ matrix comprised of coefficients of the linear differential forms

$$\begin{aligned}\omega^1(d)&=\omega_1^1\,dx_1+\cdots+\omega_n^1dx_n,\\ &\ldots\ldots\ldots\ldots\\ \omega^{n-m}(d)&=\omega_1^{n-m}dx_1+\cdots+\omega_n^{n-m}dx_n.\end{aligned} \tag{6.21}$$

The Pfaff forms coefficients (6.21) are the functions of x and may be found from the condition that each of the linear forms, $\omega^1,\ldots,\omega^{n-m}$ is brought to zero being multiplied by each vector column of matrix $B(x)$ [120], i.e.

$$W(x)B(x)=0 \tag{6.22}$$

The values of the Pfaff forms coefficients may be obtained by decoupling the matrices:

$$W(x)=[W_1(x)\,\vdots\,W_2(x)],\quad B(x)=\begin{bmatrix}B_1(x)\\ \cdots\\ B_2(x)\end{bmatrix},$$

where B_2 is the base minor matrix of matrix $B(x)$, i.e. it is assumed that $\det B_2\neq 0$ in (6.14). Rewriting Eq. (6.22) as

$$W_1(x)B_1(x)+W_2(x)B_2(x)=0 \tag{6.23}$$

and specifying matrix $W_1(x)$ in an arbitrary way, for example, $W_1=I_{n-m}$, we find that

$$W_2(x)=-B_1(x)B_2^{-1}(x). \tag{6.24}$$

As a result, the Pfaff system (6.20) is found unambiguously and it is equivalent to the system

$$A(x)dx^2+dx^1=0. \tag{6.25}$$

where

$$\begin{aligned}A&=-B_1B_2^{-1}=(a_j^i),\quad i=1,\ldots,n-m;\\ &\qquad\qquad\qquad\qquad j=n-m+1,\ldots,n;\\ dx^{1\mathrm{T}}&=(dx_1,\ldots,dx_{n-m}),\\ dx^{2\mathrm{T}}&=(dx_{n-m+1},\ldots,dx_n).\end{aligned}$$

The integral surfaces of the Pfaff system (6.25), provided they exist, i.e. that system (6.25) is completely integrable, may be found in an explicit form [120]

$$x^1=\varphi^0(x^2,c), \tag{6.26}$$

where $c^{\mathrm{T}}=(c_1,\ldots,c_{n-m})$ are the integration constants, and $\varphi^0\in\mathbb{R}^{n-m}$. Solving Eq. (6.26) with respect to c we obtain $n-m$ independent integral surfaces $\varphi(x)=c$.

Thus, if the Pfaff system (6.20) is completely integrable, then the coefficients of Pfaff forms $\omega^1,\ldots,\omega^{n-m}$, multiplied by a scalar function (integrating factor), may be regarded as the gradient vector components of functions $\varphi_i(x)$, $i=1,\ldots,n-m$, which means that functions $\varphi_i(x)$ satisfy condition (6.22) and, therefore, condition (6.19) as well. Thus, the use of functions $\varphi_i(x)$ as the transformation functions (6.17) solves the first part of our problem.

The transformation (6.17) introduced above is locally unique, since

$$\det\left[\frac{\partial y}{\partial x}\right]=\det\begin{bmatrix}\dfrac{\partial\varphi}{\partial x^1} & \dfrac{\partial\varphi}{\partial x^2}\\ 0 & I_m\end{bmatrix}\not\equiv 0.$$

Geometric interpretation. In each point of space $\mathbb{R}^n$, the vector columns $b^1,\ldots,b^m$ of the control matrix $B(x)$ form a hyperplane tangential to the above obtained surfaces $\varphi(x)=c$ and thus orthogonal to the chosen coordinate axes of subspace y^1. This gives us a new system of equations equivalent to the initial one, whose $n-m$ rows lack the control vector u which actually proves that the equations of system (6.14) may be presented in space y in the regular form of (6.16).

The applicability of the above technique, however, is limited by the requirement of the complete integrability of the Pfaff system. In compliance with the Frobenious theorem, this requirement is identical to zeroing the outer derivatives of its forms by virtue of the system equation. This makes the necessary and sufficient condition of the Pfaff system to be completely integrable [120].

In our case we have taken $W_1(x)=I_{n-m}$, therefore the outer derivatives of the Pfaff system

$$\begin{aligned}\omega^1(d)&=a^1_{n-m+1}dx_{n-m+1}+\cdots+a^1_n dx_n+dx_1,\\ &\cdots\\ \omega^{n-m}(d)&=a^{n-m}_{n-m+1}dx_{n-m+1}+\cdots+a^{n-m}_n dx_n+dx_{n-m},\end{aligned}\tag{6.27}$$

corresponding to system (6.25) are defined as the sums of the outer products

$$d\omega^i(d)=\sum_{j=n-m+1}^{n} da^i_j\wedge dx_j,\quad i=1,\ldots,n-m.\tag{6.28}$$

The $\wedge$ symbol in (6.28) denotes the outer product characterized by the following properties:

$$dx_j\wedge dx_i=-dx_i\wedge dx_j,\quad dx_i\wedge dx_i=0,$$

and $da^i_j=\sum_{\alpha=1}^{n}\frac{\partial a^i_j}{\partial x_\alpha}dx_\alpha$ is a complete differential of the function $a^i_j(x)$.

Substituting the complete differentials of the functions $da^i_j(x)$ into (6.28) and presenting the elements $dx_1,\ldots,dx_{n-m}$ as linear functions of $dx_{n-m+1},\ldots,dx_n$ by virtue of (6.25), obtain the conditions of complete integrability of system (6.25) in the form of the equalities

$$\frac{\partial a^i_\alpha}{\partial x_\beta}-\sum_{j=1}^{n-m}\alpha^j_\beta\frac{\partial a^i_\alpha}{\partial x_j}=\frac{\partial a^i_\beta}{\partial x_\alpha}-\sum_{j=1}^{n-m}a^j_\alpha\frac{\partial a^i_\beta}{\partial x_j},\tag{6.29}$$

where $i = 1, \ldots, n-m$ and α and β denote various pairwise combinations from the set of numbers $(n-m+1, \ldots, n)$. It is evident that the number of such combinations for each Pfaff's form will be C_m^2.

The design problem regularization algorithm is as follows:

1. Matrix $B_2(x)$ of the base minor of matrix $B(x)$ is found defining the coordinates of vector x^2 and the "dependent" coordinates of vector x^1.
2. A Pfaff system is constructed in the form (6.25).
3. The coefficients of the Pfaff forms are found in compliance with (6.24).
4. The conditions of integrability (6.29) are written out and if these conditions are satisfied, the integral surfaces $\varphi(x) = C$ are found.
5. A new state vector in the form (6.17) is introduced.

Differentiation and substitution of variables allows the following regular form system of equations to be obtained:

$$\dot{y}^1 = f_1(y^1, y^2),$$
$$\dot{y}^1 = f_2(y^1, y^2) + B_2(y^1, y^2)u.$$

Thus the problem of the reduction to the regular form is solvable only for one class of systems which recognizes the conditions of the Frobenious theorem. But even in this case the analytical solution is a real challenge since there are no standard ways available for obtaining integral surfaces.

Considered below are scalar control systems and some particular cases of vector control systems for which a coordinate transformation of the type (6.17) may be found reducing system (6.14) to its regular form.

4.1 Single-Input Systems

1. Let a system be described by the equations

$$\dot{x} = f(x) + b(x)u, \tag{6.30}$$

where $x \in \mathbb{R}^n$ and $u \in \mathbb{R}^1$, $b(x)$ is a vector function with components $b_i(x)$, and an assumption is made that $b_n(x) \neq 0$. Let $x^2 = x_n$, then the corresponding Pfaff system is of the form

$$\begin{aligned} \omega^1(d) &= -\frac{b_1(x)}{b_n(x)} dx_n + dx_1 = 0 \\ &\cdots\cdots\cdots \\ \omega^{n-1}(d) &= -\frac{b_{n-1}(x)}{b_n(x)} dx_n + dx_{n-1} = 0. \end{aligned} \tag{6.31}$$

Let us demonstrate that scalar control systems always meet the requirement of the complete integrability of the corresponding Pfaff equations $\omega^i(d) = 0$. The outer derivative of the Pfaff form $\omega^i(d), i = 1, \ldots, n-1$, is

$$d\omega^i(d) = \frac{\partial a_n^i(x)}{\partial x_1} dx_1 \wedge dx_n + \cdots + \frac{\partial a_n^i(x)}{\partial x_{n-1}} dx_{n-1} \wedge dx_n,$$

where $a_n^i(x) = -b_i(x)/b_n(x)$. Replacing the elements $dx_1, \ldots, dx_{n-1}$ with dx_n, by virtue of (6.31) we have

$$d\omega^i(d) = \sum_{j=1}^{n-1} a_n^j(x) \frac{\partial a_n^i(x)}{\partial x_j} dx_n \wedge dx_n$$

As follows from the outer product properties, $dx_i \wedge dx_i = 0$ and therefore $d\omega^i(d) = 0$. Consequently, any of the Eqs. (6.31) is integrable.

It may be interesting to see that the same conclusion can be reached in a different way. Some simple manipulations bring the system of Pfaff's equations (6.31) to the equivalent form

$$\frac{dx_1}{b_1(x)} = \frac{dx_2}{b_2(x)} = \cdots = \frac{dx_n}{b_n(x)}$$

which is nothing but a symmetric form of the system of ordinary differential equations $\dot{x} = b(x)$; the first integrals of such a system are known to exist if functions $b_i(x)$ are continuous and if they satisfy the Lipshitz condition. This fact is used in [58] to find singular optimal controls in scalar control systems.

2. Assume that vector function $b(x)$ in (6.30) depends on just a single coordinate, for instance, x_n, i.e.

$$\dot{x} = f(x) + b(x_n)u, \tag{6.32}$$

where $b_n(x_n) \neq 0$. In this case the equations of the integral surfaces $\varphi(x) = c$ of the system (6.31) or of the system $\dot{x} = b(x_n)$ have the form

$$\varphi_i(x) = -x_i + \int_0^{x_n} \frac{b_i(\gamma)}{b_n(\gamma)} d\gamma + c_i = 0, \quad i = 1, \ldots, n-1. \tag{6.33}$$

Then the coordinate transformation

$$y_i = \varphi_i(x), \quad y_n = x_n, \quad i = 1, \ldots, n-1$$

reduces system (6.32) to the form

$$\begin{aligned} \dot{y}_i &= v_i(y_1, \ldots, y_n), \quad i = 1, \ldots, n-1, \\ \dot{y}_n &= v_n(y_1, \ldots, y_n) + b_n(y_n)u \end{aligned} \tag{6.34}$$

and the problem of choosing the required sliding modes is reduced to a selection of a "control" $s_0(y_1, \ldots, y_{n-1})$ such that brings the desired properties to the system

$$\dot{y}_i = v_i(y_1, \ldots, y_{n-1}, s_0), \quad i = 1, \ldots, n-1.$$

The equivalency of (6.32) to (6.34) follows from the fact that the nonlinear transformation functional matrix

$$\frac{\partial y}{\partial x}=\left[\begin{array}{cc:cc} -1 & 0 & 0 & b_1/b_n \\ 0 & -1 & 0 & b_2/b_n \\ \cdots & \cdots & \cdots & \cdots \\ 0 & 0 & -1 & b_{n-1}/b_n \\ 0 & 0 & 0 & 1 \end{array}\right]$$

has a full rank.

4.2 Multiple-Input Systems

1. Consider a time-varying system

$$\dot{x}=f(x,t)+B(t)u, \tag{6.35}$$

where $x\in\mathbb{R}^n$, $u\in\mathbb{R}^m$ and rank $B=m$. Assuming $x_0=t$ and $\dot{x}_0=1$, rewrite system (6.35) as

$$\begin{bmatrix}\dot{x}_0\\ \dot{x}^1\\ \dot{x}^2\end{bmatrix}=\begin{bmatrix}1\\ \bar{f}_1(x,x_0)\\ \bar{f}_2(x,x_0)\end{bmatrix}+\begin{bmatrix}0\\ B_1(x_0)\\ B_2(x_0)\end{bmatrix}u, \quad \det B_2\neq 0$$

and make sure that the corresponding Pfaff equations

$$dx_0=0, \quad dx^1+A(x_0)sx^2=0, \tag{6.36}$$

where $A(x_0)=-B_1(x_0)B_2^{-1}(x_0)$, are completely integrable. The integration of (6.36) yields integral surfaces in the form

$$x_0=c_0, \quad x^1+A(x_0)x^2=c, \quad c^{\mathrm{T}}=(c_1,\ldots,c_{n-m}).$$

Substituting the variables

$$t=x_0, \quad y^1=x^1+A(x_0)x^2, \quad y^2=x^2$$

we reduce the initial system to its regular form

$$\dot{y}^1=\bar{f}_1(y^1,y^2,t),$$
$$\dot{y}^2=\bar{f}_2(y^1,y^2,t)+B_2(t)u.$$

This transformation is nonsingular because

$$\det\left(\frac{\partial y}{\partial x}\right)=\det\begin{pmatrix}I_{n-m} & A(t)\\ 0 & I_m\end{pmatrix}\neq 0.$$

A similar transformation reducing the system to its regular form will be employed in the subsequent chapters of the book in treating various linear system design problems.

2. Let a system be described by the equations

$$\dot{x}^k = f_k(x), \quad \dot{x}^{n-k} = f_{n-k}(x) + B(x^k)u \tag{6.37}$$

in a nontrivial case of $k < n - m$. This class of dynamic systems is characterized by the fact that the Pfaff equation system designed in compliance with the above procedure may be presented as two subsystems

$$\begin{aligned}&\omega^i(d) \equiv dx_j = 0, \quad j = 1, \dots, k,\\ &\omega^i(d) \equiv a^i_{k+1}(x^k)dx_{k+1} + \cdots + a^i_{k+m}dx_{k+m} + dx_i = 0, \quad i = k+m+1, \dots, n\end{aligned} \tag{6.38}$$

The solutions to the first subsystem, $x_j = c_j$, $j = 1, \dots, k$, mean that $a^i(x^k) = \text{const}$ which makes the second subsystem completely integrable, with the equations of its $n - m + k$ integral surfaces being of the form

$$a^i_{k+1}(x^k)x_{k+1} + \cdots + a^i_{k+m}(x^k) + x_i = c_i, \quad i = k+m+1, \dots, n$$

The nonsingular coordinate transformation

$$\begin{aligned}&y^k = x^k, \quad y^{n-(m+k)} = A(x^k)x^m + x^{n-(m+k)}, \quad y^m = x^m;\\ &A(x^k) = -B_1(x^k)B_2^{-1}(x^k),\\ &x^{\mathrm{T}} = ((x^k)^{\mathrm{T}}, (x^{n-(m+k)})^{\mathrm{T}}, (x^m)^{\mathrm{T}}), \quad B^{\mathrm{T}} = (B_1^{\mathrm{T}}, B_2^{\mathrm{T}}), \quad \det B_2 \neq 0\end{aligned} \tag{6.39}$$

results in a reduction of system (6.37) to its regular form

$$\begin{aligned}&\dot{y}^{n-m} = v_{n-m}(y), \quad (y^{n-m})^{\mathrm{T}} = ((y^k)^{\mathrm{T}}, (y^{n-(m+k)})^{\mathrm{T}}),\\ &\dot{y}^m = v_m(y) + D(y)u, \quad \det D \neq 0\end{aligned}$$

This equation has been derived after differentiating the variables $y_1, \dots, y_n$

$$\begin{aligned}&\dot{y}_k = f_k(x),\\ &\dot{y}^{n-(m+k)} = \frac{\partial A(x^k)}{\partial x^k} f_k(x)x^m + A(x^k)f_m(x) + f_{n-(m+k)}(x),\\ &\dot{y}^m = f_m(x) + B_2(x^k)u, \quad f^{\mathrm{T}}_{n-k} = (f^{\mathrm{T}}_{n-(m+k)}, f^{\mathrm{T}}_m)\end{aligned}$$

and substituting the solutions of (6.39) with respect to x into the system.

3. It may turn out that the condition of the complete integrability in system (6.14) is satisfied not for the entire matrix $B(x)$, but rather for just a part of it which depends on some group of vector x coordinates, i.e.

$$\begin{aligned}&\dot{x}^1 = f_1(x) + B_1(x^1)u^{m-l}\\ &\dot{x}^2 = f_2(x) + B_2(x)u^{m-l} + B_3(x)u^l,\end{aligned}$$

where $x^{1\mathrm{T}} = (x_1, \dots, x_{n-l})$, $x^{2\mathrm{T}} = (x_{n-l+1}, \dots, x_n)$, $(u^l)^{\mathrm{T}} = (u_1, \dots, u_l)$, $(u^{m-l})^{\mathrm{T}} = (u_{l+1}, \dots, u_m)$, rank $B_3(x) = l$ and $B_1(x^1)$ is the $(n-l) \times (m-l)$ matrix which may be used as a basis for the design of the completely integrable Pfaff system. In this case we may "spend" the control vector, u^l, to design a sliding mode on the intersection of discontinuity surfaces $s_l = x^2 - s_{0l}(x^1) = 0$ where $s_l, s_{0l} \in \mathbb{R}^l$. The equation of sliding over this manifold is written as

$$\dot{x}^1 = f_1(x^1, s_{0l}(x^1)) + B_1(x^1)u^{m-l}.$$

Under the above assumption, this system may be reduced to the regular form with the use of the algorithm described.

Example. Consider a third order system

$$\dot{x}_1 = f_1(x) + (x_2^3 + 2x_1x_3)u_1 + \left(2\frac{x_1}{x_2} + 4x_1\frac{x_3}{x_2} + 2x_2^2\right)u_2,$$

$$\dot{x}^2 = f_2(x) + x_2x_3u_1 + (1 + 2x_3)u_2,$$

$$\dot{x}_3 = f_3(x) + x_2u_1 + 2u_2,$$

where $x^{\mathrm{T}} = (x_1, x_2, x_3)$. At the first step of the procedure we establish that the base minor of matrix $B(x)$ may be

$$B_2(x) = \begin{bmatrix} x_2x_3 & 1 + 2x_3 \\ x_2 & 2 \end{bmatrix}$$

because $\det B_2 = -x_2 \neq 0$. The components of vector x^2, correspondingly, will be variables x_2 and x_3 and the Pfaff system (6.25) in this case will be the equation

$$\omega^1(d) \equiv a_2^1 dx_2 + a_3^1 dx_3 + dx_1 = 0$$

$$[a_2^1 a_3^1] = -B_1B_2^{-1} = \left[-\frac{2x_1}{x_2} - x_2^2\right].$$

The condition of integrability of (6.29)

$$\frac{\partial a_3^1}{\partial x_2} - a_2^1\frac{\partial a_3^1}{\partial x_1} = \frac{\partial a_2^1}{\partial x_3} - a_3^1\frac{\partial a_2^1}{\partial x_1}, \quad -2x_2 = -(-x_2^2)\left(-\frac{2}{x_2}\right)$$

holds, therefore the coefficient of the Pfaff equation presented as

$$dx_1 = \frac{2x_1}{x_2}dx_2 + x_2^2dx_3$$

are the partial derivatives

$$\frac{\partial x_1}{\partial x_3} = x_2^2, \quad \frac{\partial x_1}{\partial x_2} = \frac{2x_1}{x_2}.$$

Substituting the solution to the first equation $x_1 = x_2^2x_3 + q(x_2)$ into the second one we obtain

$$\frac{dq(x_2)}{dx_2} = \frac{2q(x_2)}{x_2}, \quad q(x_2) = cx_2^2.$$

Consequently, the equation of the integral surface of the Pfaff system has the form

$$x_1 = x_2^2 x_3 + c x_2^2$$

or

$$\varphi(x) = \frac{x_1}{x_2^2} - x_3 = c.$$

Introduce a vector of new variables:

$$y_1 = \frac{x_1}{x_2} - x_3, \quad y_2 = x_2, \quad y_3 = x_3.$$

Differentiating these and applying the inverse transformation, we obtain

$$\dot{y}_1 = f_1(y),$$
$$\dot{y}_2 = f_2(y) + y_2 y_3 u_1 + (1 + 2y_3)u_2,$$
$$\dot{y}_3 = f_3(y) + y_3 u_1 + 2u_2.$$

The equation of sliding over the intersection of two surfaces on which the control components u_1 and u_2 are subjected to discontinuity may be written for this system in the regular form of (6.15) as

$$\dot{y}_1 = f_1(y_1, s_{01}(y_1), s_{02}(y_1)).$$

The functions $s_{01}(y_1)$ and $s_{02}(y_1)$ are to be chosen in accordance with the control system functioning criterion.

It should be noted in conclusion that if the Pfaff system of the control system to be designed is not completely integrable, then parametric optimization may be used as a possible technique for solving the design problem. For this purpose the function $s = c^{\mathrm{T}} \Psi(x)$ may be chosen in the form of a finite series of some preselected system of the functions $\Psi^{\mathrm{T}}(x) = (\Psi_1(x), \ldots, \Psi_N(x))$ and the series expansion coefficients $c_1, \ldots, c_N$ may be found from the condition of minimization of the given performance functional.

Chapter 7

Eigenvalue Allocation

In this chapter, we make use of the decomposition method for finding the desired allocation of eigenvalues in time-invariant linear systems of the form

$$\dot{x} = Ax + Bu, \quad x \in \mathbb{R}^n, \quad u \in \mathbb{R}^m, \tag{7.1}$$

with constant matrices A and B. The discontinuity surfaces are assumed to be linear, i.e.

$$s = Cx \tag{7.2}$$

with C being a $(m \times n)$-dimensional matrix.

It will be recalled that if in (7.1) sliding mode occurs on the manifold $s = 0$, the system's behaviour is described by an $(n - m)$-th-order differential equation. It is to this motion of dimensionality lower than the original one that we apply the notions of controllability and stabilizability with the aim of solving the eigenvalue allocation problem.

1 Controllability of Stationary Linear Systems

The discontinuous control for the system (7.1), (7.2) will be designed through some basic notions and relations of the modern linear system theory that are presented below. Methodologically, it would be more convenient to discuss the results of the linear theory in this and the following sections according to a systematic presentation of the monographs [6, 62, 92] but not to the "primary source".

The first question to be answered before proceeding to the problem of control is that of controllability. According to Kalman, the linear system (7.1) is *controllable* if it may be transferred from any initial state $x(t_0)$ to any final state $x(t_1)$ in finite time $t_1 - t_0$. The necessary and sufficient controlability condition is as follows

$$\operatorname{rank} P = n, \quad P = (B, AB, A^2B, \ldots, A^{n-1}B), \tag{7.3}$$

i.e. the column vectors of the controllability matrix P should generate an n-dimensional space. The pair $\{A, B\}$ is referred to as *controllable* if the condition (7.3) is fulfilled.

Now we present another treatment of the controllability that will be used below for the desired eigenvalue allocation. The matrix A transforms the linear space $\mathbb{R}^n$ into a space of the same dimensionality, i.e.

$$A\{\mathbb{R}^n\} \subset \mathbb{R}^n. \tag{7.4}$$

The subspace $\mathscr{K} \subset \mathbb{R}^n$ is called invariant for the matrix A if

$$A\{\mathscr{K}\} \subset \mathscr{K}. \tag{7.5}$$

From now on, we shall discuss non-trivial invariant subspace[1] i.e. subspaces whose dimensionality is less than that of the system:

$$\dim\{\mathscr{K}\} < n. \tag{7.6}$$

Introduce the so-called *control subspace* $\mathscr{B}$ formed by the system of base column vectors of the matrix B. The necessary and sufficient controllability condition of the linear system (7.1) has in terms of the invariant subspaces (7.5), (7.6) and controll subspace the following form

$$\mathscr{B} \not\subset \mathscr{K}. \tag{7.7}$$

Sufficiency. If the control space does not belong to any of the invariant subspaces, there will be a non-zero m-dimensional constant vector α such that

$$b = B\alpha, \quad b \notin \mathscr{K} \tag{7.8}$$

Consider the controllability matrix for the pair $\{A, b\}$

$$P_\alpha = (b, Ab, \ldots, A^{n-1}b). \tag{7.9}$$

If the columns of matrix P_α are linearly dependent, their base generates a non-trivial invariant subspace $\mathscr{K}$. Indeed, it follows from the linear dependence

[1] The one-dimensional non-trivial invariant subspaces, in particular, are defined by the eigenvectors of the matrix A.

that for some $l < m$

$$A^l b = \sum_{i=0}^{l-1} A^i b \lambda_i,$$

where λ_i are constant coefficients. Vector

$$A^{l+1} b = A^l b \lambda_{l-1} + \sum_{i=1}^{l-1} A^i b \lambda_{i-1} \tag{7.10}$$

like vectors $A^k b$, $k > l+1$, is, obviously, linear combinations of vectors $b, \ldots, A^{l-1} b$. The matrix A may be shown to transform into itself the linear subspace of a dimensionality lower than n generated by the same vectors. This implies that the vector b as its element belongs to the invariant subspace which contradicts the condition (7.8). Consequently, the assumption of matrix P_α singluarity is not correct, i.e. its rank is n, and the pair $\{A, b\}$ is controllable. Since control in the form of

$$u = \alpha u_0 \tag{7.11}$$

leads to a controllable system with scalar control u_0, the original system or pair $\{A, B\}$ is controllable as well. The sufficiency of (7.7) is proved.

Necessity. Assume that the control subspace $\mathscr{B}$ belongs to a non-trivial invariant subspace $\mathscr{K}$ whose dimensionality is less than n. Therefore, there always will be a $(1 \times n)$-dimensional matrix l transforming the n-dimensional space into a one-dimensional space such that

$$l\{\mathscr{K}\} = 0. \tag{7.12}$$

Since we have assumed that $\mathscr{B} \subset \mathscr{K}$, then, according to (7.5) and (7.12) $A^i\{\mathscr{B}\} \subset \mathscr{K}$ and $lA^i B = 0$, $i = 0, 1, 2, \ldots$. Correspondingly, any time derivative of the scalar linear state function $y = lx$ is independent of control

$$\frac{d^i y}{dt^i} = lA^i x, \quad i = 0, 1, 2, \ldots. \tag{7.13}$$

For some number $k \leqq n$, the linear forms in the right sides of the Eqs. (7.13) are linearly dependent, therefore

$$\frac{d^k y}{dt^k} + \sum_{i=0}^{k-1} \gamma_i \frac{d^i y}{dt^i} = 0, \tag{7.14}$$

where γ_i are constant coefficients.

The solution of differential Eq. (7.14) with respect to the coordinate y is control-independent. Thus, the fact that the control subspace belongs to a non-trivial invariant subspace implies the non-controllability of the original system which proves the necessity of the condition (7.7).

2 Canonical Controllability Form

In the case of non-controllable linear system (7.1), it seems reasonable to decompose the entire state space into a subspace where controllability exists, and another subspace where motion is independent of control. This decomposition is good in visualizing the class of design problems solvable within the framework of the system under consideration.

Let us demonstrate in a way similar to that of matrix P_α (7.9) that if the pair $\{A, B\}$ is not controllable, the column vectors of the controllability matrix $P = \{B, AB, \ldots, A^{n-1}B\}$ generate a non-trivial invariant subspace of the matrix A.

Let vector x belong to the subspace $\mathscr{K}$ of dimensionality smaller than n and having a base composed of the linearly independent columns of matrix P. The vector Ax, then, belongs to the subspace generated by the column vectors of the matrices $AB, A^2B, \ldots, A^nB$. One can demonstrate at the same time that

$$A^nB = \sum_{i=0}^{n-1} A^iB\Lambda_i$$

where Λ_i are constant matrices. (To this end, one must apply the reasoning used for the matrix P_α of Sect. 1 to each column of the matrix B). Consequently, the vector Ax is a linear combination of the base vectors of matrix P, i.e.

$$A\{\mathscr{K}\} \subset \mathscr{K} \tag{7.15}$$

which proves the above assertion.

Let us now proceed to determining a non-singular transformation that would allow us to find the controllable and non-controllable subspaces of the system (7.1). Let the matrix T_1 consist of a set of base column vectors of matrix P or subspace $\mathscr{K}$, and the matrix T_2 consist of column vectors such that the square matrix

$$T = (T_1, T_2) \tag{7.16}$$

is non-singular. The equation of motion with respect to the new state vector y defined by

$$Ty = x \tag{7.17}$$

has the following form

$$\dot{y} = T^{-1}ATy + T^{-1}Bu. \tag{7.18}$$

Decompose the matrix T^{-1} into two submatrices

$$T^{-1} = \begin{pmatrix} U_1 \\ U_2 \end{pmatrix} \tag{7.19}$$

so that the submatrix U_1 has the same dimensionality as the transposed matrix

T_1. Since

$$T^{-1}T=\begin{pmatrix}U_1\\U_2\end{pmatrix}(T_1,T_2)=\begin{pmatrix}U_1T_1 & U_1T_2\\U_2T_1 & U_2T_2\end{pmatrix}=I_n \tag{7.20}$$

where I_n is an n-dimensional identity matrix, $U_2T_1=0$, therefore, for any $x\in\mathscr{K}$

$$U_2x=0. \tag{7.21}$$

According to the decomposition (7.16) and to (7.19), the matrices in the Eq. (7.18) are representable as

$$T^{-1}AT=\begin{pmatrix}U_1\\U_2\end{pmatrix}A(T_1,T_2)=\begin{pmatrix}U_1AT_1 & U_1AT_2\\U_2AT_1 & U_2AT_2\end{pmatrix},$$

$$T^{-1}B=\begin{pmatrix}U_1\\U_2\end{pmatrix}B=\begin{pmatrix}U_1B\\U_2B\end{pmatrix}. \tag{7.22}$$

The columns of matrix T_1 belong to $\mathscr{K}$, hence, the columns of matrix AT_1 as well as those of B belong to $\mathscr{K}$. For such vectors, the condition (7.21) is satisfied, or

$$U_2AT_1=0,\quad U_2B=0. \tag{7.23}$$

Thus, the Eq. (7.18) of motion with respect to the state vector y related to the original vector x by the non-singular transformation (7.17) are representable as

$$y=\begin{pmatrix}y_1\\y_2\end{pmatrix},\quad \begin{aligned}\dot{y}_1&=A_{11}y_1+A_{12}y_2+B_1u,\\ \dot{y}_2&=A_{22}y_2.\end{aligned} \tag{7.24}$$

The values of all the matrices in (7.24) are determined through (7.22) and (7.23). The vector y_2 is obviously independent of the control vector u, and its behaviour is defined by the eigenvalues of matrix A_{22}.

Now we show that the pair $\{A_{11},B_1\}$ is controllable, i.e. that the rank of matrix

$$P_{y_1}=(B_1,A_{11}B_1,\ldots,A_{11}^{q-1}B_1) \tag{7.25}$$

is q if q is the dimensionality of the base of matrix P. Notably, $U_1T_1=I_q$ according to (7.20) and there exists a $(q\times mn)$-dimensional matrix Λ such that $P=T_1\Lambda$, $\operatorname{rank}\Lambda=q$. Taking into consideration these relations and the condition $U_2T_1=0$, the controllability matrix of (7.18) may be reduced to

$$P_y=T^{-1}P=\begin{pmatrix}U_1\\U_2\end{pmatrix}T_1\Lambda=\begin{pmatrix}I_q\\0\end{pmatrix}\Lambda=\begin{pmatrix}\Lambda\\0\end{pmatrix}.$$

The rank of matrix P_y is, obviously, equal to q. On the other hand, the elements of matrix P_y are computable through the notation (7.24):

$$\begin{pmatrix}A_{11} & A_{12}\\0 & A_{22}\end{pmatrix}^i\begin{pmatrix}B_1\\0\end{pmatrix}=\begin{pmatrix}A_{11}^iB_1\\0\end{pmatrix},\quad i=0,1,\ldots,n-1 \tag{7.26}$$

Comparison of (7.25) and (7.26) reveals that

$$P_y = \begin{pmatrix} P_{y_1} \\ 0 \end{pmatrix}$$

Since the rank of matrix P_y is q, that of matrix P_{y_1} is also q, and it follows that the pair $\{A_{11}, B_1\}$ is controllable.

According to the decomposition of vector (7.24), the subspace of all the vectors y_1 is referred to as controllable, that of y_2 as non-controllable, and the system notation (7.24) is called *canonical controllability form.*

3 Eigenvalue Allocation in Linear Systems. Stabilizability

We begin our discussion of the desired eigenvalue allocation of characteristic equation with the case of scalar control in (7.1). The linear control of these systems may be conveniently designed in the so called *canonical space or canonical variable space.* Each subsequent coordinate of the canonical space is the time derivative of the preceding one, and the system equations are, therefore, as follows

$$\dot{y} = \begin{pmatrix} 0 & 1 & 0 \cdots 0 & 0 \\ 0 & 0 & 1 \cdots 0 & 0 \\ \cdots & \cdots & \cdots & \cdots \\ 0 & 0 & 0 \cdots 0 & 1 \\ -a_1 & -a_2 & -a_3 \cdots -a_{n-1} & -a_n \end{pmatrix} y + \begin{pmatrix} 0 \\ 0 \\ \vdots \\ 0 \\ 1 \end{pmatrix} u, \tag{7.27}$$

where $y \in \mathbb{R}^n$, $u \in \mathbb{R}^1$. It will be readily seen that the elements of the lowermost row are the coefficients of the characteristic equation

$$p^n + a_n p^{n-1} + \cdots + a_1 = 0. \tag{7.28}$$

It follows from (7.27) and (7.28) that the determination of a linear control providing the desired values of characteristic equation roots (or, which is the same, coefficients) presents no special problem for systems written with respect to the canonical variables.

In this connection, determine the conditions under which for the system

$$\dot{x} = Ax + bu, \quad x \in \mathbb{R}^n, \quad u \in \mathbb{R}^1 \tag{7.29}$$

there will be a non-singular transformation $Ty = x$ such that the equation with respect to the new state vector

$$\dot{y} = T^{-1}ATy + T^{-1}bu \tag{7.30}$$

has the form of (7.27). Let us take the column vectors $t_1, t_2, \ldots, t_n$ of the matrix

T in accordance with the following relations

$$t_n = b, \tag{7.31}$$

$$t_{k-1} = At_k + a_k t_n, \quad k = n, \dots, 2, \tag{7.32}$$

where a_k are the coefficients of the characteristic equation of the system (7.29) representable in the form of (7.28). From (7.32) obtain

$$t_k = A^{n-k} t_n + a_n A^{n-k-1} t_n + \cdots + a_{k+2} At^n + a_{k+1} t_n, \tag{7.33}$$

$$t_1 = A^{n-1} t_n + a_n A^{n-2} t_n + \cdots + a_3 A t_n + a_2 t_n. \tag{7.34}$$

Following the recurrent relations (7.32), compute the values of vector $T^{-1}ATy$ in (7.30):

$$T^{-1}ATy = \sum_{k=1}^{n} T^{-1}At_k y_k = \sum_{k=1}^{n} (T^{-1}t_{k-1} - a_k T^{-1} t_n) y_k \tag{7.35}$$

It follows from the condition $T^{-1}T = I_n$ that the k-th element of vector $T^{-1}t_k$ is equal to one, the remaining elements being zero, and, therefore,

$$(T^{-1}t_{k-1} - a_k T^{-1}t_n) y_k = \begin{matrix} k-1 \left\{ \begin{matrix} \\ \\ \\ \end{matrix} \right. \\ \\ n-k+1 \left\{ \begin{matrix} \\ \\ \\ \end{matrix} \right. \end{matrix} \begin{bmatrix} 0 \\ \vdots \\ 0 \\ 1 \\ 0 \\ \vdots \\ -a_k \end{bmatrix} y_k, \quad k = 2, \dots, n. \tag{7.36}$$

Taking into consideration (7.34), obtain for the first term in (7.35)

$$T^{-1}ATy_1 = T^{-1}(A^n + a_n A^{n-1} + \cdots + a_2 A + a_1) t_n y_1 - a_1 T^{-1} t_n y_1.$$

According to the Kelley–Hamilton theorem, any matrix satisfies its characteristic equation, i.e.

$$T^{-1}At_1 y_1 = -a_1 T^{-1} t_n = \begin{bmatrix} 0 \\ \vdots \\ 0 \\ -a_1 \end{bmatrix} y_1 \tag{7.37}$$

The vector preceding control in (7.30) is found in a similar manner:

$$T^{-1}b = T^{-1}t_n = \begin{bmatrix} 0 \\ \vdots \\ 0 \\ 1 \end{bmatrix} \tag{7.38}$$

It follows from the comparison of (7.27), (7.30), (7.36), (7.37) and (7.38) that the transformation matrix T with columns computed through (7.31) and (7.32) reduces the original system (7.29) to the canonical variables. It rests to determine the conditions under which the transformation matrix T is non-singular. By replacing the column vectors in T by their values in (7.33) and t_n by b, represent it as the product of two matrices

$$T=(b, Ab, \ldots, A^{n-1}b)\begin{bmatrix} a_2 & a_3 & \cdots & a_n & 1 \\ a_3 & a_4 & \cdots & 1 & 0 \\ \cdot & \cdot & \cdot & \cdot & \cdot \\ a_n & 1 & \cdots & 0 & 0 \\ 1 & 0 & \cdots & 0 & 0 \end{bmatrix}. \tag{7.39}$$

The first matrix is the controllability matrix of the system (7.27), the determinant of the second matrix is equal to one. Therefore, the transformation $Ty=x$ is non-singular if the original system or pair $\{A, b\}$ is controllable.

The above reasoning leads to the following procedure for designing scalar linear control for the controlable system (7.29) so as to obtain the desired eigenvalues:

1. The coefficients $a_1, \ldots, a_n$ of the characteristic polynomial $\det(pI_n - A)=0$ of the open-loop system are computed.
2. By means of the non-singular transformation $Ty=x$ the system (7.29) is reduced to the Eq. (7.27) with respect to the canonical variables. The transformation matrix is found through the recurrent formulas (7.31), (7.32) or through (7.39).
3. The characteristic polynomial of the feedback system is found through the desired eigenvalues λ_i

   $$(p-\lambda_1)(p-\lambda_2)\cdots(p-\lambda_n)=p^n+a_n^* p^{n-1}+\cdots+a_1^*.$$

   For the coefficients a_i^* to be real, the complex eigenvalues must be, of course, pairwise conjugate.
4. Control is chosen in the form of a function of canonical variables $u=\Delta a y$ where Δa is row vector with elements $a_i^* - a_i$.
5. Control in the form of a linear function of the original state vector is determined

   $$u=\Delta a T^{-1} x.$$

This procedure is also applicable to the controllable systems with vector control. Indeed, as it was established in the discussion of controllability conditions (Sect. 7.1), there always will be a vector α such that the control of the form of $u=\alpha u_0$ (7.11) reduces the problem to controllable system with scalar control.

To conclude, let us discuss from the viewpoint of eigenvalue allocation the linear systems where the controllability condition is not obeyed. Their properties may be conveniently studied if they are represented in the canonical controllability form (7.24). Under any linear control, the spectrum of matrix in the feedback system equation, obviously, will always contain all the eigenvalues of matrix A_{22}. The non-controllable system, therefore, cannot be represented by an equation with respect to the canonical variables because this representation would amount to the possibility of "shifting" all the characteristic equation roots. Since the pair $\{A_{11}, B_1\}$ in (7.24) is controllable, only a part of the eigenvalues may be allocated arbitrarily by taking the control in the form of linear function of y_1. In the cases where the controlled variables are the components of vector y_1 constituting the controllable subspace, one can provide their desired variation.

The control system, however, may be recognized as working if it is asymptotically stable. This fact is the cause for considering a wider class as compared with that of controllable systems where one can also advantageously solve the problem of control with respect to some operation criteria. We mean the class of stabilizable linear systems: the pair $\{A, B\}$ is referred to as *stabilizable* if there is a linear control such that the feedback system is asymptotically stable. As a matter of fact, we have already formulated the necessary and sufficient stabilizability condition: the eigenvalues of matrix A_{22} must have negative real parts.

It should be remarked that the controllable system is stabilizable because there is no non-controllable subspace in its state space.

4 Design of Discontinuity Surfaces

Consideration is given to control in the multidimensional linear system

$$\dot{x} = Ax + Bu, \quad x \in \mathbb{R}^n, \quad u \in \mathbb{R}^m, \tag{7.40}$$

and without loss of generality it is assumed that the rank of matrix B or control space dimensionality coincides with the control dimensionality. Indeed, if the column vectors of matrix B are linearly dependent, then

$$Bu = B'\Lambda u = B'u', \quad u' = \Lambda u,$$

where B' is $(n \times m')$-dimensional matrix, $m' < m$, consisting of the base vectors of control space, Λ is constant $(m' \times m)$-dimensional matrix whose column elements are the coefficients of expansion of the columns of matrix B with respect to base vectors. As the result, there is a system with new control u' having lower dimensionality coinciding with that of the control space. It will be noted that it follows from the condition $\operatorname{rank} B = m'$ that $\operatorname{rank} \Lambda = m'$ and, therefore, the original control u may be determined after designing the control u'.

Thus, in the problem under consideration

$$\operatorname{rank} B = m. \tag{7.41}$$

The control vector components have discontinuities on some planes $s_i(x) = 0$, $i = 1, \ldots, m$, that should be chosen so that sliding mode motion on their intersection

$$s = Cx = 0, \quad s^T = (s_1(x), \ldots, s_m(x)), \quad C = \text{const} \tag{7.42}$$

is described by differential equations having the desired allocation of characteristic equation roots.

This problem will be solved in the space of new variables related to the original ones by a non-singular linear transformation

$$x' = Mx, \quad x' \in \mathbb{R}^n, \tag{7.43}$$

such that

$$MB = \begin{pmatrix} 0 \\ B_2 \end{pmatrix}, \tag{7.44}$$

where B_2 is quadratic $(m \times m)$-dimensional matrix. In order to satisfy the condition (7.44), the first $n - m$ rows of matrix M should be composed of the base of $(n - m)$-dimensional subspace orthogonal to the control subspace. The remaining m rows are chosen so that $\operatorname{rank} M = n$ and the matrix B_2 is non-singular. For example, B^T can be taken as these rows, then $B_2 = B^T B$ and $\det B_2 \neq 0$.

The behaviour of (7.40) in the space x' is described by

$$\dot{x}' = MAM^{-1}x' + MBu$$

or

$$\dot{x}_1 = A_{11}x_1 + A_{12}x_2 \tag{7.45}$$

$$\dot{x}_2 = A_{21}x_1 + A_{22}x_2 + B_2 u, \tag{7.46}$$

where x_1 and x_2 are vectors consisting of $n - m$ and m components of vector x',

$$MAM^{-1} = \begin{pmatrix} A_{11} & A_{12} \\ A_{21} & A_{22} \end{pmatrix}.$$

The equation of the manifold $s = 0$ with respect to the new variables, correspondingly, has the form

$$s = C_1 x_1 + C_2 x_2,$$

where $CM^{-1} = (C_1, C_2)$, C_1 and C_2 are $(m \times n)$- and $(m \times m)$-dimensional matrices, respectively.

Below, consideration will be given only to the discontinuity surfaces for which $\det CB \neq 0$, i.e. the sliding equations may be written unambiguously. Since $CB = C_2 B_2$, this requirement amounts to the condition $\det C_2 \neq 0$. Without loss of generality, one may confine oneself to the case of $C_2 = I_m$ which is due to the invariance property to linear transformation of the vector as was found in Sect. 6.2.

In order to obtain the sliding mode motion equation one has to solve the set of equations $s = 0$ and $\dot{s} = 0$ with respect to x_2 and u, substitute the results into the system (7.45), (7.46), and skip the last m equations in it. This procedure results in the equations

$$\dot{x}_1 = A_1 x_1 + A_{12} x_2 \tag{7.47}$$

$$x_2 = -C_1 x_1. \tag{7.48}$$

The system (7.47) may be regarded as an open-loop system with $(n-m)$-dimensional state vector x_1 and m-dimensional control x_2. Thus, we come across the common problem of designing a feedback linear system whose dimensionality is m less than that of the original system.

The problem of the desired eigenvalue allocation for the characteristic equation defining sliding mode motion is solved by the following lemma.

Lemma. *If the pair $\{A, B\}$ in (7.40) is controllable, the pair $\{A_{11}, A_{12}\}$ in (7.47) is also controllable.*

Proof. Assume that the pair $\{A_{11}, A_{12}\}$ is not controllable. By means of the non-singular transformation $x_1 = T y_1$ ($y_1 \in \mathbb{R}^{n-m}$) reduce the system (7.47) to the canonical controllability form (7.24):

$$\dot{y}'_1 = A'_{11} y'_1 + A'_{12} y'_2 + B' x_2, \quad \dot{y}'_2 = A'_{22} y'_2, \tag{7.49}$$

where $y_1^T = (y_1'^T, y_2'^T)$, the pair $\{A'_{11}, B'\}$ is controllable. The system (7.45), (7.46) may be rewritten in compliance with this transformation and the subspace of all the vectors y'_2 will be non-controllable. Since the system (7.45), (7.46) results from the non-singular transformation of the system (7.40) which does not violate controllability, the original system must be non-controlable which contradicts the condition. The lemma is proved.

From the lemma follows

Theorem. *If the pair $\{A, B\}$ is controllable, then, by an appropriate choice of the matrix C_1, the eigenvalues of matrix in the sliding mode equation*

$$\dot{x}_1 = (A_{11} - A_{12} C_1) x_1$$

may be arbitrarily placed (the complex eigenvalues must be pairwise conjugate).

Proof. If follows from the lemma that the pair $\{A_{11}, A_{12}\}$ is controllable. As it has been established in Sect. 7.3, linear control enables any eigenvalue allocation for controllable systems.

The theorem allows us to formulate for controllable systems a discontinuity surface design procedure:

(1) the matrix M in the transformation (7.43) is determined;
(2) in accordance with the procedure of Sect. 7.3, a matrix C_1 is found such that the eigenvalues $\lambda_1, \ldots, \lambda_{n-m}$ of matrix $A_{11} - A_{12} C_1$ characterizing sliding mode motion be the desired ones;

(3) the equations of discontinuity surfaces in the space of original variables are chosen as $s=(C_1, I_m)Mx=0$.

It must be noted in conclusion that the problem of asymptotic stability of sliding mode motions is also solvable by means of the canonical controllability form. It follows directly from (7.49) that the sliding mode motion is stabilizable if the pair $\{A, B\}$ in (7.40) is stabilizable.

Example. Consider a third-order system with scalar control

$$\dot{x} = Ax + bu, \quad x \in \mathbb{R}^3, \quad u \in \mathbb{R}^1, \tag{7.50}$$

$$A = \begin{pmatrix} -10 & -5 & -2 \\ 26 & 15 & 7 \\ -27 & -18 & -10 \end{pmatrix}, \quad b = \begin{pmatrix} -1 \\ 2 \\ -1 \end{pmatrix}.$$

It is assumed that the discontinuous control induces sliding mode motion on the surface $s=cx=0$ where c is (1×3)-dimensional row vector. One has to choose the coefficients of the plane equation so that the second-order differential equation $(n-m=2)$ describing sliding mode motion has eigenvalues

$$\lambda_1 = -5, \quad \lambda_2 = -10. \tag{7.51}$$

Prior to proceeding to the design, let us make sure that the system in question is controllable:

$$\det(b, Ab, A^2b) = \det \begin{pmatrix} -1 & 2 & -7 \\ 2 & -3 & 14 \\ -1 & 1 & 10 \end{pmatrix} = 3 \neq 0.$$

According to the above design procedure, define the matrix M so that its first two rows be orthogonal to the vector b and rank $M=3$:

$$M = \begin{pmatrix} 2 & 1 & 0 \\ 3 & 2 & 1 \\ 2 & 2 & 1 \end{pmatrix}, \quad Mb = \begin{pmatrix} 0 \\ 0 \\ 1 \end{pmatrix}, \quad \det M = -1.$$

As the result of coordinate transformation $x' = Mx$ obtain

$$\dot{x}_1 = A_{11}x_1 + A_{12}x_2, \tag{7.52}$$

$$\dot{x}_2 = A_{21}x_1 + A_{22}x_2 + B_2 u, \tag{7.53}$$

where x_1 and x_2 are, respectively, two- and one-dimensional vectors

$$A_{11} = \begin{pmatrix} -1 & 2 \\ 1 & -3 \end{pmatrix}, \quad A_{12} = \begin{pmatrix} 1 \\ 1 \end{pmatrix}, \quad A_{21} = (2, 1), \quad A_{22} = 1, \quad B_2 = 1.$$

The equation of a discontinuity plane with respect to the new variables has the form of $s = c_1 x_1 + x_2$ where c_1 is a two-dimensional row vector.

As might be expected, the pair $\{A_{11}, A_{12}\}$ is controllable

$$\det(A_{12}, A_{11}A_{12}) = \det\begin{pmatrix} 1 & 1 \\ 1 & -2 \end{pmatrix} \neq 0.$$

(Note that it is more convenient to verify the controllability condition for the subsystem (7.52), because its dimensionality is lower than that of the original system). The next step is the reduction of the subsystem (7.52) to the space of canonical variables in compliance with the procedure of Sect. 7.3. To this end, one needs the coefficients of the characteristic polynomial of matrix A_{11}

$$\det(A_{11} - pI_2) = \det\begin{pmatrix} -1-p & 2 \\ 1 & -3-p \end{pmatrix} = p^2 + 4p + 1,$$

$$a_1 = 1, \quad a_2 = 4.$$

According to (7.31), (7.32),

$$t_2 = b = \begin{pmatrix} 1 \\ 1 \end{pmatrix}, \quad t_1 = A_{11}t_2 + a_2 t_2 = \begin{pmatrix} 5 \\ 2 \end{pmatrix}$$

are the columns of the transformation matrix T.

The transformation

$$Ty = x, \quad T = \begin{pmatrix} 5 & 1 \\ 2 & 1 \end{pmatrix}$$

as applied to the problem under consideration leads to the system

$$\dot{y} = T^{-1}A_{11}Ty + T^{-1}A_{12}x_2,$$

$$T^{-1}AT = \begin{pmatrix} 0 & 1 \\ -1 & -4 \end{pmatrix}, \quad T^{-1}A_{12} = \begin{pmatrix} 0 \\ 1 \end{pmatrix} \tag{7.54}$$

i.e. the Eq. (7.54) is written with respect to the canonical variables. For the eigenvalues of sliding mode motion equation be equal to the given values (7.51), the characteristic polynomial of feedback system should be as follows:

$$(p+5)(p+10) = p^2 + 15p + 50, \quad \text{i.e.} \quad a_1^* = 50, \quad a_2^* = 15.$$

It will be recalled that when taken with the opposite sign, the coefficients of the last equation with respect to the canonical variables coincide with those of the characteristic polynomial. Hence, in order to obtain their given eigenvalues, the coordinate x_2 considered as control must be taken in the form of $x_2 = -\, c_1^* y_1$, $c_1^* = (c_{11}, c_{12})$, $c_{11} = a_1^* - a_1 = 49$, $c_{12} = a_2^* - a_1 = 11$. Control x_2 as the function of vector x_1 is defined, correspondingly, by the relation

$$x_2 = -c_1^* T^{-1} x_1 = -c_1 x_1, \quad c_1 = c_1^* T^{-1} = (12\tfrac{1}{3}, 2),$$

i.e. the discontinuity plane equation is

$$s = (c_1, 1)x' = c_1 x_1 + x_2 = 0.$$

The design procedure is completed by the passage to the original state space x:

$$s=(c_1, 1)Mx=0, \quad s=(32\tfrac{2}{3}, 18\tfrac{1}{3}, 3)x=0. \tag{7.55}$$

At the occurrence of sliding mode on the plane (7.55), the eigenvalues corresponding to the differential equation of this motion assume the given values.

5 Stability of Sliding Modes

The first stage of design, allocation of the eigenvalues of the characteristic equation of sliding mode motion on the manifold $s=0$, is folllowed by the second stage, concerned with stability of sliding modes themselves or, stated differently, with stability of motion in a subspace $(s_1, \dots, s_m)$,

$$\dot{s}=CAx+CBu. \tag{7.56}$$

To solve this problem, we make use of the two versions of the Lyapunov function having the quadratic form as described in Sect. 4.4, and also of the Lyapunov function in the form of sum of absolute values (4.13).

As it was shown in Chap. 4, control in all cases should be sought in the form of (4.20) if the scalar function $F(x, t)$ is the upper estimate for any of the equivalent control components. To determine it, one must solve the equation $\dot{s}=0$ obtained from (7.56) with respect to u. Since

$$CB=(CM^{-1}(MB)=(C_1, I_m)\begin{pmatrix}0\\ B_2\end{pmatrix} \quad \text{and} \quad \det B_2 \neq 0,$$

it follows that $\det CB \neq 0$ and u_{eq} is a linear function of the state vector with bounded coefficients. Therefore, there exists a positive parameter α such that the function

$$\alpha|x|, \quad \left(|x|=\sum_{i=1}^{n}|x_i|\right)$$

will be the upper estimate of the components of vector u_{eq}. This function will be used below for determination of discontinuous control providing sliding mode stability "in the large" over all the manifold of discontinuity surfaces.

Remembering that $\det CB=\det B_2 \neq 0$, one may use various invariant transformations of discontinuity surfaces or control vector (6.2), (6.5). (These transformations are applicable in the case of $\det GB \neq 0$, and in the case of linear discontinuity surfaces the gradient matrix G for functions $s_i(x)$ is equal to the matrix C). If D is an $(m \times m)$-dimensional constant diagonal matrix with elements $d_i \neq 0$, and $\Omega=D(CB)^{-1}$, $\Omega=(CB)^{-1}D$ in the transformations (6.2) and (6.5), respectively, the motion equations in the respective subspaces s^* and

s are as follows

$$\dot{s}^* = Hx + Du, \tag{7.57}$$

$$\dot{s} = Hx + Du^*, \tag{7.58}$$

where $H = D(CB)^{-1}CA$ in the first case (7.57), where the components of control vector u have discontinuities on the planes $s_i^* = 0$, and $H = CA$ in the second case (7.58), where the planes $s_i = 0$ are discontinuity surfaces of the components of u^*. Since (7.57) and (7.58) are formally identical, we confine our discussion to the first case.

Remembering that the function $\alpha|x|$ is the upper estimate of the equivalent control components, by analogy with (4.20) take control u as a piecewise linear function of the system state vector

$$u = -\alpha|x| \operatorname{sign} Ds^*, \tag{7.59}$$

where $\operatorname{sign} Ds^*$ is m-dimensional column vector with components $\operatorname{sign} d_i s_i^*$, α is a positive constant. The positive definite quadratic form

$$v = \tfrac{1}{2} s^{*T} s^* \tag{7.60}$$

is used as the Lyapunov function. Its time derivative on the system trajectories is computed through (7.57) and (7.60)

$$\frac{dv}{dt} = s^{*T}(Hx - \alpha|x| D \operatorname{sign} Ds^*). \tag{7.61}$$

One may readily see that $dv/dt < 0$ everywhere outside of the discontinuity surface points if

$$|d_i|\alpha > h_i, \quad i = 1, \dots, m, \tag{7.62}$$

where $h_i = \max\limits_j |h_{ij}|$, h_{ij} are the elements of matrix H. Then $\dfrac{dv}{dt}$ is estimated by the inequality

$$\frac{dv}{dt} < -|x||s^*| r_0, \tag{7.63}$$

where the positive value

$$r_0 = \min_i (\alpha|d_i| - h_i), \quad |s^*| = \sum_{i=1}^{m} |s_i^*|.$$

According to the stability theorem of Sect. 4.2, if the inequalities (7.62) are satisfied, the origin of the subspace $(s_1^*, \dots, s_m^*)$ is the point of stable equilibrium in the large, i.e. sliding mode occurs always on the manifold $s^* = 0$ (or, which is the same, on the manifold $s = 0$).

In this algorithm all the components of the control vector (7.59) are proportional to the norm of the state vector with the same proportionality factor α that should obey (7.62). These conditions may be relaxed and, thus, the value of control may be diminished if u is taken in the form of a piecewise linear function of all the coordinates x_i of the state vector with different discontinuous coefficients

$$u = -\Psi x. \tag{7.64}$$

The elements ψ_{ij} of the $(m \times n)$-dimensional matrix Ψ in (7.64) obey the following logical law:

$$\psi_{ij} = \begin{cases} \alpha_{ij} & \text{for} \quad d_i s_i^* x_j > 0, \quad i = 1, \ldots, m, \\ \beta_{ij} & \text{for} \quad d_i s_i^* x_j < 0, \quad j = 1, \ldots, n, \end{cases}$$

where α_{ij} and β_{ij} are constant coefficients. The time derivative of the positive definite quadratic form (7.60) is then

$$\frac{dv}{dt} = s^{*T}(Hx - D\Psi x) = \sum_{i=1}^{m} s_i^* \sum_{j=1}^{n} (h_{ij} - d_i \psi_{ij}) x_j.$$

If

$$|d_i| \alpha_{ij} > h_{ij}, \quad |d_i| \beta_{ij} < h_{ij}, \tag{7.65}$$

$dv/dt < 0$ everywhere outside of the discontinuity surfaces, i.e. the inequalities (7.65) also provide sliding mode on the intersection of discontinuity surfaces, $s^* = 0$ coinciding with the manifold $s = 0$. The time derivative of the Lyapunov function may be also estimated through the inequality (7.63) if

$$r_0 = \min_i \min_j |h_{ij} - d_i \psi_{ij}| \tag{7.66}$$

A similar reasoning may be applied to the case where the matrix preceding control in the equation of motion projection on the subspace s and s^* is transformed to the symmetrical positive definite one by means of (6.2) and (6.5). As the result, obtain in (7.57) and (7.58) a matrix D of the form $D = D^T$, $D > 0$. By taking the positive definite quadratic form

$$v = \tfrac{1}{2} s^* D^{-1} s^* \tag{7.67}$$

as the Lyapunov function, compute its derivative on the trajectories of the system (7.57) (and again confine ourselves only to this case):

$$\frac{dv}{dt} = s^{*T}(H^* x + u), \tag{7.68}$$

where $H^* = D^{-1} H$, $H^* = \| h_{ij}^* \|$. The function dv/dt, obviously, will be strictly negative everywhere outside of the discontinuity surfaces if

$$u = -\alpha |x| \operatorname{sign} s^*,$$

$$\alpha > h_i^*, \quad h_i^* = \max_j |h_{ij}^*|, \quad i = 1, \ldots, m, \tag{7.69}$$

and its upper bound coincides with (7.63) for

$$r_0 = \min_i (\alpha - h_i^*). \tag{7.70}$$

For the algorithm

$$u = -\Psi x, \quad \psi_{ij} = \begin{cases} \alpha_{ij} & \text{for} \quad s_i x_j > 0 \\ \beta_{ij} & \text{for} \quad s_i x_j < 0, \end{cases}$$

according to (7.68) dv/dt will be negative if

$$\alpha_{ij} > h_{ij}^*, \quad \beta_{ij} < h_{ij}^*.$$

In the estimate (7.63) for this case

$$r_0 = \min_i \min_j |h_{ij}^* - \psi_{ij}|. \tag{7.71}$$

Notably, in the case of diagonal matrix D there is no need in the Lyapunov functions approach because if (7.62) or (7.65) are fulfilled, sliding mode stability is provided for each of the discontinuity surfaces $s_i^* = 0$ taken separately according to the condition (1.9). If the matrix D is taken symmetrical and positive definite, sliding mode motion may not occur on each of the surfaces of $s_i^* = 0$ taken separately in spite of sliding mode stability on the manifold $s^* = 0$.

Two more cases where the question about sliding mode stability may be answered by means of Lyapunov functions conclude this section.

The first case is a generalization of systems with diagonal matrix D in (7.57) or (7.58). The Lyapunov function (7.60) may be used if D turns out to be Hadamard matrix, i.e. the inequalities

$$|d_{ii}| - \sum_{\substack{j=1 \\ j \neq i}}^{m} |d_{ij}| = \Delta d_i > 0, \quad i = 1, \ldots, m$$

are fulfilled for its elements. Denote the diagonal matrix with elements d_{ii} by D_0. The time derivative of the positive definite function (7.60) on the trajectories of the system (7.57) with control $u = -\alpha \operatorname{sign} D_0 s^*$ then has the form

$$\frac{dv}{dt} = s^{*T}(Hx - \alpha |x| D \operatorname{sign} D_0 s^*)$$

This function will be negative definite if $\alpha \Delta d_i > h_i$. In the estimate (7.63) that may be used for the systems with Hadamard matrix,

$$r_0 = \min_i (\alpha \Delta d_i - h_i).$$

Finally, the second case where the matrix D in (7.57) or (7.58) satisfies $D + D^T > 0$. This case was discussed in Sect. 4.3 by means of the piecewise smooth Lyapunov function (4.12). In virtue of (7.57) with control (7.69), an estimate similar to (4.25) can be obtained for the derivative of (4.13):

$$\frac{dv}{dt} = -\alpha |x| \cdot \| z \| \left(\lambda_0 - \frac{\| H \|}{\alpha} \right),$$

where λ_0 is the minimal eigenvalue of matrix $D+D^T$, and $\|H\| = \max_{|x|=1} \|Hx\|$. If the inequality $\alpha > \|H\|/\lambda_0$ is obeyed, the function $\dot{v}$ is, obviously, negative definite and, according to (4.24),

$$\frac{dv}{dt} \leqq -\alpha|x|r_0, \tag{7.72}$$

where $r_0 = \lambda_0 - \|H/\alpha > 0$.

Thus, for the four types of matrices D piecewise linear controls enable stability "in the large" of sliding modes on all the manifold $s=0$. The particular choice of D and reduction to it by means of transformations (6.2) or (6.5) are dictated by the simplicity of transformation realization.

6 Estimation of Convergence to Sliding Manifold

If sliding mode motion for the system (7.40), (7.42) is completely defined by the eigenvalues chosen at our will, the question about motion in the subspace s (or s^*) still has only a qualitative answer – the existence of the quadratic-form Lyapunov function ensures only stability of sliding motions. In order to estimate the rate of convergence of the state vector to the manifold $s=0$ we take advantage of the comparison principle as presented in Sect. 4.5. For the piecewise linear control functions (7.59) and (7.64) in question, the finiteness condition of positive definite quadratic-form Lyapunov function is not satisfied. Indeed, the Lyapunov function vanishes in finite time if the estimate (4.49) holds for its time derivative. Since $|x| \geqq \|x\|$, $|s^*| \geqq \|s^*\| \geqq (2v/\lambda_M)^{1/2}$ according to (7.63)

$$\frac{dv}{dt} < -\|x\|r_0\|s^*\| \leqq -\|x\|r_0\sqrt{\frac{2v}{\lambda_M}}, \tag{7.73}$$

i.e. the estimate (4.49) is always violated for $\|x\| \to 0$.

In all the cases, the Lyapunov function was taken as quadratic form $v = s^{*T}Ws^*$ (matrix W being equal to I_m (7.60) or D^{-1} (7.67)). The parameter r_0 in (7.73) assumes one of the values (7.63), (7.66), (7.70) or (7.71).

Let us see which estimate of the function v will be obtained if the comparison principle is applied to the inequality (7.73). The Lyapunov function is positive definite quadratic form, therefore, $v \leqq \lambda_M\|s^*\|^2$ or $\|s^*\| \geqq \lambda_M^{-1/2}v^{1/2}$ where λ_M is the maximal eigenvalue of matrix W. The estimate (7.73) is, correspondingly, rewritten as

$$\frac{dv}{dt} \leqq -\mu\|x\|v^{1/2}, \quad \mu = r_0\lambda_M^{-1/2}. \tag{7.74}$$

Let at initial time $\|x(t_0)\| = x_0$. Then, with regard to $s^* = \Omega Cx$ obtain

$$v(t_0) = v_0 = \lambda_c x_0^2, \tag{7.75}$$

where λ_c is the maximal eigenvalue of matrix $(\Omega C)^T W(\Omega C)$. Assume that

$$\|x(t)\| > \xi x_0 \tag{7.76}$$

for some number $0<\xi<1$ and $t>t_0$. Then, for motion in the domain (7.76) from (7.73), (7.75), (7.76) obtain by analogy with (4.50) the equation of the estimate of v [128]:

$$\dot{\rho} = -\mu\xi x_0 \rho^{1/2}, \quad \rho(t_0)=\rho_0=v_0. \tag{7.77}$$

The solution of (7.77)

$$\rho(t,t_0,\rho_0)=(\rho_0^{1/2}-\mu\xi x_0(t-t_0))^2 \tag{7.78}$$

vanishes for any initial ρ_0 if

$$t_\rho = t_0 + \rho_0^{1/2}\frac{1}{\mu\xi x_0}.$$

Since

$$v(t,t_0,v_0) \leqq \rho(t,t_0,\rho_0), \quad \rho_0=v_0 \leqq \lambda_c x_0^2$$

then

$$t_v \leqq t_\rho = t_0 + \frac{\lambda_c^{1/2}}{\mu\xi},$$

$$v(t_v,t_0,v_0)=0,$$

which implies that within the framework of our assumption (7.76) the function v will be finite and sliding motion will start in it in finite time. If the inequality (7.76) is not fulfilled over the interval $t_0 \leqq t \leqq t_v$, then

$$\|x(t_\xi)\| = \xi x_0, \quad t_0 < t_\xi \leqq t_v. \tag{7.79}$$

For this half-interval there exists a number L independent of x_0 and t_ξ such that[1]

$$\|x(t)\| \leqq x_0 e^{L(t_\xi - t_0)} \leqq x_0 M, \quad M=e^{L(t_\rho - t_0)}, \quad t_0<t<t_\xi.$$

Regarding the time instant t_ξ as the initial one, we see in a similar manner that the state vector will get on the manifold $s^*=0$ at time t_{v_1}, $t_{v_1} < t_\xi + \lambda_c^{1/2}/\mu\xi$, or

$$\|x(t_{\xi_1})\| \leqq \xi^2 x_0, \quad \|x(t)\| \leqq \xi^2 x_0 M,$$

$$t_{\xi_1} < t_{v_1}, \quad t_\xi < t < t_{\xi_1}$$

---

[1] For solution of the Eq. (7.40)

$$x(t) = x_0 + \int_0^t (Ax+Bu)d\tau$$

with piecewise linear control there exists a positive number L such that

$$\|x(t)\| \leqq x_0 + L\int_0^t \|x(\tau)\|\, d\tau.$$

According to the Bellman–Gronwall lemma [9], the estimate presented below follows from this inequality.

The repeated application of this reasoning to the subsequent time intervals with initial conditions $\xi^i x_0$ $(i=2,3,\ldots,)$ leads to the following conclusions:

the system state vector gets on the manifold $s^*=0$ in finite time, or

there exists a decaying exponential estimate for the solution of the system (7.40)

$$\|x(t)\| \leqq M x_0 e^{-(t-t_0)/T} \tag{7.80}$$

where

$$T=-(t_\rho-t_0)(\ln \xi)^{-1}.$$

Notably, if the manifold $s^*=0$ is reached in a finite time, the solution $x(t)$ before the instant of getting on the manifold is bounded by the exponent (7.80) as well.

Thus, in the discontinuous control system under study sliding mode occurs in finite time or the system is asymptotically stable. The value $t_\rho-t_0$ defining the time of reaching $s=0$ or the convergence rate in the case of asymptotic stability, is inversely proportional to the parameter r_0, and, therefore, it follows from (7.63), (7.66), (7.70) and (7.71) that the processes going on in the subspace s (or s^*) and ending on the manifold $s=0$ can be speeded up by just increasing the coefficients α or $|\alpha_{ij}|$ and $|\beta_{ij}|$ in (7.59) and (7.64).

Demonstrate now that the same conclusion about the convergence to the intersection of discontinuity surfaces may be drawn in the case of $D+D^T>0$ where sliding mode stability was checked by means of the Lyapunov function having the form of a sum of magnitudes. According to (7.72), the condition (4.48) is never obeyed for $|x|\to 0$. Hence, for the system in question one cannot use the result of Sect. 4.6 asserting that the state vector get on the manifold $s=0$ in finite time.

Assume that for some $0<\xi<1$ and $t>t_0$ motion within the domain

$$|x|>\xi|x(t_0)| \tag{7.81}$$

occurs. For the Lyapunov function (4.13), the inequality $v=|s|\leqq|C|\cdot|x|$ holds where $|C|=\max\limits_{|x|=1}|Cx|$. According to (7.72), the function v is decreasing, and therefore

$$v\leqq|C||X(t_0)|. \tag{7.82}$$

It follows from the comparison of (7.72), (7.81) and (7.82) that

$$v(t_v)=0 \quad \text{or} \quad s(t_v)=0, \quad t_v<\frac{|C|}{\alpha\xi r_0}.$$

If the inequality (7.81) is not satisfied over the interval $t_0\leqq t\leqq t_v$ obtain, by analogy with (7.79), $|x(t_\xi)|=\xi x(t_0)$, $t_0\leqq t_\xi<t_v$. Further, the above reasoning as applied to the quadratic-form Lyapunov functions leads to the conclusion that in this case sliding modes also occur in finite time, or the system is asymptotically stable. In this case also reduction of the time of getting on the manifold $s=0$ and speeding up convergence to the state space origin also is attained through increasing the coefficient α in the control algorithm (7.69).

Chapter 8

Systems with Scalar Control

1 Design of Locally Stable Sliding Modes

The methods of this chapter for allocation of the eigenvalues of the characteristic equation of sliding mode motion may be, of course, applied to the special case to systems with one-dimensional control

$$\dot{x} = Ax + bu, \tag{8.1}$$

where $u \in \mathbb{R}^1, b$ is $(n \times 1)$-dimensional column vector, and the pair $\{A, b\}$ is controllable. For the system (8.1) we present a design procedure that differs from the procedure of Chap. 7 and allows one to realize stable sliding modes with desirable eigenvalue allocation over all the discontinuity plane by means of control as a piecewise linear function only of a part of the state vector components rather than all of them.

According to the equivalent control method, the sliding mode motion is described in the case of linear discontinuity surface by $(n-1)$-st order linear equation. As was shown above, for a controllable system the roots of the characteristic equation of sliding mode motion may be placed in an arbitrary manner, and the problem reduces to that of providing sliding mode stability. The procedure for finding control in (8.1) will be regarded in this sequence.

Assume that for (8.1) there exists a linear control

$$u_l = \gamma x, \quad \gamma = (\gamma_1, \ldots, \gamma_k, 0, \ldots, 0), \tag{8.2}$$

depending only on k (e.g. first) coordinates of the state vector $x_1,\ldots,x_k$ for which $n-1$ eigenvalues of the linear system $\lambda_1,\ldots,\lambda_{n-1}$ correspond to the desired allocation of sliding mode motion eigenvalues, and the last eigenvalue λ_n may be an arbitrary real number. We now demonstrate that in this case by means of a control having the form of a piecewise linear function of the same coordinates stable sliding modes may be provided over all the discontinuity surface, the roots of the characteristic equation of sliding mode motion being $\lambda_1,\ldots,\lambda_{n-1}$.

Write (8.1) as

$$\dot{x} = A^*x + b(u - u_l), \tag{8.3}$$

where $A^* = A + b\gamma$, the eigenvalues of matrix A^* are $\lambda_1,\ldots,\lambda_n$. Let c be the eigen row vector of the matrix A^* corresponding to the eigenvalue λ_n, i.e.

$$cA^* = \lambda_n c. \tag{8.4}$$

The Eq. (8.4) has a non-trivial solution with respect to c, therefore, there is a non-singular transformation enabling representation of (8.3) with $u = u_l$ in the following form:

$$\dot{x}^1 = A_{11}x^1 + a_s s, \quad \dot{s} = \lambda_n s, \tag{8.5}$$

where $s = cx$, the components of the $(n-1)$-dimensional vector x^1 coincide with $(n-1)$ component of the initial state vector x, A_{11} is $(n-1)\times(n-1)$- dimensional matrix, and a_s is $(n-1)\times 1$-dimensional vector. Since at $u = u_l$ (8.3) is equivalent to (8.5), the eigenvalues of A_{11} are equal to $\lambda_1,\ldots,\lambda_{n-1}$.

In the general case where control u does not coincide with u_l, the system motion Eq. (8.1) or (8.3) in the space x^1, s will be

$$\begin{aligned} \dot{x}^1 &= A_{11}x^1 + a_s s + b_1(u - u_l), \\ \dot{s} &= \lambda_n s + cb(u - u_l), \end{aligned} \tag{8.6}$$

where b_1 is $(n-1)$-dimensional column vector,

$$cb \neq 0. \tag{8.7}$$

(Had the conditions (8.7) been violated, the scalar function s would have been independent of the control which contradicts the assumption about the controllability of (8.1).)

Take the plane $s = 0$ as the discontinuity surface of control

$$u = \begin{cases} u^+(x) & \text{for} \quad s > 0, \\ u^-(x) & \text{for} \quad s < 0, \end{cases} \tag{8.8}$$

where $u^+(x)$ and $u^-(x)$ are continuous state functions. According to the method of equivalent control, in order to obtain the sliding mode equations one must substitute u_{eq}, the solution of $\dot{s} = 0$ with respect to control, into (8.6) and assume that $s = 0$. Obviously $u_{\text{eq}} = u_l$ and $\dot{x}^1 = A_{11}x^1$ is the sliding equation. The eigenvalues of the matrix A_{11} are $\lambda_1,\ldots,\lambda_{n-1}$, i.e. we have obtained the desired eigenvalue allocation for sliding mode motion.

The inequalities (1.9) allow us to derive the conditions for stable sliding modes in the system (8.6), (8.7), (8.8) over all the plane $s=0$ to exist:

$$cbu^{+} < cbu_l, \quad cbu^{-} < cbu^{+}. \tag{8.9}$$

Since u_l is a linear combination of k coordinates of the state vector, the inequalities (8.9) may be fulfilled by using piecewise linear control u with respect to the same coordinates

$$u = -\Psi x - \delta, \tag{8.10}$$

with

$$\Psi = (\psi_1, \ldots, \psi_k, 0, \ldots, 0),$$

$$\psi_i = \begin{cases} \alpha_i & \text{for} \quad (cb)x_i s > 0, \\ \beta_i & \text{for} \quad (cb)x_i s < 0, \end{cases} \qquad i = 1, \ldots, k,$$

$$\delta = \delta_0 \operatorname{sign}(cb)s,$$

δ_0 is an arbitrarily small constant or system-state-dependent positive value that is non-zero everywhere but, possibly, in the point $x=0$,

$$\alpha_i \geqq -\gamma_i, \quad \beta_i \leqq -\gamma_i. \tag{8.11}$$

Thus, if the set $\lambda_1, \ldots, \lambda_{n-1}$ is the desired eigenvalue allocation for sliding mode and there exists a linear control (8.2) allowing one to obtain these eigenvalues for $k<n$ in the linear system (8.1), all the discontinuity surface may be converted into a sliding domain with the desired sliding motions by means of control algorithms of the (8.10) type that are simpler as compared with (7.64).

The inverse assertion holds as well: if over all the discontinuity plane in the system (8.1) with discontinuous control (8.10) there exists a stable sliding mode defined by the roots $\lambda_1, \ldots, \lambda_{n-1}$ of characteristic equation, then there is a linear control of the form of (8.2) for which the linear system (8.1) has the same eigenvalues. To substantiate this assertion, let us discuss the behaviour of the system in the space x^1, s if the $(n-1)$-dimensional vector x^1 again consists of the components of the original state vector x. If for some l in the vector c

$$c_l \neq 0, k < l \leqq n \tag{8.12}$$

then the coordinate x_l will not be included into x^1; if

$$c = (c_1, \ldots, c_k, 0, \ldots, 0), \tag{8.13}$$

the vector x^1 is independent of one of the coordinates $x_l, 1 \leqq l \leqq k$. After this non-singular transformation, the motion equations are

$$\dot{x}^1 = \bar{A}_{11} x^1 + a_s s + b_1 u \tag{8.14}$$

$$\dot{s} = \lambda s + r x^1 + cbu \tag{8.15}$$

where λ is scalar coefficient, and r is $(n-1)$-dimensional row vector.

In (8.15) $cb \neq 0$, because otherwise $\dot{s}$ is continuous and sliding mode cannot occur over all the plane $s=0$. The existence of stable sliding mode in any point of the plane $s=0$ implies that in the case of (8.12)

$$r=(r_1,\dots,r_k,0,\dots,0) \tag{8.16}$$

and in the case of (8.13)

$$r=(r_1,\dots,r_{l-1},r_{l+1},\dots,r_k,0,\dots,0). \tag{8.17}$$

Indeed, if it turns out that $r_p \neq 0$ for $p>k$, the condition (1.9) will be violated in the point of discontinuity surface with the coordinates $x_i=0, i=1,\dots,k$, $x_p \neq 0, |x_p| > |cb|\delta_0$.

Let us write the equation of sliding mode on the plane $s=0$ by using the equivalent control method. To this end, substitute the solution of the equation $\dot{s}=0$ (8.15) at $s=0$

$$u_{\mathrm{eq}} = -(cb)^{-1}rx^1 \tag{8.18}$$

into (8.14)

$$\dot{x}^1 = (\bar{A}_{11} - (cb)^{-1}b_1 r)x^1 + a_s s, \quad \dot{s} = \lambda s \tag{8.19}$$

By convention, the eigenvalues of (8.19) are equal to $\lambda_1,\dots,\lambda_{n-1}$. On the other hand, the substitution of the linear control

$$u_l = -(cb)^{-1}rx^1 \tag{8.20}$$

into the system (8.14), (8.15) leads to the following system of equations

$$\dot{x}^1 = (\bar{A}_{11} - (cb)^{-1}b_1 r)x^1 + a_s s, \quad \dot{s} = \lambda s$$

where $\lambda_1,\dots,\lambda_{n-1}$ are also the eigenvalues. According to (8.16), (8.17), in linear control all the coefficients $r_i (i>k)$ are zero which substantiates the above assertion.

Let us formulate the major result of this study of linear scalar-control system in the form of the following

Theorem. *For the existence of control in the form of (8.10) providing in system (8.1) the eigenvalue allocation $\lambda_1,\dots,\lambda_{n-1}$ desired for sliding mode motion and the sliding mode stability over all the discontinuity plane $s=0$, it is necessary and sufficient that there exists a linear control of the form of (8.2) such that the spectrum of matrix A^* in (8.3) contains the same eigenvalues.*

In particular, if the asymptotic stability of the origin in the state space is investigated, it is necessary to find a linear control such that there be $n-1$ roots of the characteristic equation in the left semi-plane of the complex plane of eigenvalues rather than n roots as it is the case with linear systems. Then, a plane may be chosen with stable sliding modes in each point, and sliding mode motion described by the $(n-1)$-st order equation may be made asymptotically stable. For this purpose, the discontinuous control must be composed of the same coordinate as the obtained linear control, and the coefficients in the control must be changed stepwise according to the switching logic (8.10), (8.11).

Let us present another version of the algorithm with piecewise linear control that solves the same problem as the control (8.10) but in contrast to it has one discontinuous coefficient only. The sliding mode stability condition (8.9) may be satisfied also if

$$u = -\psi_l u_l - \delta, \tag{8.21}$$

where

$$\psi_l = \begin{cases} \alpha_l & \text{for} \quad u_l s > 0 \\ \beta_l & \text{for} \quad u_l s < 0, \end{cases}$$

α_l and β_l are constant coefficients,

$$\alpha_l \geqq -1, \quad \beta_l \leqq -1 \tag{8.22}$$

The control component δ is taken in the same form as in (8.10).

2 Conditions of Sliding Mode Stability "in the Large"

Unlike the algorithms (7.59) and (7.64) of Sect. 7.5, the controls (8.10) and (8.21) are piecewise linear functions of only a part of the state vector components rather than of all of them and, therefore, they are simpler to realize. However, the problem of designing with them a system having the desired eigenvalue allocation for sliding mode motion has not yet been solved because the design procedures of Sect. 7.5 were oriented to sliding modes stable "in the large" whereas the conditions (8.9), (8.11) ensure only the local stability. Let us, therefore, consider the conditions that should be obeyed by the eigenvalues of each of the linear subsystems of the discontinuous system (8.1) with controls (8.10) or (8.21) in order to provide the convergence of the state vector from any initial state to the discontinuity plane $s = 0$. Below, we are going to present the results of works [12, 13] where the above formulated problem was solved.

The system behaviour will be considered in the space of variables $y_1, \ldots, y_{n-1}, s$ related to x^1 and s by a non-singular transformation

$$y = x^1 - \frac{1}{cb} b_1 s, \tag{8.23}$$

where $y^{\mathrm{T}} = (y_1, \ldots, y_{n-1})$. The motion equations in the space of new coordinates are found from (8.23) and (8.6)

$$\dot{y} = A_{11} y + as, \tag{8.24}$$

$$\dot{s} = \lambda_n s + cb(u - u_l), \tag{8.25}$$

with

$$a = a_s + \frac{1}{cb} A_{11} b_1 - \frac{\lambda_n}{cb} b_1.$$

Consider first the case where control u is taken in the form of (8.21) and $\delta = 0$, or

$$u = -\frac{\alpha_l + \beta_l}{2} + \frac{\alpha_l - \beta_l}{2}|u_l| \operatorname{sign} cbs. \tag{8.26}$$

Now we determine the conditions under which the state vector in a system with control (8.26) always gets into the plane $s = 0$, and the conditions are valid if the small component δ is non-zero.

After the substitution of the control (8.26) into (8.25) obtain

$$\dot{s} = \lambda_n s + \mu u_l - k|u_l| \operatorname{sign} s, \tag{8.27}$$

with

$$\mu = -cb\frac{\alpha_l + \beta_l}{2} - cb,$$

$$k = |cb|\frac{\alpha_l - \beta_l}{2},$$

$$u_l = qy + q_0 s, \quad q = (q_1, \dots, q_{n-1})$$

the constant coefficients $q_0, q_1, \dots, q_{n-1}$ are found from (8.2) and (8.23). Depending on the sign of λ_n, two cases may be identified in the problem of state vector convergence to the plane $s = 0$. In the first case where $\lambda_n < 0$, the condition of sliding mode stability on the plane $s = 0$ is evident: if

$$k > |\mu| \quad \text{or} \quad \alpha_l > -1, \quad \beta_l < -1, \tag{8.28}$$

the positive definite function $\frac{1}{2}s^2$ and its time derivative along the trajectories of the system (8.27) satisfy the conditions of the second theorem on stability "in the large" (Sect. 4.2). Only a special case is disregarded where the plane $u_l = 0$ contains whole trajectories of one of the linear subsystems because in this case the plane $s = 0$ also contains whole trajectories. According to the definition of Ch. 4, this motion is not sliding mode.

The inequalities (8.28) are at the same time the necessary stability condition "in the small" for sliding mode on the entire plane $s = 0$ because if the sign of at least one of them is changed, $\dot{s}$ at $u_l \neq 0$ will not change the sign on the plane $s = 0$ and sliding mode canot occur. For the boundary case where one or both inequalities in (8.28) become equalities, the plane $s = 0$ contains whole trajectories but, although stable, these motions are not sliding modes. Notably, for $\lambda_n < 0$ the conditions of stability "in the small" (8.22) as obtained in the last section for the control (8.21) with small in the amplitude discontinuous function δ are at the same time the conditions of stability "in the large".

Consider now a non-trivial case where $\lambda_n \geqq 0$. It will be assumed below that A_{11} defining sliding mode motion is Hurwitz matrix and that the necessary condition of sliding mode stability "in the small" (8.28) is obeyed.

For the linear part of the system (8.24), (8.27), let us determine the transfer function from the input s to the output $-u_l$

$$\chi(p) = q(A_{11} - pI_{n-1})^{-1}a - q_0. \tag{8.29}$$

In the case where the system (8.24), (8.27) has real solutions in the form of

$$y(y) = y(0)e^{pt}, \quad s(t) = s(0)e^{pt}, \tag{8.30}$$

the following equation should be used in order to find them:

$$\frac{p - \lambda_n}{k} = -\left(\frac{\mu}{k}\operatorname{sign}\chi(p) + 1\right)|\chi(p)|. \tag{8.31}$$

Theorem. *For one of the two conditions*

$$\lim_{t\to\infty} s = 0 \quad \text{or} \quad s(t_s) = 0, \quad t_s > 0 \tag{8.32}$$

to be fulfilled, it is necessary and sufficient that the Eq. (8.31) does not have non-negative real eigenvalues.

The theorem formulates the condition for the state vector to reach the plane $s = 0$ in finite time or asymptotically from any initial points $s(0) \neq 0$. If one assumes that the Eq. (8.31) has a non-negative eigenvalue p_0, it is possible to choose initial conditions such that the solution will be in the form of (8.30) for $p = p_0 \geqq 0$. In this case none of the conditions (8.32) for state vector convergence to the plane $s = 0$ will be fulfilled which proves the necessity of the theorem condition.

Without loss of generality, consider the system behaviour in the domain $s > 0$ and substantiate the sufficiency of this condition. Introduce an auxiliary function

$$W_\lambda = s + h_\lambda y + \delta_1 v^{1/2} \tag{8.33}$$

where h_λ is a constant row vector depending on the scalar parameter λ, δ_1 is a positive number, the positive definite quadratic form $v = y^T Hy$ satisfies the condition $A_{11}H + HA_{11} < 0$. Since A_{11} is Hurwitz matrix, there will be always such a matrix H. The time derivative of function v on the trajectories of (8.24) satisfies the inequality

$$\frac{dv}{dt} = y^T(A_{11}^T H + HA_{11})y + 2aHys \leqq -\gamma_1 v + 2\gamma_2 v^{1/2} s, \tag{8.34}$$

where $\gamma_1 > 0$, $\gamma_2 > 0$ are numbers. With due regard to the inequality (8.34), the derivative of W_λ calculated in virtue of the system (8.24), (8.25) is

$$\frac{dW_\lambda}{dt} = \lambda_n s + \mu u_l - k|u_l| + h_\lambda(A_{11}y + as) + (-\tfrac{1}{2}\delta_1\gamma_1 v^{1/2} + \delta_1\gamma_2 s). \tag{8.35}$$

Introduce the notations

$$h_\lambda = (\nu_\lambda - \mu) q (A_{11} - \lambda I_{n-1})^{-1},$$

$$\nu_\lambda = \mu - \frac{\lambda_n - \lambda + \gamma_2 \delta_1 + \varepsilon}{\chi(\lambda)}, \tag{8.36}$$

where $\varepsilon > 0$ is a positive number.

First, calculate $h_\lambda(A_{11} y + as)$ with allowance for the notations (8.36):

$$h_\lambda(A_{11} y + as) = \frac{\lambda_n - \lambda + \gamma_2 \delta_1 + \varepsilon}{\chi(\lambda)} (qy + \lambda q (A_{11} - \lambda I_{n-1})^{-1} y$$
$$+ q(A_{11} - \lambda I_{n-1})^{-1} as).$$

By substituting, respectively, $u_l - q_0 s$ for qy and $\chi(\lambda)$ for $q(A_{11} - \lambda I_{n-1})^{-1} a - q_0$, and $\nu_\lambda - \mu$ for $-(\lambda_n - \lambda - \gamma_2 \delta_1 + \varepsilon)/\chi(\lambda)$, and h_λ for $(\nu_\lambda - \mu) q (A_{11} - \lambda I_{n-1})^{-1}$ obtain

$$h_\lambda(A_{11} + as) = (\nu_\lambda - \mu) u_l + \lambda h_\lambda y - (\lambda_n - \lambda + \gamma_2 \delta_1 + \varepsilon) s.$$

Substitute the resulting relation into (8.35)

$$\frac{dW_\lambda}{dt} \leqq \lambda W_\lambda + \nu_\lambda u_l - k |u_l| - (\lambda + \tfrac{1}{2}\gamma_1) \delta_1 v^{1/2} - \varepsilon s. \tag{8.37}$$

The proof framework is as follows. Depending on s, the state vector y belongs to some surface $W_\lambda = 0$. It will be demonstrated below that the solution of the system (8.24), (8.25) tends to the origin, or for $s > 0$ the parameter λ must assume a value such that its corresponding surface will lie under the plane $s = 0$ and, as a result, the state vector will get at some instant on the discontinuity plane $s = 0$.

Denote by $\lambda_{cr} = \lambda_n + \gamma_2 \delta + \varepsilon$ and show that

$$k > |\nu_\lambda| \quad \text{with} \quad \lambda \in [0, \lambda_{cr}]. \tag{8.38}$$

The inequality (8.38) reduces to two inequalities

$$k |\chi(\lambda)| > \mu \chi(\lambda) + \lambda - \lambda_{cr} \quad \text{for} \quad \mu \chi(\lambda) + \lambda - \lambda_{cr} > 0 \tag{8.39}$$

$$k |\chi(\lambda)| > -\mu \chi(\lambda) - \lambda + \lambda_{cr} \quad \text{for} \quad \mu \chi(\lambda) + \lambda - \lambda_{cr} < 0. \tag{8.40}$$

Rewrite (8.40) as

$$-|\chi(\lambda)| \left(1 + \frac{\mu}{k} \operatorname{sign} \chi(\lambda) \right) > (\lambda - \lambda_{cr})/k. \tag{8.41}$$

The inequality (8.41) cannot become equality under any $\lambda \in [0, \lambda_{cr}]$ and $\chi(\lambda_{cr}) \neq 0$, otherwise, the Eq. (8.31) would have a non-negative eigenvalue. Since $k > \mu$ (8.28) and $\chi(\lambda_{cr}) \neq 0$, the inequality (8.41) is obeyed in the point $\lambda = \lambda_{cr}$. The function $\chi(\lambda)$ is continuous for $\lambda \in [0, \lambda_{cr}]$ (it will be recalled that A_{11} is Hurwitz matrix), therefore, the inequality (8.41) or (8.40) is satisfied over this interval. The validity of the inequality (8.39) follows from $k > \mu$ and $\chi(\lambda_{cr}) \neq 0$.

Let us assume that for $\lambda = 0$ at the initial time $W_0(0) > 0$. If the condition (8.38) is obeyed, the derivative (8.37) of this function is negative. By means of the technique used in Sect. 6.6 for stydying the convergence to the manifold $s = 0$, one can show for solution of the system (8.24), (8.27) that there is an upper-estimate decaying exponent, or that the solution in finite time will get on the surface $W_0 = 0$. The former case means that the system is asymptotically stable and that the condition $\lim s = 0$ for $t \to \infty$ is fulfilled. Consider now the latter case where the motion from the surface $W_0(t_1) = 0$ begins at some time t_1. The motion outside the plane $s = 0$ may be represented as the motion of the end of state vector over the surface $W_\lambda = 0$ with a time-varrying parameter $\lambda(t)$. Variation of s is defined then by

$$s(y, \lambda) = -h_\lambda y - \delta_1 v^{1/2}, \tag{8.42}$$

and the system state is defined by the $(n-1)$-dimensional vector y and scalar parameter λ.

For fixed vector y and $\lambda \in [0, \lambda_{cr}]$, the function $s(y, \lambda)$ is continuous because A_{11} is a Hurwitz matrix. It follows from the condition $h_{\lambda_{cr}} = 0$ and (8.42) that $s(y, \lambda_{cr}) < 0$. The function $s(y, \lambda)$ is plotted in Fig. 16 under fixed y. Notably, if several of λ may correspond to the same s, the greatest of them will be always chosen in what follows, and for this value

$$\frac{\partial s}{\partial \lambda} = -\frac{dh_\lambda}{d\lambda} y \leqq 0. \tag{8.43}$$

(Thereby we assume that in some points $\lambda(t)$ undergoes stepwise changes, e.g. for $s = s_1$ the passage from λ_1 to λ_2 occurs).

As the state vector is always on one of the surfaces $W_\lambda = 0$,

$$\left.\frac{dW_\lambda}{dt}\right|_{\lambda = \text{const}} + \frac{dW_\lambda}{d\lambda}\frac{d\lambda}{dt} = 0$$

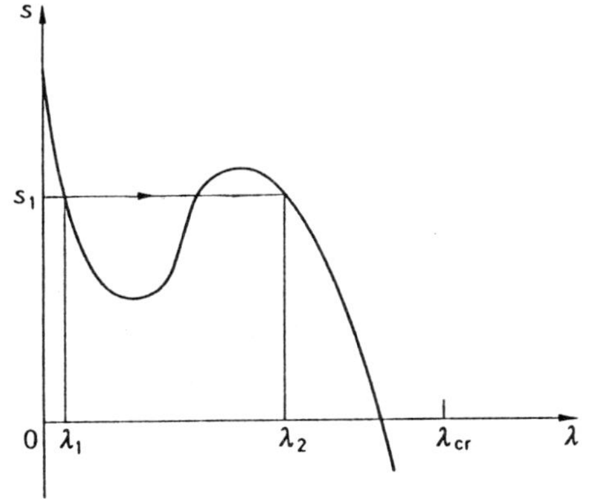

Fig. 16

and, according to (8.33), (8.37) with $W_\lambda = 0$

$$\left(\frac{dh_\lambda}{d\lambda} y\right)\frac{d\lambda}{dt} \geqq (\lambda + \tfrac{1}{2}\gamma_1)\delta_1 v^{1/2}. \tag{8.44}$$

The function h_λ is continuous and continuously differentiable over the interval $\lambda \in [0, \lambda_{cr}]$; therefore, there is a positive number γ_3 such that

$$\left\| \frac{dh_\lambda}{d\lambda} y \right\| \leqq \gamma_3 v^{1/2}. \tag{8.45}$$

From (8.43), (8.44), (8.45) obtain the lower bound of the speed of parameter λ for $\lambda \in [0, \lambda_{cr}]$

$$\frac{d\lambda}{dt} \geqq \frac{1}{2}\frac{\delta_1 \gamma_1}{\gamma_3}.$$

Consequently, at a finite time instant $t_2 > t_1$

$$\lambda(t_2) = \lambda_{cr}, \quad 0 < t_1 < t_2 \leqq \frac{2\lambda_{cr}\gamma_3}{\delta_1 \gamma_1}.$$

As it was noted above, the surface $W_{\lambda_{cr}} = 0$ is completely under the plane $s = 0$ or $s(y, \lambda_{cr}) < 0$; therefore, any trajectory beginning on the surface $W_0(0)$ will cross the plane $s = 0$ in finite time in virtue of the continuity of $s(\lambda)$. The theorem is proved.

In the case where the control (8.10) is used with $\delta = 0$, the theorem condition holds if (8.31) is rewritten as

$$p - \lambda_n = -\sum_{i=1}^{k} (\mu \operatorname{sign} \chi_i(p) - k_i)|\chi_i(p)|,$$

where

$$\chi_i(p) = q_i(A_{11} - pI_{n-1})^{-1}a - q_{0i},$$

$$x_i = q_i y + q_{0i}s, \quad q_i = (q_{1i}, \ldots, q_{n-1,i}),$$

$$\mu_i = -cb\left(\frac{\alpha_i + \beta_i}{2} - \gamma_i\right), \quad k_i = |cb|\frac{\alpha_i - \beta_i}{2}$$

Inequalities similar to (8.28) also must be obeyed:

$$k_i > |\mu_i|, \quad i = 1, \ldots, k. \tag{8.46}$$

This implies that if the theorem conditions are fulfilled and sliding modes are stable "in the small", they will be also stable "in the large". We have established that it suffices for this purpose to add to the control the component $\delta = \delta_0 \operatorname{sign}(cb)s$. If δ_0 is taken constant or having the form of $\delta_0 = \delta_0^1|y|$ for sufficiently small δ_0 or δ_0', the proof of the theorem about state vector getting into the plane $s = 0$ is completely the same.

3 Design Procedure: An Example

It follows from the last two sections that the design procedure should involve the following sequence of steps. First, the choice of linear control (8.2) for which $n-1$ roots of the characteristic equation have the desired allocation. Second, the coefficients of the discontinuity plane equation from the condition (8.4) and piecewise linear control (8.10) or (8.21) are determined. Finally, parameters of linear subsystems are determined so as to satisfy the conditions of the theorem on sliding mode stability "in the large" together with the conditions (8.28) or (8.46). Obviously, there always is a number k in (8.2) for which the conditions of each design stage will be satisfied by means of control (8.10) because the problem of the desired eigenvalue allocation for sliding modes and their stability "in the large" has been solved in Sect. 7.4, 5 for $k=n$.

Let us consider by way of example a 3-rd order system described by the equations

$$\dot{x}_1 = x_2, \quad \dot{x}_2 = x_3, \quad \dot{x}_3 = u. \tag{8.47}$$

Assume that it is desired to provide a given degree of stability η for sliding mode motion, i.e.

$$\operatorname{Re}\lambda_{1,2} \leqq -\eta \tag{8.48}$$

with λ_1 and λ_2 being the roots of the characteristic equation defining the sliding mode motion. Let us take linear control in the form of

$$u_l = \gamma x_1, \tag{8.49}$$

where γ is scalar coefficient. For $\gamma > 0$, two of the three roots of the characteristic equation of (8.47) with control (8.49)

$$\lambda_{1,2} = -\gamma^{1/3}\left(\frac{1}{2} \pm j\frac{\sqrt{3}}{2}\right), \quad \lambda_3 = \gamma^{1/3}$$

have negative real parts, and for $\gamma^{1/3} \geqq 2\eta$ they satisfy (8.48).

As it was found in Sect. 8.1, two versions of control are possible for these systems: (8.10) with k discontinuous coefficients or (8.21) with a single one. In our example $k=1$ and the versions differ only in the form of presentation. Let us dwell upon the version (8.21) which for the system (8.47), (8.49) is as follows

$$u = -\psi_l u_l - \delta, \tag{8.50}$$

with

$$\psi_l = \begin{cases} \alpha_l & \text{for} \quad u_l s > 0, \\ \beta_l & \text{for} \quad u_l s < 0, \end{cases}$$

$$s = c_1 x_1 + c_2 x_2 + x_3, \quad \delta = \delta_0 \operatorname{sign} s$$

The coefficients c_1, c_2 and 1 are the components of the eigenvector c of matrix

$$A^* = A + b\gamma = \begin{pmatrix} 0 & 1 & 0 \\ 0 & 0 & 1 \\ \gamma & 0 & 0 \end{pmatrix}$$

corresponding to the eigenvalue $\lambda_3 = \gamma^{1/3}$ or

$$cA^* = \gamma^{1/3} c. \tag{8.51}$$

The solution of this equation

$$c = (\gamma^{2/3}, \gamma^{1/3}, 1) \tag{8.52}$$

enables determination of the discontinuity plane $s = 0$. The equation for sliding on this plane may be derived by substituting $x_3 = -c_1 x_1 - c_2 x_2$ into (8.47)

$$\dot{x}_1 = x_2, \quad \dot{x}_2 = -\gamma^{2/3} x_1 - \gamma^{1/3} x_2$$

As could be expected, the numbers

$$\lambda_{1,2} = \gamma^{1/3}\left(\frac{1}{2} \pm j\frac{\sqrt{3}}{2}\right)$$

are the roots of the characteristic equation of the resulting second-order system. Represent its equations in the form of (8.24), (8.27)

$$\begin{aligned} \dot{x}_1 &= x_2, \\ \dot{x}_2 &= -\gamma^{2/3} x_1 - \gamma^{1/3} x_1 + s, \\ \dot{s} &= \gamma^{1/3} s - \mu u_l - k|u_l| \operatorname{sign} s. \end{aligned} \tag{8.53}$$

with

$$\mu = -\frac{\alpha_l + \beta_l}{2} - 1, \quad k = \frac{\alpha_l - \beta_l}{2}, \quad u_l = \gamma x_1.$$

The coefficients α_l and β_l satisfy the condition (8.28). It follows from the comparison of (8.24), (8.27) and (8.53) that for the case in question

$$A_{11} = \begin{pmatrix} 0 & 1 \\ -\gamma^{2/3} & -\gamma^{1/3} \end{pmatrix}, \quad a = \begin{pmatrix} 0 \\ 1 \end{pmatrix}, \quad q = (\gamma, 0), \quad q_0 = 0.$$

and the transfer function (8.29) and Eq. (8.31) are, respectively, as follows

$$\chi(p) = -\frac{\gamma}{p^2 + \gamma^{1/3} p + \gamma^{2/3}},$$

$$\frac{p - \gamma^{1/3}}{k} = \left(-\frac{\mu}{k} \operatorname{sign} \frac{\gamma}{p^2 + \gamma^{1/3} p + \gamma^{2/3}} + 1\right) \left|\frac{\gamma}{p^2 + \gamma^{1/3} p + \gamma^{2/3}}\right|. \tag{8.54}$$

Since $\lambda_3 = \gamma > 0$, according to the theorem conditions we have just to make sure that only real non-negative eigenvalues are missing in (8.54). Consequently, $p^2 + \gamma^{1/3} p + \gamma^{2/3} > 0$ and after a transformation we obtain $p^3 = \gamma + \gamma\mu - \gamma k$.

The theorem condition will be fulfilled if $k > \mu + 1$ or $\alpha_l > 0$.

As a result, we have found all the coefficients in the control (8.49), (8.50): $\gamma = 8\eta^3$, $\alpha_l > 0$, $\beta_l < -1$, $c_1 = 4\eta^2$, $c_2 = 2\eta$. Notably, the desired stability degree of (8.47) at the occurrence of sliding mode motion is provided by two unstable continuous linear subsystems, and the control itself is a piecewise linear function of only one system coordinate. In the class of linear systems, control in a similar problem must be a linear combination of all the three coordinates, otherwise, the system will not be asymptotically stable.

Here we have discussed design methods for systems with scalar controls that are piecewise linear functions of only some of the state vector coordinates rather than all of them. Ref. [143] generalizes the methods of systems with vector controls.

4 Systems in the Canonical Form

As it was noted in the discussion of control algorithms for the class of variable-structure systems with scalar control and scalar output (Sect. 1.3) the desired control quality was attained through inducing sliding modes in the space of error coordinate and its derivatives $x, \dot{x}, \ldots, x^{(n-1)}$ [130]. This choice of the canonic space is due to the fact that sliding modes over a plane

$$s = c_1 x + c_2 \dot{x} + \cdots + c_{n-1} x^{(n-2)} + x^{(n-1)} = 0$$

in this space are invariant to parameteric perturbations and external disturbances. Indeed, the plane equation is at the same time the differential equation of sliding mode (this fact follows also from the equivalent control method) whose solution depends only on the coefficients $c_1, \ldots, c_{n-1}$ as it is the case (1.11) with the second-order systems.

But it should be remembered that as a rule the original equations are written with respect to variables characterizing physical processes in individual system elements rather than to canonical variables which implies that the system behaviour is described by a general equation

$$\dot{x} = Ax + bu \tag{8.55}$$

and some coordinate

$$y_1 = c_0 x, \quad c_0 = (c_{01}, c_{02}, \ldots, c_{0n}) \tag{8.56}$$

is the controlled variable. It is assumed that in (8.55) the pair $\{A, b\}$ is controllable and $\{c_0, A\}$ is observable. In the following chapter we shall discuss in more detail why observability is required in the design, but here we just give the condition under which this requirement fulfilled:

$$\operatorname{rank}\begin{pmatrix} c_0 \\ c_0 A \\ \vdots \\ c_0 A^{n-1} \end{pmatrix} = n. \tag{8.57}$$

It is important to note that the standard technique of reducing controlled systems to the canonical space (Sect. 7.3) may be unacceptable for two reasons. First, the coordinate y_1 in (7.21) after the transformation (7.39) is not, generally speaking, equal to the controlled variable (8.56). Second, exact parameters of the matrix A of the original system (8.55) are required for realization of (7.39).

In this connection, it would be reasonable to consider control system design in the space of coordinate y_1 (8.56) and its derivatives under the assumption of direct differentiability of the coordinate. With this problem formulation, the study of system (8.55) reducibility to the canonical space and the conditions under which sliding mode motions may have the desired properties is required.

To reduce (8.55) to the canonical form, let us determine the derivatives of the output y_1 (8.56). Since (8.55) is controllable i.e.

$$\operatorname{rank}(b, Ab, \ldots, A^{n-1}b) = n,$$

all the numbers $c_0 A^i b$ cannot be simultaneously equal to zero, otherwise the vector c_0 will be zero which contradicts the problem formulation. Consequently, there will be $l \leq n$ such that

$$\begin{aligned} y_1^{(i)} &= c_0 A^i x, \quad i = 0, 1, \ldots, l-1, \\ y_1^{(l)} &= c_0 A^l x + c_0 A^{l-1} bu, \quad c_0 A^{l-1} b \neq 0. \end{aligned} \tag{8.58}$$

Introduce notations

$$\dot{y}_i = y_{i+1}, \quad i = 1, \ldots, l-1 \tag{8.59}$$

and then the last equation in (8.58) has the following form

$$\dot{y}_l = c_0 A^l x + c_0 A^{l-1} bu. \tag{8.60}$$

If sliding motion is induced in an l-dimensional subspace on the plane

$$s = cy^l = 0, \tag{8.61}$$

where $c = (c_1, \ldots, c_{l-1}, 1), (y^l)^{\mathrm{T}} = (y_1, \ldots, y_l)$, then its equation may be derived by substitution of $-\sum_{i=1}^{l-1} c_i y_i$ for y_l in (8.59):

$$\begin{aligned} \dot{y}_i &= y_{i+1}, \quad i = 1, \ldots, l-2, \\ \dot{y}_{l-1} &= -\sum_{i=1}^{l-1} c_i y_i. \end{aligned} \tag{8.62}$$

The sliding mode motion in the canonical subspace $y_1, \ldots, y_l$ is, obviously, dependent only on the coefficients $c_1, \ldots, c_{l-1}$ of the equation of plane (8.61), i.e. is invariant.

In order to obtain a control algorithm providing stability "in the large", let us write a differential equation with respect to the coordinate s by means of (8.58) through (8.61):

$$\dot{s} = dx + c_0 A^{l-1} bu, \quad d = \sum_{i=1}^{l} c_i c_0 A^i, \quad c_l = 1 \tag{8.63}$$

One can readily see that the derivative of the positive definite function $v=\frac{1}{2}s^2$ of the scalar argument s as computed with regard to (8.63) is negative definite if

$$u=-\psi x, \tag{8.64}$$

$$\psi=(\psi_1,\ldots,\psi_n),$$

$$\psi_i=\begin{cases}\alpha_i & \text{for} \quad c_0A^{l-1}bx_is>0,\\ \beta_i & \text{for} \quad c_0A^{l-1}bx_is<0,\end{cases}$$

$$\alpha_i>|c_0A^{l-1}b|^{-1}d_i, \quad \beta_i<|c_0A^{l-1}b|^{-1}d_i, \quad d=(d_1,\ldots,d_n).$$

According to Sect. 7.6, for the solution of the system in question with control (8.64) there exists an upper bound as a decreasing exponent, or the state vector in finite time gets on the plane $s=0$ over which stable sliding modes exist everywhere. The decay rate of the exponent and, as a result, the decay rate of the controlled variable $y_1(t)$ or the convergence rate of vector x to the plane $s=0$ grow with the magnitude of coefficients α_i and β_i. An appropriate choice of the coefficients $c_1,\ldots,c_{l-1}$ enables the desired dynamics of sliding mode motion described by (8.62) with respect to the controlled variable and its derivatives.

The control (8.64), thus, allows one to solve the problem of controlling the output coordinate by deliberately inducing sliding modes in the canonical l-dimensional subspace. But this system may be regarded as working only if the general motion of the n-th-order system is asymptotically stable. Below, a class of systems featuring this property will be identified.

Since the system under consideration is observable, according to (8.57)

$$\operatorname{rank}\begin{pmatrix}c_0\\ c_0A\\ \vdots\\ c_0A^{l-1}\end{pmatrix}=l,$$

and, therefore, the system consisting of the first l equations in (8.58) may be solved with respect to l components of the vector x. Enumerate x so as to make these l components constitute a vector $(x^l)^{\mathrm{T}}=(x_{n-l+1},\ldots,x_n)$. Then

$$x^l=Rx^{n-l}+R_yy^l, \tag{8.65}$$

where $x^{n-l}=(x_1,\ldots,x_{n-l})$, and the constant matrices R and R_y are defined as the result of solving the system of l equations of (8.58) with respect of x^l. By substituting the system (8.59), (8.60) for last l equations in (8.55) and linear functions (8.65) for x^l in all the equations, obtain equations describing the system behaviour in the space x^{n-l}, y^l:

$$\begin{aligned}&\dot{x}^{n-l}=A_{n-l}x^{n-l}+Dy^l+b^{n-l}u,\\ &\quad \dot{y}_i=y_{i+1}, \quad i=1,\ldots,l-1,\\ &\quad \dot{y}_l=r^{n-l}x^{n-l}+d^ly^l+c_0A^{l-1}bu,\end{aligned} \tag{8.66}$$

where $A_{n-l}, D, b^{n-l}, r^{n-l}$ and d^l are constant matrices, vector and rows of corresponding dimensionality. (This transformation of the coordinates, obviously, is non-singular).

Let us determine the sliding mode equations in the system (8.66) on the plane $s=0$ (8.61). To this end, one must substitute into the first equation of the system (8.66) the equivalent control as found from

$$\dot{s} = \sum_{i=1}^{l-1} c_i y_{i+1} + r^{n-l} x^{n-l} + d^l y^l + c_0 A^{l-1} b u = 0$$

$$u_{\text{eq}} = -(c_0 A^{l-1} b)^{-1} \left(\sum_{i=1}^{l-1} c_i y_{i+1} + r^{n-l} x^{n-l} + d^l y^l \right).$$

Further, $-\sum_{i=1}^{l-1} c_i y_i$ must be substituted for y_l, and sliding mode equations in the canonical subspace y^l (8.62) should be added to this $(n-1)$-th-order equation. Bearing in mind that u_{eq} is a linear function of the state vector coordinates, obtain as the result of the above procedure $(n-1)$-st-order linear system

$$\dot{x}^{n-1} = \bar{A} x^{n-1} + \bar{D} y^{l-1},$$

$$\dot{y}_i = y_{i+1}, \quad i = 1, \ldots, l-2,$$

$$\dot{y}_{l-1} = -\sum_{i=1}^{l-1} c_i y_i \tag{8.67}$$

where $\bar{A}$ and $\bar{D}$ are constant matrices, and $(y^{l-1})^{\mathrm{T}} = (y_1, \ldots, y_{l-1})$.

The system (8.62) describes the projection of sliding mode motion on the subspace y^l. The coefficients c_i defining this motion were chosen in terms of the requirements to variations of the coordinate y_1. It is, therefore, natural to assume that (8.62) is asymptotically stable. And then the asymptotic stability of the general motion (8.67) is completely defined by the eigenvalues of matrix $\bar{A}$. It follows from the above transformations that the matrix $\bar{A}$ is independent of the control parameters (8.61), (8.64) and is defined only by the coefficients of matrix A, vector b and row c_0 in the original system (8.55). Consequently, the proposed method of control is confined to the class of systems for which

$$\operatorname{Re} \lambda_i(\bar{A}) < 0, \quad i = 1, \ldots, n-l, \tag{8.68}$$

where $\lambda_i(\bar{A})$ are the eigenvalues of matrix $\bar{A}$.

In order to take the constraint (8.68) into account directly in the design, it is recommendable to formulate it in terms of the original system (8.55). Regarding the scalars u and y as input and output, respectively, write the transfer function of the system (8.55), (8.56),

$$W(p) = c_0 (A - p I_n)^{-1} b = \frac{P(p)}{Q(p)},$$

where $Q(p)$ is the system characteristic polynomial, $P(p)$ is a polynomial of degree $n-l$ (this fact may be easily established from (8.60): the coordinate $\dot{y}_l$ dependent on u is equal to $y_1^{(l)}$, and, therefore, $y_1^{(n)}$ is the function of $u, \dot{u}, \ldots, u^{(n-l)}$).

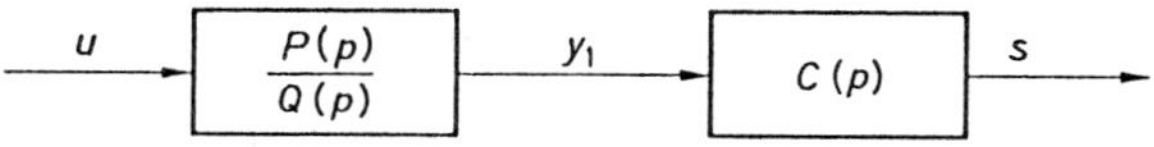

Fig. 17

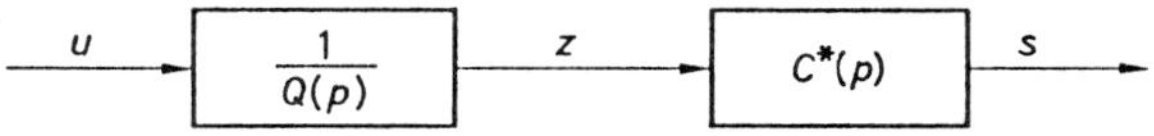

Fig. 18

The block-diagram of an open-loop system whose polynomial $C(p)$ is defined by the equation of plane $s=0$:

$$C(p)=p^{l-1}+c_{l-1}p^{l-2}+\cdots+c_2p+c_1.$$

is shown in Fig. 17. Since $s=0$ in sliding mode, then $s=C(p)y_1=0(p\equiv d/dt)$, i.e. as could be expected the behaviour of output y_1 is defined by the roots $\lambda_1,\ldots,\lambda_{l-1}$ of the characteristic equation of the system (8.62) $C(p)=0$.

The block-diagram of open-loop system is also representable as in Fig. 18 if $C^*(p)=P(p)C(p)$. At the occurrence of sliding mode, $s=C^*(p)z=0$ and the behaviour of coordinate z will be defined by the $(l-1)$ roots of the polynomial $C(p)$ and $n-l$ roots of the polynomial $P(p)$. These $n-1$ roots completely define the sliding mode motion because of the order of its differential Eq. (8.67) is also $n-1$.

This conclusion means that the design method relying upon inducing invariant sliding modes in canonical subspaces is applicable to time-invariant linear controlled plants whose transfer function zeros have negative real parts. Only if there is no zero in the transfer function (or $l=n$ in (8.58)), the dimensionalities of the canonical variable space and state space x coincide and system behaviour is completely defined by the sliding Eq. (8.62).

In conclusion we present different versions of control algorithms providing sliding modes in the canonical subspace y^l. Control is always generated by means of the coordinates $y_1,\ldots,y_l$ because the function s defining the discontinuity surface is their linear combination. In the control (8.64), these coordinates may be used instead of the components of vector x^l:

$$u=-\psi x^{n-l}-\psi^l y^l \tag{8.69}$$

with

$$\psi=(\psi_1,\ldots,\psi_{n-l}),\quad \psi^l=(\psi_1^l,\ldots,\psi_l^l),$$

$$\psi_i=\begin{cases}\alpha_i & \text{for}\quad c_0A^{l-1}bx_is>0,\\ \beta_i & \text{for}\quad c_0A^{l-1}bx_is<0,\end{cases}\qquad i=1,\ldots,n-l,$$

$$\psi_i^l=\begin{cases}\alpha_i^l & \text{for}\quad c_0A^{l-1}by_is>0,\\ \beta_i^l & \text{for}\quad c_0A^{l-1}by_is<0,\end{cases}\qquad i=1,\ldots,l.$$

The control (8.69) in the form of piecewise linear function of all the components of system state vector (8.64) may be used for inducing invariant sliding modes in the subspace y^l (this conclusion directly follows from the results of Sect. 7.6 like in the case of system (8.55) with control (8.64)).

Now, we make use of the results of Sect. 8.1 to simplify the control algorithm (8.69). Assume that there is a linear control

$$u_l = \gamma y^l, \quad \gamma = (\gamma_1, \ldots, \gamma_k, 0, \ldots, 0), \quad k < l$$

such that the linear system

$$\dot{y}_i = y_{i+1}, \quad i = 1, \ldots, l-1,$$
$$\dot{y}_l = d^l y^l + c_0 A^{l-1} b u_l$$

has the desired allocation of $l-1$ eigenvalue. Then, like in the case of (8.10) in order to have in space y^l sliding modes defined by the same eigenvalue allocation and stable "in the small" over all the plane $s = 0$, it suffices to take control in the form of

$$u = -\psi x^{n-l} - \bar{\psi}^l y^l - \delta$$

with

$$\bar{\psi}^l = (\psi_1^l, \ldots, \psi_k^l, 0, \ldots, 0)$$
$$\delta = \delta_0 \operatorname{sign} c_0 A^{l-1} b u_l,$$

ψ and $\bar{\psi}^l$ are defined in accordance with (8.69). The justification of the exclusion of $l-k$ components from the control (8.69) is exactly similar to the proof of the theorem in Sect. 8.1. In order to provide sliding mode stability "in the large", the conditions of the theorem in Sect. 8.2 are to be used.

Finally, we consider the case where the components of state vector x are not measurable, but in addition to the controlled coordinate y_1 it is possible to obtain its $n-1$ derivatives. Let us discuss the system motion in the space $y_1, \dot{y}_1, \ldots, y_1^{(n-1)}$ described by

$$\dot{y}_i = y_{i+1}, \quad i = 1, \ldots, n-1$$
$$\dot{y}_n = -\sum_{i=1}^{n} a_i y_i + \sum_{i=0}^{n-l-1} h_{i+1} u^{(i)} + c_0 A^l b u^{(n-l)}, \tag{8.70}$$

where a_i are the coefficients of characteristic polynomial

$$Q(p) = p^n + a_n p^{n-1} + \cdots + a_1,$$

h_i are the coefficients of the polynomial

$$P(p) = c_0 A^{l-1} b p^{n-l} + a_n p^{n-1} + \cdots + a_1.$$

The equations in the space of canonical variables $y_1, \ldots, y^{(n-1)}$ contain the derivatives of control action. Discontinuous controls lead in these systems to discontinuous trajectories in the state space, and their behaviour may be analyzed by means of generalized functions [2]. In order to be still able to realize invariant sliding modes, one should choose control so as to have continuous function s with sign opposite to the sign of rate $\dot{s}$ that has first-order discontinuities on the surface $s=0$. These conditions are satisfied if the output of dynamic system

$$u=u_1 \quad \dot{u}_i=u_{i+1}, \quad i=1,\dots,n-l-1$$

$$\dot{u}_{n-l}=\sum_{i=1}^{n-l}\lambda_i u_i+v, \tag{8.71}$$

is used as control where v is the discontinuous function of vector y^l and state coordinates $u_1,\dots,u_{n-l}$ that are assumed to be measurable, the coefficients λ_i may be arbitrary and chosen so as to facilitate realization. Bearing in mind that $u^{(i)}=u_{i+1}$ $(i=1,\dots,n-l-1)$ and $u^{(n-l)}=\dot{u}_{n-l}$, represent (8.70) as

$$\dot{y}_i=y_{i+1}, \quad i=1,\dots,n-1,$$

$$\dot{y}_n=-\sum_{i=1}^{n}a_i y_i+\sum_{i=1}^{n-l}(h_i+c_0A^{l-1}b\lambda_i)u_i+c_0A^{l-1}bv. \tag{8.72}$$

If v is considered as control action, one can readily see that the dynamic systems (8.71), (8.72) and (8.66) differ only in dimensionality and the form of matrix A_{n-m}, D and column b^{n-l}. Therefore, one may avail of any of the design procedures described in this section in order to realize invariant sliding modes on the plane

$$s=\sum_{i=1}^{n}c_i y_i=0, \quad c_n=1$$

in the space of $y_1,\dots,y_n$ described by equations

$$\dot{y}_i=y_{i+1}, \quad i=1,\dots,n-2,$$

$$\dot{y}_{n-1}=-\sum_{i=1}^{n-1}c_i y_i. \tag{8.73}$$

In order to determine sliding equations in all the $(2n-1)$-dimensional space $(y_1,\dots,y_n,u_1,\dots,u_{n-l})$, find the equivalent control from the equation $\dot{s}=0$ provided that $s=0$:

$$u_{\text{eq}}=-(c_0A^{l-1}b)^{-1}\left(\sum_{i=1}^{n-1}(c_{i-1}-a_i-c_ic_{n-1}+c_ia_n)y_i\right.$$
$$\left.+\sum_{i=1}^{n-l}(h_i+c_0A^{l-1}b\lambda_i)u_i\right.$$

and substitute it into the last equation of (8.71):

$$\dot{u}_i=u_{i+1}, \quad i=1,\dots,n-l-1$$

$$\dot{u}_{n-l-1}=-(c_0A^{l-1}b)^{-1}\sum_{i=1}^{n-l}h_iu_i$$

$$-(c_0A^{l-1}b)\sum_{i=1}^{n-1}(c_{i-1}-a_i-c_ic_{n-1}+c_ia_n)y_i. \tag{8.74}$$

Reasoning by the analogy with the condition (8.68) for the system (8.67), the asymptotic stability of general motion in the system (8.73) (rather than of motion in the canonical subspace only) is defined by the eigenvalues of the characteristic equation of (8.74) for $y_i = 0$ $(i = 1, \ldots, n-1)$. Since this characteristic equation has the form $P(p) = 0$, for the technique under consideration oriented to designing sliding modes in the canonical space, the system can be regarded as able to control only the plants having transfer function zeros with negative real parts.

Chapter 9
Dynamic Optimization

1 Problem Statement

Optimization of linear systems of the type (7.1) usually relies upon minimization of the quadratic criterion

$$I = \int_0^{\infty} (x^{\mathrm{T}} Q x + u^{\mathrm{T}} R u) dt, \tag{9.1}$$

where Q is a positive semi-definite symmetric matrix, and R is a positive definite symmetric matrix. If the controlled variable is of the form $y = Dx$ where $y \in \mathbb{R}^k$, $k < n$, and D is constant $k \times n$ matrix, then taking

$$Q = D^{\mathrm{T}} D \geqq 0 \tag{9.2}$$

one obtains that the first term in the criterion (9.1) is $y^{\mathrm{T}} y$ and characterizes the degree of deviation of the controlled variable from the zero state. The second term in the criterion defines the penalty for controll "expenses". The relation between the weight matrices Q and R defines the tradeoff between the two contradictory desires such as to have a rapidly decaying control process and to reduce power consumption for its realization.

Optimal systems of this kind are of interest, on the one hand, because criterion (9.1) formalizes the control problem sufficiently well, and, on the other hand, because the optimal control is linear [6, 23, 71, 92, 95], which is advantageous in terms of implementation. Existence conditions and methods for finding of optimal control will be discussed below in Sect. 3. Here it will be only noted that the latter problem involves solution of an n-th order matrix equation, and that one might run into significant analytical and computational difficulties in the case of high-dimensionality problem.

It is suggested in this chapter to solve the $(n-m)$-dimensional problem of dynamic optimization of sliding mode motion by means of the decoupling principle as described in the beginning of the present part, and next to provide stability of this motion "in the large" by means of the discontinuous control. Like in the eigenvalue allocation problem, the increase in norm of the control vector speeds up the beginning of the optimal sliding mode.

All at once the question arises about the form of optimality criterion. Obviously, functional (9.1) is not suited to this aim because sliding mode motion is control-independent and defined by the equation of discontinuity surfaces. Therefore, we consider the functional

$$I=\int_0^\infty x^{\mathrm{T}}Qx\,dt \tag{9.3}$$

as such a criterion and try to determine the discontinuity surfaces resulting in the minimum of the functional. Should sliding mode on their intersection occur, the instant when the sliding mode begins is regarded as the initial point in function (9.3).

Apart from functional (9.3), this chapter will discuss also on optimality criterion dependent on u_{eq}. This function defines the average value of control required for realization of sliding modes and, therefore, characterizes control "expenses". Last, consideration will be given to the optimization of control processes in time-varying systems and to cases of finite time of control.

Like in Chap. 7, let us start from the basic concepts and methods of multivariable linear system theory, that will be used below for solution of optimal problems with quadratic criteria.

2 Observability, Detectability

Let us discuss the possibility of reconstructing the state vector of (7.1) through measured values of the components of the output vector

$$y=Cx \tag{9.4}$$

where $y\in\mathbb{R}^m$, C is a constant matrix. Let $y(t,t_0,x_0,u)$ by the values of vector y on the system trajectories with initial condition $x(t_0)=x_0$ and control $u(t)$. The system (7.1), (9.4) is referred to as *reconstructable* if for all t_1 there exists an instant t_0, $t_0<t_1$ such that it follows from

$$y(t,t_0,x_0,u)=y(t,t_0,x_0',u) \tag{9.5}$$

that $x_0 = x'_0$ for all $u(t)$, $t_0 \leqq t \leqq t_1$. It stems the definition that for reconstructable systems the state of system x at time t_1 may be determined through previous observations of vector y for $t_0 \leqq t \leqq t_1$.

The concept of "reconstructability" complements that of "observability". A system is referred to as *observable* if for all t_0 there will be $t_1 > t_0$ such that $x_0 = x'_0$ follows from

$$y(t, t_0, x_0, u) = y(t, t_0, x'_0, u), \quad t_0 \leqq t \leqq t_1$$

for all $u(t)$. For observable systems, state vector x at time t_0 may be determined through future values of vector y over the interval $t_0 \leqq t \leqq t_1$. For linear systems with constant parameters, the conditions of observability and reconstructability coincide and have a simpler form as compared with (9.5) [92]: system (7.1), (9.4) is reconstructable/observable if the equality

$$y(t, t_0, x_0, 0) = 0 \quad t_0 \leqq t \leqq t_1.$$

implies that $x_0 = 0$.

Let us use the geometrical representation of motions in order to explain why (7.1), (9.4) may be unobservable. In Sect. 7.1 we have introduced the notion of non-trivial invariant subspace $\mathscr{K}$ of matrix A. One may readily see that if in the linear system (7.1) with $u = 0$

$$x_0 \in \mathscr{K}, \tag{9.6}$$

then

$$x(x_0, t_0, t) \in \mathscr{K}, \quad t \geqq t_0 \tag{9.7}$$

Since the dimensionality of subspace $\mathscr{K}$ is smaller than n, there exists matrix R such that

$$R\{\mathscr{K}\} = 0. \tag{9.8}$$

If matrix C of (9.4) coincides with at least one matrix R of the (9.8) type, then the initial condition of (9.6) means that $y(t_0) = 0$ and $y(x_0, t_0, t) = 0$, $t \geqq t_0$, in virtue of (9.7). This equality holds for all x from the set $\mathscr{K}$. Consequently, vector x cannot be determined through observations of vector y, and system (7.1), (9.4) is unobservable.

Similar to the formulation of controllability conditions in terms of invariant subspaces in Sect. 7.1, one may demonstrate that if matrix C does not coincide with any of matrices R of the (9.8) type then

$$\operatorname{rank} F = n, \tag{9.9}$$

with

$$F = \begin{pmatrix} C \\ CA \\ \vdots \\ CA^{n-1} \end{pmatrix}.$$

Matrix F in (9.3) is called *observability matrix*. Condition (9.9) is necessary and sufficient for observability of system (7.1), (9.4) [92]. If this condition is met, pair $\{A, C\}$ is accordingly referred to as *observable*.

As it was seen above, the state vector cannot be reconstructed for unobservable systems and the initial position of state vector on non-trivial manifold $\mathscr{K}$. It would be of interest, in this connection, to determine which components of the state vector are reconstructable for arbitrary initial state vector rather than for the state vector in manifold. Assume that for the unobservable system (7.1)

$$\operatorname{rank} F = q$$

Define an $(n-q)$-dimensional subspace $\mathscr{K}$ of n-dimensional column vectors such that

$$F\{\mathscr{K}\} = 0. \tag{9.10}$$

Let us demonstrate that this subspace is a non-trivial invariant subspace for matrix A. Exactly as in Sect. 7.2, it was shown that any column vector of matrix AP may be presented in terms of base vectors of matrix P, one may demonstrate in this case that the row vectors of matrix FA are linear combinations of the base vectors of matrix F. Since any element of the subspace $\mathscr{K}$ is orthogonal to base row vectors of matrix F then $FA\{\mathscr{K}\} = 0$, and

$$A\{\mathscr{K}\} \subset \mathscr{K} \tag{9.11}$$

by the definition of subspace $\mathscr{K}$ (9.10). Condition (9.11) means that for matrix A subspace $\mathscr{K}$ is invariant.

Let the rows of $(q \times n)$-dimensional matrix U_1 be the base vectors of matrix F, and the columns of matrix U_2^T constitute the base of subspace $\mathscr{K}$. Since the rows of U_1 are orthogonal to the columns of U_2^T (9.10), matrix

$$U = \begin{pmatrix} U_1 \\ U_2 \end{pmatrix}$$

is non-singular, and linear transformation $z = Ux$ reduces system (7.1), (9.4) to the form

$$\dot{z} = UAU^{-1}z + UBu \tag{9.12}$$

$$y = CU^{-1}z. \tag{9.13}$$

Let us divide matrix U^{-1} into two submatrices T_1 and T_2 of the dimensionalities $n \times q$ and $n \times (n-q)$, respectively: $U^{-1} = (T_1, T_2)$. For this decomposition

$$UU^{-1} = \begin{pmatrix} U_1 \\ U_2 \end{pmatrix}(T_1, T_2) = \begin{pmatrix} U_1T_1 & U_1T_2 \\ U_2T_1 & U_2T_2 \end{pmatrix} = \begin{pmatrix} I_q & 0 \\ 0 & I_{n-q} \end{pmatrix}, \tag{9.14}$$

$$UAU^{-1} = \begin{pmatrix} U_1AT_1 & U_1AT_2 \\ U_2AT_1 & U_2AT_2 \end{pmatrix}, \tag{9.15}$$

$$CU^{-1} = (CT_1, CT_2). \tag{9.16}$$

Since the rows of U_1 constitute the basis of matrix F and space $\mathscr{K}$ contains all the vectors orthogonal to F, it follows according to (9.10), (9.11) from condition $U_1 T_2 = 0$ (9.14) that $T_2 \in \mathscr{K}$, $AT_2 \in \mathscr{K}$ and

$$U_1 A T_2 = 0. \tag{9.17}$$

As matrix C is submatrix of F, condition (9.10) means that

$$CT_2 = 0 \tag{9.18}$$

Comparison of (9.12) through (9.18) suggests that (9.12) and (9.13) are representable with respect to the new state vector z as

$$\dot{z} = \begin{pmatrix} A_{11} & 0 \\ A_{21} & A_{22} \end{pmatrix} z + \begin{pmatrix} B_1 \\ B_2 \end{pmatrix} u, \tag{9.19}$$

$$y = (C_1, 0) z. \tag{9.20}$$

Since for linear systems observability is defined at the zero control, the behaviour of vector z_1 consisting of the first q components of vector z is defined at $u = 0$ only by matrix A_{11}, and output y depends only on these components. Following the reasoning used in proving controllability of pair A_{11} and B_1 in the canonical form (7.25), observability of pair $\{A_{11}, C_1\}$ may be proved in the case as well.

Representation of the system in the form of (9.19), (9.20) is referred to as *canonical observability form*, and subspaces z_1 and z_2 are referred to as *subspaces of observable* and *non-observable states*, respectively.

Note another feature elucidating the concept of observability which is important from the viewpoint of the theory of linear optimal systems. It follows from the consideration of the canonical observability from (9.19), (9.20) that for non-observable systems with any kind of control using information only about the output variable vector y the system may be made asymptotically stable only in the case of asymptotically stable motion in the subspace of non-observable states. These systems are called *detectable*. Consideration of the equation system represented in the canonical observability form (9.19), (9.20) reveals the necessary and sufficient detectability condition: matrix A_{22} should be asymptotically stable. Pair $\{A, C\}$ in this case is accordingly, called *detectable*. It follows directly from the definition of detectable system that all the asymptotically stable systems and observable systems are detectable.

3 Optimal Control in Linear Systems with Quadratic Criterion

The goal of any design of optimal linear system (7.1) with criterion (9.1) is to construct a feedback system, i.e. to obtain a control as a function of the state vector. From an engineering standpoint, it is reasonable to consider only situations where controls minimizing functional (9.1) at the same time provide asymptotic stability of the equilibrium $x=0$. In this connection, let us find the class of matrices A, B, Q and R where precisely these situations occur.

The first evident condition is stabilizability of pair $\{A, B\}$ because, otherwise, asymptotic stability cannot be provided. The second condition is obtained by representing (7.1) in the canonical observability form (9.19), (9.20) if vector $y=Dx$ defined by (9.2) is regarded as the output. Denoting by z_1 and z_2 the vectors of observable and nonobservable states, obtain as before

$$\dot{z}_1 = A_{11}z_1 + B_1 u, \tag{9.21}$$

$$\dot{z}_2 = A_{21}z_1 + A_{22}z_2 + B_2 u, \quad y=(D_1, 0)\begin{pmatrix} z_1 \\ z_2 \end{pmatrix}, \tag{9.22}$$

with pair $\{A_{11}, D_1\}$ being observable. Criterion (9.1) may be correspondingly rewritten as

$$I = \int_0^\infty (z_1^{\mathrm{T}} D_1^{\mathrm{T}} D_1 z_1 + u^{\mathrm{T}} R u)dt. \tag{9.23}$$

One may easily see that the problem has been reduced to determination of the optimal control for subsystem (9.21) with the optimality criterion (9.23). This control, evidently, will be the function only of the coordinates of observable subspace z_1 (It is to be noted that, according to the first condition, pair $\{A, B\}$ is stabilizable and, therefore, pair $\{A_{11}, B_1\}$ is also stabilizable, thus ensuring the asymptotic stability of subsystem (9.21)). The asymptotic stability of matrix A_{22} is the necessary condition of asymptotic stability of (9.21), (9.22) with any control which is dependent only z_1. According to the definition of the last section, this requirement is equivalent to the condition of detectability of pair $\{A, D\}$. Thus, stabilizability of pair $\{A, B\}$ and detectability of pair $\{A, D\}$ are necessary conditions for asymptotic stability of (7.1) with criterion (9.1).

To determine the optimal control, let us resort to the method of dynamic programming [46]. The Bellman equation being applied to (7.1), (9.1) will have the form of

$$\min_u (x^{\mathrm{T}} Q x + u^{\mathrm{T}} R u + (\operatorname{grad} S)^{\mathrm{T}}(Ax + Bu)) = 0 \tag{9.24}$$

where $S(x)$ is the Bellman function, $\operatorname{grad} S$ is a column vector. In the point of minimum, the gradient of the function in (9.24) with respect to control is zero, i.e.

$$2Ru + B^{\mathrm{T}} \operatorname{grad} S = 0$$

and the optimal control u^* is

$$u^* = -\tfrac{1}{2} R^{-1} B^{\mathrm{T}} \operatorname{grad} S. \tag{9.25}$$

After substitution of u^* into (9.24), obtain

$$x^{\mathrm{T}} Q x - \tfrac{1}{4}(\operatorname{grad} S)^{\mathrm{T}} B R^{-1} B^{\mathrm{T}} \operatorname{grad} S + (\operatorname{grad} S)^{\mathrm{T}} A x = 0 \tag{9.26}$$

For this partial differential equation, we shall seek solution with respect to $S(x)$ in terms of quadratic form $S(x) = x^T P x$ where P is constant symmetrical matrix. Then, $\operatorname{grad} S = 2Px$ and it follows from (9.26) which should hold at all x that

$$Q - PBR^{-1}B^T P + PA + A^T P = 0. \tag{9.27}$$

The optimal control (9.25) is, correspondingly, determined from

$$u^* = -R^{-1}B^T Px \tag{9.28}$$

where matrix P meets (9.27) which is called *Riccati equation*. This equation is the necessary optimally condition as well as Bellman equation.

All the mentioned above necessary conditions are sufficient at the same time: if in (7.1), (9.1) pair $\{A, B\}$ is stabilizable and pair $\{A, D\}$ is detectable, $Q \geqq 0$, $R \geqq 0$, the Riccati equation has a unique solution defining the optimal control (9.28), the feedback system (7.1), (9.28) being asymptotically stable [92].

4 Optimal Sliding Modes

The problem of sliding motion optimization already was formulated in Sect. 1: for system (7.1), it is necessary to design discontinuity surfaces for which the optimality criterion (9.3) is minimal at motion in sliding mode. Notably, for system (7.1) criterion (9.3) leads to the so called *singular* optimal problems [28], [52], [58] where control in the system state space is not limited anywhere with the exception of some subspace of smaller dimensionality.

References [67], [100] apply the apparatus of singularly perturbed systems to these singular problems. The approach resides in regularization via introduction of a small penalty for control in functional (9.3) (whence their name-"cheap control" systems)

$$I = \int_0^\infty (x^T Q x + \mu^2 u^T R u)dt, \quad R > 0$$

and then tending parameter μ to zero. As a result, obtain also a linear control with coefficients inverse to μ; motion in the optimal system consists of a fast and slow components, the slow motion trajectories being on a manifold of smaller dimensionality.

A similar situation occurs when matrix $R \geqq 0$ rather than being strictly positive. This can be easily accounted for by the fact that the term $u^T R u$ in functional (9.1) is the penalty for control, and, therefore, at $R = 0$ or $R \geqq 0$ all the control components or some of their combinations may be infinitely increased "free of charge".

Let us particularize the problem by first transforming the system in question (7.1) to the form (7.45), (7.46) by means of non-singular transformation (7.43).

Sliding mode motion and, consequently, discontinuity surfaces are assumed to be defined by

$$s = s_0(x_1) + x_2 = 0, \quad s \in \mathbb{R}^m, \tag{9.29}$$

where $s_0(x_1)$ is an arbitrary function of vector x_1, i.e. these surfaces are not in advance chosen linear. The design consists in choosing function $s_0(x_1)$ such that functional (9.3) reaches its minimum at motion over manifold $s = 0$. It follows from the problem statement that at the zero time only the components of the $(n-m)$-diminisional vector x_1 are to be defined, the rest of the state vector components or m-dimensional vector x_2 is determined from (9.29).

As the result, we have to determine the optimal m-dimensional control in system (7.45) with vector $x_2 = -s_0(x)$ as control and criterion (9.3) representable as

$$I = \int_0^\infty (x_1^T Q_{11} x_1 + 2x_1 Q_{12} x_2 + x_2^T Q_{22} x_2) dt \tag{9.30}$$

and

$$(M^{-1})^T Q M^{-1} = \begin{pmatrix} Q_{11} & Q_{12} \\ Q_{21} & Q_{22} \end{pmatrix}, \quad Q_{21}^T = Q_{12}$$

Below, consideration will be given to the case of $Q_{22} > 0$. Introduce a new variable v related to x_1 and x_2 as follows

$$v = x_2 + Q_{22}^{-1} Q_{12} x_1 \tag{9.31}$$

and rewrite (7.45) and (9.30) in compliance with (9.31):

$$\dot{x}_1 = (A_{11} - A_{22} Q_{22}^{-1} Q_{12}^T) x_1 + A_{12} v, \tag{9.32}$$

$$I = \int_0^\infty (x_1 (Q_{11} - Q_{12} Q_{22}^{-1} Q_{12}^T) x_1 + v^T Q_{22} v) dt. \tag{9.33}$$

If in the case of $Q_{22} > 0$ x_1 is considered as the state vector and v as control for sliding mode motion, obtain a usual statement of the optimal control problem with quadratic criterion of the (7.1), (9.1) type.

For problem (7.45), (9.30), we formulate the optimality conditions based upon the Riccati Eq. (9.27) as a theorem.

Theorem. *If $Q_{22} > 0$, pair $\{A, B\}$ is stabilizable and pair $\{A_{11} - A_{12} Q_{22}^{-1} A_{12}^T, D\}$ is detectable, where*

$$D^T D = Q_{11} - Q_{12} Q_{22}^{-1} Q_{12}^T \tag{9.34}$$

then the optimal vector x_2 playing in (7.45) the part of control has the following form

$$x_2 = -s_0(x_1) = -(Q_{22}^{-1} A_{12} P + Q_{22}^{-1} Q_{12}^T) x_1, \tag{9.35}$$

where P is the unique solution to the matrix Riccati equation

$$\begin{aligned} &P(A_{11} - A_{12} Q_{22}^{-1} Q_{12}^T) + (A_{11} - A_{12} Q_{22}^{-1} Q_{12}^T)^T P \\ &\quad - P A_{12} Q_{22}^{-1} A_{12}^T P + (Q_{11} - Q_{12} Q_{22}^{-1} Q_{12}^T) = 0 \end{aligned} \tag{9.36}$$

The proof of the theorem emphasizes the equivalence of systems (7.35) and (9.32) from the viewpoint of controllability. Indeed, all the non-controllable (non-stabilizable) coordinates of (7.35) are control-independent in (9.32) as well, and, therefore, generation of control x_2 in (7.45) in the form of a sum of new control v and action $-Q_{22}Q_{12}^{\mathrm{T}}x_1$ cannot affect the system controllability (stabilizability). On the other hand, it may be concluded through similar reasoning that if system (9.32) or pair $\{A_{11}-A_{12}Q_{22}^{-1}A_{12}^{\mathrm{T}}, D\}$ (D is found from (9.34)) is not stabilizable, the original system (7.1) or pair $\{A, B\}$ are not stabilizable as well which contradicts the statement of theorem. Thus, all the necessary and sufficient conditions of optimality as formulated in Sect. 3 hold for system (9.32) with criterion (9.33). The Riccati Eq. (9.27) will be written for this system in the form of (9.36), and optimal control is found through (9.28):

$$v^* = -Q_{22}^{-1}A_{12}^{\mathrm{T}}Px_1.$$

Since vectors x_1, x_2 and v are related by (9.31), (9.35) defines the discontinuity surfaces over which sliding mode is to be set up. Q.E.D.

This conclusion implies that the optimality is realized in sliding on linear discontinuity surfaces and is, therefore, described by the linear differential equations (7.45), (9.35).

Let us formulate for the discontinuity planes design procedure for (7.1) with quadratic criterion (9.3):

(1) determine matrix M of transformation (7.43);
(2) bring system (7.1) and criterion (9.3) into the form of (7.45), (7.46) and (9.30), respectively;
(3) solve the matrix Riccati Eq. (9.36) with respect to P; and
(4) determine the equations of discontinuity surfaces (9.29) or (9.35) in the form of a function of the original variables

$$s = (Q_{22}^{-1}(A_{22}^{\mathrm{T}}P + Q_{12}^{\mathrm{T}}), I_m)Mx = 0.$$

Since the discontinuity surfaces have turned out to be linear, there is no need to focus our attention on the stability of sliding modes themselves (or the stability of motion in subspace s) and estimation of convergence to manifold $s = 0$; all the methods used in Sect. 7.5, 6 for the design of control u and upper estimates are directly applicable also to setting up the stable optimal sliding modes.

5 Parametric Optimization

As was already noted, the absence of a penalty function for control $u^{\mathrm{T}}Ru$ in criterion (9.3) for the linear system (7.1) results in a singular optimal problem whose solution is a linear state function with coefficients tending to infinity. The problem of dynamic optimization of sliding mode motions with respect to the same criterion is regular, and the coefficients of equations of discontinuity planes

$$s = C_1 x_1 + x_2 = 0 \tag{9.37}$$

are finite. Therefore, the coefficients of piecewise linear controls (7.60) and (7.65) providing the stability of sliding modes at the intersection of planes chosen in this manner will be finite as well. It should be also noted that, nevertheless, criterion (9.3) indirectly allows for control "expenses". The values of coefficient α in control (7.59) or the coefficients of matrix Ψ in control (7.64) are dependent on matrix C_1 which in its turn defines vector x_2 that plays a part of control in subsystem (7.45) with optimality criterion (9.30).

Consider now a criterion that directly depends on the control vector. Sliding mode motion is defined by equivalent control

$$u_{eq} = -B_2^{-1}(C_1 A_{11} - C_1 A_{12} C_1 + A_{21} - A_{22} C_1) x_1 \tag{9.38}$$

that makes function $\dot{s}$ computed on the trajectories of system (7.45), (7.46) vanish over the manifold $s = 0$ (9.37). Let us consider u_{eq} as a measure of control "expenses" and, correspondingly, write the optimality criterion as

$$I = \int_0^\infty (x^T Q x + u_{eq}^T R u_{eq}) dt, \tag{9.39}$$

where $Q \geqq 0$, $R \geqq 0$ are symmetrical matrices, and the zero time coincides with the beginning of sliding mode. Since in sliding mode $x_2 = -C_1 x_1$, criterion (9.39) becomes with allowance for (9.38) as follows

$$I = \int_0^\infty x_1^T W x_1 dt, \tag{9.40}$$

with

$$W = Q_{11} + \Gamma^T (B_2^{-1})^T R B_2^{-1} \Gamma - 2 Q_{12} C_1 + C_1^T Q_{22} C_1,$$
$$\Gamma = C_1 A_{11} - A_{22} C_1 - C_1 A_{12} C_1 + A_{21}.$$

As the result, the problem has been reduced to the minimization of criterion (9.40) with respect to C_1 on the trajectories of (7.46) for $x_2 = -C_1 x_1$, i.e.

$$\dot{x}_1 = (A_{11} - A_{12} C_1) x_1. \tag{9.41}$$

The optimization problem in question is classified among the problems of designing linear systems with constant feedback coefficients. The optimal solution in this case, generally, depends on the initial conditions [96], [105], [171]. Since the initial vector $x_1(0)$ usually is not known, it is desirable to obviate the dependence of solution on $x_1(0)$.

The most common way to do this is to consider vector $x_1(0)$ as a random variable with zero expectation and covariance matrix [96]

$$M\{x_1(0) x_1(0)^T\} = I_{n-m}. \tag{9.42}$$

The task of optimization consists in determining matrix C_1 such that the expectation of criterion (9.40)

$$\bar{I} = M\{I\} \tag{9.43}$$

becomes minimal. Let us compute this criterion as a function of the initial conditions. Take a quadratic form $v = -x_1^T K x_1$, $K = \text{const}$, such that its time derivative over solutions of (9.41) is equal to the integrand in (9.40), i.e.

$$\frac{dv}{dt} = x_1^T W x_1 \tag{9.44}$$

$$K(A_{11} - A_{12}C_1) + (A_{11} - A_{12}C_1)^T K = -W. \tag{9.45}$$

Then, assuming that for the optimal matrix C_1 system (9.41) is asymptotically stable, obtain from (9.40) and (9.44) criterion I:

$$I = x_1(0)^T K x_1(0). \tag{9.46}$$

For the chosen covariance matrix (9.42), criterion (9.43) is evaluated by (9.46) as $\bar{I} = tr\, K$. The parametric optimization problem, thus, has been reduced to the usual determination of conditional extremum: for the optimal matrix C_1, the trace of matrix K must be minimal provided that K is the solution to (9.45) which is called the Lyapunov equation.

The interested reader is referred to [15], [30], [61], [96], [154] for discussions of the existence and uniqueness of the solution to this problem, as well as computational procedures enabling its determination.

Since the discontinuity surfaces are chosen to be linear, once again may avail of the methods providing stability of sliding modes as described in Sect. 7.5.

6 Optimization in Time-Varying Systems

Now, let us consider a more general formulation of the dynamic optimization problem where the control system and weight coefficients in the criterion are time-varying and the control time is finite:

$$\dot{x} = A(t)x + B(t)u, \tag{9.47}$$

$$\int_0^{t_1} (x^T Q(t) x + u^T R(t) u)\, dt, \quad t_1 < \infty, \tag{9.48}$$

where time-varying matrices $A(t)$, $B(t)$, $Q(t)$ and $R(t)$ meet all the above conditions for systems with time-invariant parameters. Moreover, the elements of the matrices and their time derivatives are assumed to be limited. The Bellman equation as applied to this class of systems will be in the following form [46]

$$-\frac{\partial S}{\partial t} = \min_u (x^T Q x + u^T R u + (\text{grad}\, S)^T (Ax + Bu)) \tag{9.49}$$

where the Bellman function $S(x, t)$ depends not only on the state vector, but on time as well. Like in the case of time-invariant systems (Sect. 3), one has to determine here the optimal control $u^*(\operatorname{grad} S, x, t)$ providing the minimum in the right side of (9.49) and, after substitution of u^* into (9.49), to solve the resulting partial differential equation (with respect to $S(x, t)$. Assuming that this solution is a quadratic form with time-varying coefficients

$$S(x, t) = x^{\mathrm{T}} P(t) x, \tag{9.50}$$

obtain

$$u^* = -B^{\mathrm{T}}(t) R^{-1}(t) P(t) x \tag{9.51}$$

and the matrix differential Riccati equation with respect to $P(t)$:

$$-\frac{dP(t)}{dt} = Q(t) - P(t)B(t)R^{-1}(t)B^{\mathrm{T}}(t)P(t) + A^{\mathrm{T}}(t)P(t) + P(t)A(t). \tag{9.52}$$

By definition, the Bellman function function $S(x, t)$ is equal to the value of functional (9.48) computed over the interval $[t, t_1]$ with the initial condition x for $u = u^*$ [46]. Obviously,

$$S(x, t_1) = 0 \text{ or, according to (9.50),}$$

$$x^{\mathrm{T}}(t_1) P(t_1) x(t_1) = 0$$

for all x; consequently,

$$P(t_1) = 0. \tag{9.53}$$

which will be the boundary condition enabling one to determine the solution of the Riccati Eq. (9.52) and then also the optimal control (9.51). Note that at $Q(t) \geqq 0$ and $R(t) > 0$ for $0 \leqq t \leqq t_1$ the Riccati Eq. (9.52) with the boundary condition (9.53) has a unique solution [23], [71].

Now we determine the equation of manifold over which the sliding mode motion is optimal with respect to the criterion

$$I = \int_0^{t_1} x^{\mathrm{T}} Q(t) x(t) dt, \quad Q(t) \geqq 0. \tag{9.54}$$

To this end, let us again make use of the non-singular (and now also time-varying) transformation (7.44), (7.45) and find motion equations with respect to new variables x_1 and x_2 in the form of (7.46), (7.47) where

$$\begin{pmatrix} A_{11}(t) & A_{12}(t) \\ A_{21}(t) & A_{22}(t) \end{pmatrix} = M(t)A(t)M^{-1}(t) + \dot{M}(t)M^{-1}(t).$$

Represent criterion (9.54) also in the form of (9.30) with time-varying weight matrices, and finally through transformation (9.31) obtain an optimization problem of the (9.32), (9.33) type with the difference that all the matrices are time-varying and the integral in the criterion is computed over the finite time interval $[0, t_1]$.

In order to solve this problem assume that $Q_{22}(t) > 0$ and make use of the Riccati Eq. (9.52)

$$-\frac{dP}{dt} = P(A_{11} - A_{12}Q_{22}^{-1}Q_{12}^{\mathrm{T}}) + (A_{11} - A_{12}Q_{22}^{-1}Q_{12}^{\mathrm{T}})^{\mathrm{T}}P$$
$$- PA_{12}Q_{22}^{-1}A_{12}^{\mathrm{T}}P + Q_{11} - Q_{12}Q_{22}^{-1}Q_{12}^{\mathrm{T}}$$

with the boundary condition (9.53). As before, the equation of time-varying manifold of $(n-m)$-th order with optimal motions in sliding mode will be in the form of (9.35) or, after the application of design procedure of Sect. 4, in the form of $s = C(t)x = 0$, where

$$C(t) = (Q_{22}^{-1}(t)(A_{22}^{\mathrm{T}}(t)P(t) + Q_{12}^{\mathrm{T}}(t), I_m)M(t).$$

The design is completed by choosing a discontinuous control vector providing the stability of sliding modes or stability of motion in the subspace s. The peculiarity of this step lies in the fact that the discontinuity surface equation is time-varying; and therefore we obtain

$$\dot{s} = C(t)A(t)x + C(t)B(t)u + \dot{c}(t)x \tag{9.55}$$

instead of the equation of motion projection on the subspace s (7.57). Matrix $(C(t)B(t))^{-1} = (B_2(t))^{-1}$ exists and is limited over interval $[0, t_1]$. Therefore, one may recourse to the invariant transformations of vector s and u described in Sect. 7.5

$$s^* = \Omega(t)s, \quad \Omega = D(C(t)B(t))^{-1}, \tag{9.56}$$

$$u = \Omega(t)u^*, \quad \Omega = (C(t)B(t))^{-1}D, \tag{9.57}$$

that lead to equations similar to (7.58) and (7.59):

$$\dot{s}^* = H(t)x(t) + Du, \tag{9.58}$$

$$\dot{s} = H(t)x(t) + Du^*, \tag{9.59}$$

where D is constant (diagonal or symmetrical) positive-definite matrix. Matrix $H(t)$ is, according to (9.55) through (9.57), has the form

$$H(t) = D(C(t)B(t)^{-1}\left(C(t)A(t) + \frac{dC(t)}{dt}\right) + \left(\frac{d}{dt}D(C(t)B(t))^{-1}\right)C(t)$$

in the first case (9.58), and

$$H(t) = C(t)A(t) + \frac{dC(t)}{dt}$$

in the second one (9.59). All the elements of matrix H are bounded time-functions because all the matrices in system (9.47) and criteria (9.48), and, consequently, matrix H are, by convention, bounded together with the time-derivatives. The fact that functions $h_{ij}(t)$ are bounded implies the applicability of control algorithms (7.60) and (7.65) to this time-varying case as well. Indeed, there always will be numbers α or α_{ij} and β_{ij} such that the inequalities given in Sect. 7.5 hold under all the values of functions $h_{ij}(t)$ from the bounded interval. As the result, the time derivatives of corresponding Lyapunov functions will be negative which guarantees the stability of sliding modes. It is at once apparent that in order to determine the desired numerical values of coefficients α or α_{ij} and β_{ij} one needs only to know the intervals within which vary the elements and their time derivatives of the matrices of the initial system and criterion.

It will be relevant to notice here that the invariant transformation of vector $s^* = \Omega(t)s$ may be carried out with the help of the design method described in Sect. 6.2 which does not require any consideration of the system behaviour in subspace s^* but, rather, makes use of the Lyapunov function $v = \frac{1}{2}s^{\mathrm{T}}Ws$ $(W > 0)$. For example, if $W = I_m$, then

$$\dot{v} = s^{*\mathrm{T}}(\Omega^{-1})^{\mathrm{T}}(CA + \dot{C})x + s^{*\mathrm{T}}(\Omega^{-1})^{\mathrm{T}}CBu \qquad (9.60)$$

Comparing (7.61) and (9.60), we obtain an evidence of the fact that with $\Omega(t) = (C(t)B(t))^{\mathrm{T}}$ the control algorithm of the form (7.59) (or (7.64)) where $D = I_m$ may also provide the stability of sliding modes "in the large" throughout the entire manifold $s = 0$. It is essential that this control technique does not impose any constraints upon the time derivative of matrix $C(t)B(t)$.

Chapter 10

Control of Linear Plants in the Presence of Disturbances

1 Problem Statement

Consider a control system operating in the presence of disturbances, and given reference inputs which define the desired profiles of controlled variables. Accordingly, the behaviour of control system in the case of linear controlled plant is described by linear non-homogeneous differential equation

$$\dot{x} = Ax + Bu + Qf(t), \tag{10.1}$$

where $f(t) \in \mathbb{R}^l$ is the vector characterizing external actions. If the error coordinates are substituted in the state vector for the controlled variables, reference inputs will play a part of disturbances in the new space. Assume that (10.1) is written after such a substitution, vector $f(t)$ then is just a disturbing action, and it is desirable to reduce its effect upon the system behaviour or eliminate it at all. This problem is pivotal in the invariance theory. Let us discuss in short the ideas that may underlie the design of invariant systems.

In the case of measurable disturbances, their action may be compensated by means of a combined control. In the combined systems control actions consist of two components: the first component describes deviation from the desired mode, and the second one which depends on the measured disturbances offsets their effect [90, 113, 138]. The limitation of this method is that in many a practical problem disturbance measurements are impeded while for the realization of invariance conditions exact information about both disturbances and system parameters is required.

Control by deviation, in the limit, also enables one to reject disturbances if the feedback gains are infinitely increased [103]. It is the system stability conditions that define the achievable accuracy within which the invariance to disturbances may be provided.

Both of the control principles mentioned here are sufficiently universal because they may be applied to any functions $f(t)$.

If there is *a priori* information about the class of non-measurable disturbances and if one has to provide in the control system independence of motions for any representative of this class, then invariant system design becomes much simpler. A most popular approach is to assume that as a time function the disturbance is solution of homogeneous linear differential equation with known constant coefficients and unknown initial conditions. In order to provide invariance, one has to choose a controller in the form of a dynamic element described by the same equation as the disturbance model [89]. The same idea is at the root of the design procedure for multivariable linear systems with vector input, output and disturbance [33, 69]. In this class of invariant systems, disturbances may be accurately compensated if the parameters of differential equation describing the disturbance model are known.

Below, consideration will be given to design methods for control systems of the (10.1) type where the desired properties, invariance included, are provided through deliberate introduction of sliding modes. Combined systems and those with non-measurable disturbances will be given separate treatment. It will be demonstrated for both cases that the invariance conditions do not imply the use of infinitely high gains and are robust, i.e. do not require for their realization exact information about disturbances or the parameters of plant or disturbance model.

2 Sliding Mode Invariance Conditions

The possibilities of designing system with invariant sliding modes were investigated thoroughly for a special case where control was scalar and system behaviour was considered in the space of canonical variables, i.e. in the coordinate space of error and its derivatives [41, 139, 140]. However, the technical difficulties involved into determination of the time derivatives of various orders are the major obstacle to the use of such specific sliding modes (although the use of sliding modes in devices performing differentiation allows one to somewhat overcome them [43, 54, 55]).

At the same time, by means of both scalar and vector controls sliding modes can be rather easily established in the space whose coordinates may be not only derivatives but arbitrary physical variables as well. Let us set forth the invariance conditions for such systems which are linear with respect to control:

$$\dot{x} = f(x, t) + B(x, t)u + h(x, t), \tag{10.2}$$

where x, f, B and u are defined by (2.7), (1.8), and n-dimensional vector $h(x, t)$

characterizes disturbances and variations of the parameters with respect to which sliding-mode motion on manifold $s=0$ must be invariant.

Let us write now sliding equations according to the equivalent control method, assuming that $\det GB \neq 0$:

$$u_{eq} = -(GB)^{-1}(Gf + Gh),$$
$$\dot{x} = f - B(GB)^{-1}Gf + (I_n - B(GB)^{-1}G)h. \tag{10.3}$$

Let $\mathscr{B}(x,t)$ be a subspace formed for each point x, t by the base vectors of matrix $B(x,t)$. For the sliding mode motion (10.3) to be invariant with respect to vector $h(x,t)$, it suffices that

$$h(x,t) \in \mathscr{B}(x,t). \tag{10.4}$$

Condition (10.4) means that there exists m-dimensional vector $\lambda(x,t)$ such that

$$h(x,t) = B(x,t)\lambda(x,t). \tag{10.5}$$

The direct substitution of vector $h(x,t)$ in the form of (10.5) into the sliding mode Eq. (10.3) demonstrates that if condition (10.4) is obeyed, invariant motions occur in sliding mode. Condition (10.4) generalizes the invariance condition obtained for linear systems in [40].

It would be easier to investigate the invariance of sliding modes in (10.1) by considering its behaviour in the space of new variables as defined by linear transformation (7.43), (7.44) which leads to a system of equations similar to (7.45)

$$\dot{x}_1 = A_{11}x_1 + A_{12}x_2 + Q_1 f$$
$$\dot{x}_2 = A_{21}x_1 + A_{22}x_2 + B_2 u + Q_2 f \tag{10.6}$$

with

$$MQ = \begin{pmatrix} Q_1 \\ Q_2 \end{pmatrix}.$$

Vector

$$\begin{pmatrix} Q_1 \\ Q_2 \end{pmatrix} f(t) = h(x,t)$$

plays in (10.6) the part of $h(x,t)$. Matrix B_2 is non-singular, and, therefore, the invariance condition will be met if

$$Q_1 = 0.$$

In this case vector $\lambda(x,t) = B_2^{-1}Q_2 f$, and the equation of sliding on manifold $s = Cx_1 + x_2 = 0$

$$\dot{x}_1 = (A_{11} - A_{12}C)x_1 \tag{10.8}$$

is independent of disturbances[1].

[1] One can easily see that the invariance condition is equivalent to $Q \in \mathscr{B}$ of [40].

3 Combined Systems

Let us assume that the set of discontinuity surfaces stems from some requirements to sliding mode motion described by (10.8). The problem is to determine a control vector providing the stability "in the large" of sliding modes over all manifold $s=0$. Notably, the same problem may occur also when the invariance condition (10.7) is not satisfied if the sliding equation has the form of

$$\dot{x}_1=(A_{11}-A_{12}c)x_1+Q_1 f(t)$$

At first, consider the systems where all the components $f_1,\ldots, f_l$ of external action vector f are measurable. The design procedure for control of the systems is similar to that of Sect. 7.5. The linear transformation of vector s or control u gives for system (10.1) equations of motion projection on subspace s^* or s which, in contrast to (7.57) or (7.58), depend not only on the state and control vectors, but also on the disturbance vector

$$\dot{s}^*=Hx+Du+Pf, \tag{10.9}$$

$$\dot{s}=Hx+Du^*+Pf, \tag{10.10}$$

where matrix $P=D(CB)^{-1}CQ$ in (10.9), and $P=CQ$ in (10.10). Since (10.9) and (10.10) are identical, let us confine ourselves to the design problem for (10.9) only where vector is linearly transformed according to (6.2).

The equivalent control $u_{\mathrm{eq}}=-D^{-1}Hx-D^{-1}Pf$ resulting from solution of equation $\dot{s}^*=0$ is a linear function with respect to the state and disturbance vectors. Therefore, there are positive numbers α and q such that function $F(x_1t)=d|x|+q|f(t)|$ is the upper estimate for all the components of equivalent control.

If D is taken to be a diagonal matrix with non-zero diagonal elements, control will be formed similar to (7.59):

$$u=(-\alpha|x|-q|f(t)|)\operatorname{sign} Ds^*. \tag{10.11}$$

In addition to (7.62), parameter q in control (10.11) must meet the following condition in order to provide sliding mode stability

$$|d_i|q>p_i \tag{10.12}$$

where $p_i=\max\limits_j|p_{ij}|$, p_{ij} are the elements of matrix P. An analogue of the algorithm (7.64) has the form of

$$u=-\psi_x-\psi^f f, \tag{10.13}$$

where elements ψ^f_{ij} of matrix Ψ^f vary according to the logical algorithm

$$\psi^f_{ij}=\begin{cases}\alpha^f_{ij} & \text{at} \quad d_i s^*_i f>0\\ \beta^f_{ij} & \text{at} \quad d_i s^*_i f_j<0,\end{cases}$$

α^f_{ij} and β^f_{ij} being constant coefficients.

Conditions (7.65) imposed on the coefficients of matrix Ψ should be met for control (10.13), together with the inequalities

$$|d_i|\alpha_{ij}^f > p_{ij}, \quad |d_i|\beta_{ij}^f < p_{ij}. \tag{10.14}$$

For the balance of cases discussed in Sect. 7.5 where matrix D or $D+D^{\mathrm{T}}$ is symmetrical and positive definite as well as for the case where D is Hadamard matrix, the generalization of design procedures for systems with measurable disturbances is evident.

Since the control algorithms (10.11) and (10.13) has turned out to be dependent on the disturbance vector as well as on the state one, the control systems under consideration should be classified as combined. For the systems of the (10.7) type, sliding mode motion is invariant to disturbances. Importantly, conditions (7.59), (10.12) and (7.65), (10.14) for invariant sliding mode to exist are inequalities whereas in the continuous systems exact compensation of disturbance is required in order to provide invariance (e.g. the control vector in (10.6) must include the term $-B_2^{-1}Q_2 f$). This fact enables one to relax the requirements to the accuracy of disturbance measurements in combined systems with discontinuous controls.

4 Invariant Systems Without Disturbance Measurements

Assume that for the system at issue (10.1) or for its equivalent system (10.6) the invariance condition (10.7) is satisfied and the components of vector f cannot be measured. As in [33, 69], the "disturbance model" is taken in the form of linear dynamic system

$$f^{(k)} + \sum_{i=0}^{k-1} \Theta_i(t) f^{(i)} = 0 \tag{10.15}$$

but in contrast to the references, the scalar coefficients $\theta_i(t)$ in (10.15) are not constant and can vary arbitrarily over any bounded interval, i.e. there exists a set of positive numbers θ_{i0} such that

$$|\theta_i(t)| \leqq \theta_{i0}. \tag{10.16}$$

It is assumed also that neither the initial conditions in (10.15), nor function $\theta_i(t)$ are measurable, and only the ranges are known over which these function can vary. The relations (10.15), (10.16) define a rather broad class of disturbances. For example, for $k=2$ it includes exponential and harmonic functions, polynomials of any finite degree (beginning from a certain time), all kinds of the products of these functions, etc.

The control device will be chosen in the following manner. The control vector is the output of dynamic element described by

$$u^{(k)} + \sum_{i=0}^{k-1} d_i u^{(i)} = v, \tag{10.17}$$

where d_i are constant scalar coefficients whose choise is dictated only by the convenience of implementation. The input v of element (10.17) will be generated below in the form of piecewise linear function of its coordinates and the coordinates of system state vector. Each of m control channels of resulting system has a k-th order dynamic element, the total order of the system being $n + mk$. The state coordinates of the additional dynamic system $u, \dot{u}, \ldots, u^{(k-1)}$ are measurable.

Let us write the motion equations of the extended system in a space consisting of vectors $x_1, x_2, \ldots, x_{k+2}$ if

$$\dot{x}_i = x_{i+1}, \quad i = 2, \ldots, k+1. \tag{10.18}$$

Since $\dot{x}_2 = x_3$, obtain from this relation and the second equation in (10.6) that

$$u = B_2^{-1}(x_3 - A_{21}x_1 - A_{22}x_2 - Q_2 f). \tag{10.19}$$

After k differentiations of (10.19) and substitution of the right sides of (10.6), (10.18) and (10.19) for the time derivatives of vectors x_i and vector u obtain, respectively,

$$u^{(i)} = B_2^{-1}\left(x_{i+3} + \sum_{j=1}^{i+2} A_j^i x_j - Q_2 f^{(i)}\right), \quad i = 1, \ldots, k-1, \tag{10.20}$$

$$u^{(k)} = B_2^{-1}\left(\dot{x}_{k+2} + \sum_{j=1}^{k+2} A_j^k x_j - Q_2 f^{(k)}\right), \tag{10.21}$$

where A_j^i and A_j^k are constant matrices. By substituting the values of derivatives of $u^{(i)}$ from (10.20) and (10.21) into (10.17) and replacing the k-th derivative of the disturbance vector according to constraint (10.15) by a linear combination of vectors $f, \ldots, f^{(k-1)}$, obtain equation with respect to the vector

$$\dot{x}_{k+2} = \sum_{i=1}^{k+2} A_i x_i + \sum_{i=0}^{k-1} (d_i - \theta_i(t)) Q_2 f^{(i)} + B_2 v \tag{10.22}$$

where A_i are constant matrices.

Bearing in mind that vectors $Q_2 f^{(i)} (i = 0, \ldots, k-1)$ may be computed from (10.20), one may represent Eq. (10.22) as

$$\dot{x}_{k+2} = \sum_{i=1}^{k+2} \bar{A}_i(t) x_i + B_2 \sum_{i=0}^{k-1} (d_i - \theta_i(t)) u^{(i)} + B_2 v, \tag{10.23}$$

where $\bar{A}_i(t)$ are matrices dependent on $\theta_i(t)$, and, consequently, on time. Introduce the notations

$$\bar{x}^{\mathrm{T}} = (\bar{x}_1^{\mathrm{T}}, \bar{x}_2^{\mathrm{T}}), \quad \bar{x}_1^{\mathrm{T}} = (x_1^{\mathrm{T}}, x_2^{\mathrm{T}}, \ldots, x_{k+1}^{\mathrm{T}}),$$

$$\bar{x}_2 = x_{k+2}, \quad \bar{u}^{\mathrm{T}} = (u^T, \dot{u}^{\mathrm{T}}, \ldots, (u^{(k-1)})^{\mathrm{T}}),$$

$$B_2 \sum_{i=0}^{k-1} (d_i - \theta_i(t)) u^{(i)} = \bar{Q}_2(t) \bar{u}$$

and rewrite the first Eq. in (10.6), and the equations of (10.18) and (10.23) that describe system behaviour as

$$\frac{d\bar{x}_1}{dt} = \bar{A}_{11}\bar{x}_1 + \bar{A}_{12}\bar{x}_2,$$
$$\frac{d\bar{x}_2}{dt} = \bar{A}_{21}(t)\bar{x}_1 + A_{22}(t)\bar{x}_2 + B_2 v + Q_2(t)\bar{u}, \tag{10.24}$$

where $\bar{A}_{11}$ and $\bar{A}_{12}$ are constant matrices, dependent on time matrices $\bar{A}_{21}(t)$, $\bar{A}_{22}(t)$ and $\bar{Q}_2(t)$ have bounded elements by virtue of the fact that coefficients $\theta_i(t)$ (10.16) are bounded. It follows from the comparison of (10.6) at $Q_1 = 0$ and (10.24) that for (10.24) the principle of combined control (see Sect. 3 above) may be used if vector v is regarded as control and $\bar{u}$ is regarded as vector of measurable disturbances, the difference being that matrices $\bar{A}_{21}(t), \bar{A}_{22}(t)$ and $Q_2(t)$ in (10.24) are time-varying. This, however, does not require the development of special design procedures. Indeed, the sliding mode stability conditions (10.11) and (10.13) are inequalities that may be always satisfied even for varying matrices $\bar{A}_{21}(t), A_{22}(t)$ and $Q_2(t)$ because their elements have bounded ranges of variations.

Thus, in order to provide sliding mode motion over some mainfold $s = \bar{C}_1\bar{x}_1 + \bar{x}_2 = 0$ one may use algorithms similar to (10.11) and (10.13) by taking the new control v in the form of a piecewise linear function of the components of vector $\bar{x}$ and of the state vector $\bar{u}$ of additional dynamic system (10.17). The sliding mode motion is described by the homogeneous differential equation with constant coefficients

$$\frac{d\bar{x}_1}{dt} = (\bar{A}_{11} - \bar{A}_{12}\bar{C}_1)\bar{x}_1 \tag{10.25}$$

and is invariant to disturbances $f_1(t), \ldots, f_l(t)$. It is an important point that the implementation of this control algorithm requires finite gains, and in doing so one does not need to measure disturbances.

5 Eigenvalue Allocation in Invariant System with Non-Measurable Disturbances

Aside from the main problem of providing independence of disturbances for sliding mode motion, it is desirable to elucidate whether it is possible to endow the motion with desired properties by varying the equation of discontinuity surfaces or matrix $\bar{C}_1$ in (10.25). Here we confine ourselves to the problem of eigenvalue allocation.

Let us explain why the need occurred to consider this problem. As was shown in Sect. 7.4, the desired eigenvalue allocation for sliding mode motion may be always provided if the initial system is controllable. For instance, according to the theorem of Sect. 7.4 for combined system (10.1) the eigenvalues defining sliding mode motion (10.8) may be arbitrarily allocated if pair $\{A, B\}$ is controllable. In order to provide invariance in a system without disturbance measurement, an additional $(k \times m)$-th order system is included and sliding mode is provided in the extended state space. The theorem of Sect. 7.4 is not fit for the analysis of this motion because it holds only when sliding modes occur in the state space x of (10.1).

Let us write the motion equations in the space $\bar{x}$

$$\frac{d\bar{x}}{dx} = \bar{A}\bar{x} + \bar{B}\bar{v} + \bar{Q}\bar{u} \tag{10.26}$$

with

$$\bar{v} = \sum_{i=1}^{k+2} \bar{A}_i(t)x_i + B_2 \sum_{i=0}^{k-1} (d_i - \theta_i(t))u^{(i)} + B_2 v. \tag{10.27}$$

According to the first equation in (10.6) at $Q_1 = 0$ and equations (10.18), (10.23) and (10.27), matrices $\bar{A}$ and $\bar{B}$ of (10.26) will be as follows

$$\bar{A} = \begin{bmatrix} A_{11} & A_{12} & 0 & 0 & \cdots & 0 \\ 0 & 0 & I_m & 0 & \cdots & 0 \\ 0 & 0 & 0 & I_m & \cdots & 0 \\ \cdot & \cdot & \cdot & \cdot & \cdots & \cdot \\ 0 & 0 & 0 & 0 & \cdots & I_m \\ 0 & 0 & 0 & 0 & \cdots & 0 \end{bmatrix}, \quad \bar{B} = \begin{bmatrix} 0 \\ 0 \\ \vdots \\ 0 \\ I_m \end{bmatrix},$$

All the zero elements in $\bar{A}$ and $\bar{B}$ are zero matrices of dimension $(n-m)m$ (in the first row of $\bar{A}$ and the first elements in $\bar{B}$), $m(n-m)$ (in the first column of $\bar{A}$) or $m \times m$ (the rest of zero elements in $\bar{A}$ and $\bar{B}$). Vector $\bar{u}$ in (10.26) is regarded as the external disturbance vector, therefore matrix $\bar{Q}$ does not affect the eigenvalue allocation of characteristic equation in sliding mode, and therefore the relations for computation of this matrix are omitted. Since matrix B_2 is non-singular, vectors v and $\bar{v}$ are one-to-one related by (10.27), and it would be more conventient for us to solve the problem of eigenvalue allocation for the sliding mode motion described by (10.25) by means of control $\bar{v}$.

Assume that the pair $\{A, B\}$ in the original system (10.1) is controllable. Let us demonstrate that in this case all the $n + m(k-1)$ eigenvalues of characteristic Eq. (10.25) may be allocated as desired. As was found in Sect. 7.4, a problem stated in this manner always is solvable if pair $\{\bar{A}, \bar{B}\}$ in (10.26) also is controllable. Consequently, we have to prove that controllability of extended system (10.26) follows from the controllability of system (10.1).

First recall that (10.1) and (10.6) are equivalent and that according to the lemma of Sect. 7.4 the pair $\{A_{11}, A_{12}\}$ in (10.6) is controllable, i.e.

$$\operatorname{rank}\{A_{12}, A_{11}A_{12}, \ldots, A_{11}^{n-m-1}A_{12}\} = n - m. \tag{10.28}$$

For the controllability matrix $\bar{P} = \{\bar{B}, \bar{A}\bar{B}, \ldots, \bar{A}^{n+mk-1}\bar{B}\}$ of the $(n+mk)$-th order system (10.26), condition

$$\operatorname{rank} \bar{P} = n + mk \tag{10.29}$$

should be met. Compute the elements of matrix $\bar{P}$:

$$\bar{A}^i = \begin{bmatrix} A_{11}^i & A_{11}^{i-1}A_{12} & \cdots & A_{11}A_{12} & A_{12} & 0 & \cdots & 0 \\ 0 & 0 & \cdots & 0 & 0 & I_m & \cdots & 0 \\ \cdots & \cdots & \cdots & \cdots & \cdots & \cdots & \cdots & \cdots \\ 0 & 0 & \cdots & 0 & 0 & 0 & \cdots & I_m \\ 0 & 0 & \cdots & 0 & 0 & 0 & \cdots & 0 \\ \cdots & \cdots & \cdots & \cdots & \cdots & \cdots & \cdots & \cdots \\ 0 & 0 & \cdots & 0 & 0 & 0 & \cdots & 0 \end{bmatrix} \left.\vphantom{\begin{matrix}0\\0\\0\end{matrix}}\right\} i \times m, \quad i = 1, \ldots, k;$$

$$\bar{A}^i = \begin{bmatrix} A_{11}^i & A_{11}^{i-1}A_{12} & \cdots & A_{11}^{i-k-1}A_{12} \\ 0 & 0 & \cdots & 0 \\ \cdots & \cdots & \cdots & \cdots \\ 0 & 0 & \cdots & 0 \end{bmatrix}, \quad i \geqq k+1$$

$$\bar{A}^i\mathrm{B} = \begin{bmatrix} 0 \\ 0 \\ \vdots \\ 0 \\ I_m \\ 0 \\ \vdots \\ 0 \end{bmatrix} \left.\vphantom{\begin{matrix}0\\0\\0\end{matrix}}\right\} i \times m, \quad i = 0, \ldots, k; \quad \bar{A}^i\bar{B} = \begin{bmatrix} A_{11}^{i-k-1}A_{12} \\ 0 \\ \vdots \\ 0 \end{bmatrix}, \quad i \geqq k+1.$$

Consider matrix $\bar{P}'$ consisting of the $m(k+1) + (n-m)m$ first columns of matrix $\bar{P}$:

$$\bar{P}' = (\bar{B}, \bar{A}\bar{B}, \ldots, \bar{A}^{n-m-k}\bar{B}), \quad \bar{P} = (\bar{P}'\bar{P}''),$$

or

$$\bar{P}' = \begin{bmatrix} 0 & 0 & \cdots & 0 & A_{12} & A_{11}A_{12} & \cdots & A_{11}^{n-m-1}A_{12} \\ 0 & 0 & \cdots & I_m & 0 & 0 & \cdots & 0 \\ \cdots & \cdots & \cdots & \cdots & \cdots & \cdots & \cdots & \cdots \\ 0 & 0 & \cdots & 0 & 0 & 0 & \cdots & 0 \\ 0 & I_m & \cdots & 0 & 0 & 0 & \cdots & 0 \\ I_m & 0 & \cdots & 0 & 0 & 0 & \cdots & 0 \\ & \underbrace{\hphantom{0 \quad 0}}_{m \times (k+1)} & & & & & & \end{bmatrix}$$

($\bar{P}''$ consists of matrices $\bar{A}^i\bar{B}$, $i > n - m + k$; it should be noted that $\bar{P} = \bar{P}'$ only if $m = 1$). It follows from condition (10.28) that rank $\bar{P}' = m(k+1) + n - m = mk + n$, i.e. the validity of condition (10.29) is proved and the pair $\{\bar{A}, \bar{B}\}$ is controllable.

Introduction of an additional dynamic element (10.17) for provision of invariance with respect to disturbances, thus, does not violate system controllability and sliding mode motion may have any desirable eigenvalue allocation.

Chapter 11

Systems with High Gains and Discontinuous Controls

1 Decoupled Motion Systems

For complicated dynamic plants that are described by non-linear time-varying high-dimensional differential equations, control systems whose general motion may be decoupled with respect to some attribute into partial components having smaller subspaces are especially attractive. Independent investigation of smaller-dimensionality problems enables a significant simplification of analysis and design of control systems.

As was found above, such a possibility is offered by the discontinuous control systems where the final phase of control is sliding mode motion. A similar situation occurs also in singularly perturbed systems (Chap. 5) if the small parameters before the derivatives in the equations enable one to decouple the general motion into the fast and slow components that may be considered independently.

High gains, the classical tool for suppression of the effects of disturbances and parametric variations, may be used in control systems for artificial decomposition of the general motion into separate components of different rates. Reference [102] was the first to specify the class of systems where the unlimited increase of open-system gain does not result in the loss of stability, and to demonstrate that the eigenvalues of characteristic equation may be partitioned into "fast" and "slow". Some new results obtained recently by the theory of linear multi-variable control systems seem to have awakened a fresh interest in high-gain systems. The asymptotic behaviour of the eigenvalues of multi-variable control systems was studied in terms of "transmission zeros" [34, 84, 123], and procedures were specified for separation of fast and slow motions [126]. In such a system, fast motions are defined by a group of eigenvalues tending to infinity, the rest of eigenvalues defining slow motions and a manifold in the state space where the slow motions trajectories lie. This technique has enabled decomposition of the design problem into two smaller-dimensionality problems [127]. A similar procedure was proposed in reference [169] which at first establishes the correspondence between singularly perturbed systems and those with high gains.

One will readily see that the design philosophies of high-gain and discontinuous-control systems have a great deal in common: in systems of both types the general motion is decoupled into smaller-dimensionality motions designed independently, slow motions and sliding modes taking place on a manifold in the system state space. In this chapter we are going to establish a correspondence between the mathematical tools used for the investigation of decoupled-motion systems and to discuss the differences between them.

The features of high-gain system design will be elucidated for the non-linear systems that are linear with respect to control (2.7). Let us assume that matrix $B(x,t)$ in this equations has l non-zero rows (e.g. last ones). The system (2.7) is then representable as

$$\begin{aligned} \dot{x}_1 &= f_1(x_1, x_2, t), \\ \dot{x}_2 &= f_2(x_1, x_2, t) + B(x_1, x_2, t)u \end{aligned} \tag{11.1}$$

with $x_1 \in \mathbb{R}^{n-l}$, $x_2 \in \mathbb{R}^l$, $u \in \mathbb{R}^m$. Since in all the problems under consideration it is assumed that rank $B = m$, then $l \geqq m$. Let control u be generated as a continuous function of the state vector

$$u = ks(x_1, x_2), \tag{11.2}$$

where $s \in \mathbb{R}^m$, and the scalar parameter k tends to infinity. Rewrite (11.1) in the following form:

$$\begin{aligned} \dot{x}_1 &= f_1(x_1, x_2, t), \\ \mu\dot{x}_2 &= \mu f_2(x_1, x_2, t) + B(x_1, x_2, t)s(x_1, x_2), \end{aligned} \tag{11.3}$$

where $\mu = 1/k$.

For small magnitudes of parameter μ, (11.3) is a singularly perturbed system. However, the methods of Sect. 5.1 are not directly applicable to it. Indeed, the zero-approximation equation obtained from (11.3) at $\mu = 0$;

$$B(x_1, x_2, t)s(x_1, x_2) = 0 \tag{11.4}$$

is equivalent to equation

$$s(x_1, x_2) = 0. \tag{11.5}$$

Since the dimensionality l of vector x_2 is not necessarily equal to m, one cannot generally determine an unambigous solution of (11.4) with respect to x_2 which should be substituted into the first equation of (11.3).

Such cases of singularly perturbed systems are referred to as *critical.* Reference [156] describes iterative procedures enabling one to determine solutions for these cases. From the viewpoint of control system design, of particular interest are not solutions but the differential equations describing various components of motion because this allows one to design them separately through independent solution of problems of smaller dimensionality. It is the case with the design of discontinuous control systems where the equivalent control method permits one to form the equations of motion on the manifold of intersection of discontinuity surfaces. It will be shown in what follows that this method allows one to determine the slow-motion equations or those of zero approximation on manifold (11.5). for high-gain systems as well.

2 Linear Time-Invariant Systems

Let us begin the discussion of high-gain systems from the most comprehensively studied case where the system behaviour is described by linear Eq. (7.1) with constant coefficients. Control in these systems will be also taken to be linear:

$$u = ks, \quad s = Cx, \tag{11.6}$$

where C is $(m \times n)$-dimensional matrix, and k is scalar gain that may be infinitely increased. As a result, obtain an equation describing the motion in a closed system

$$\dot{x} = Ax + kBCx. \tag{11.7}$$

We confine ourselves to the discussion of the case where

$$\operatorname{rank} BC = \operatorname{rank} CB = m \tag{11.8}$$

and present for it without proof the major results of the study of asymptotic behaviour of (11.7) under the unlimited increase of gain k.

The analysis carried out in [126, 169] has revealed that under k tending to infinity the eigenvalues of the characteristic equation fall into two groups. The first group consists of $(n - m)$ eigenvalues and defines the slow motion. Its eigenvalues coincide with the system transmission zeros if control is regarded as an input action and vector s as output one. According to [126], the eigenvalues of the first group $\lambda_1, \ldots, \lambda_{n-m}$ are solutions to the equation

$$\det \begin{pmatrix} A - \lambda I_n & B \\ C & 0 \end{pmatrix} = 0, \tag{11.9}$$

the second group consists of m eigenvalues $\bar{\lambda}_1, \ldots, \bar{\lambda}_m$ defining the fast motion in the system. All the eigenvalues of the group tend to infinity with the growth of k, the asymptotes of their loci in the complex plane being dependent on matrix CB:

$$\bar{\lambda}_i = k\left(\lambda_i(CB) + O\left(\frac{1}{k}\right)\right), \quad i = 1, \ldots, m, \tag{11.10}$$

where $\lambda_i(CB)$ are the eigenvalues of matrix CB (they are different from zero because $\det CB \neq 0$ according to (11.8), $\lim_{k \to \infty} O\left(\frac{1}{k}\right) = 0$. Obviously, for fast motions to be stable the real parts of the eigenvalues of matrix CB must be negative:

$$\operatorname{Re} \lambda_i(CB) < 0, \quad i = 1, \ldots, m. \tag{11.11}$$

If condition (11.11) is met, the particular motions defined by the eigenvalues of (11.10) decay rapidly, and in the vicinity of the order of $O(1/k)$ of manifold $s = 0$ slow motion occurs in the system. This fact may be easily accounted for: at significant deviation of the state vector from manifold $s = 0$, control (11.6) grows dramatically with large k, thus resulting in high rate of vector x variation and, as a result, in the occurrence of fast motions.

Linear systems are designed in two steps [126, 127]. At first, the desired slow motion is generated by proceeding from its characteristic equation (11.9) and choosing an appropriate matrix C. Bearing in mind that the linear transformation of C

$$C^* = \Omega C \tag{11.12}$$

does not change solutions of (11.9), we choose the $(m \times m)$-dimensional matrix Ω so that the eigenvalues of matrix C^*B that define the fast motions be equal to the desired values. This procedure realizes the desired fast ans slow motions in the system with control $u = ks$, $s = C^*x$ under the infinite increase of gain k.

Compare now the continuous linear systems considered above with the discontinuous ones which are designed by inducing sliding mode on manifold $s = 0$. According to the equivalent control technique, the sliding equation is determined through

$$\dot{x} = Ax + Bu_{\text{eq}}, \quad s = Cx, \quad \dot{s} = 0. \tag{11.13}$$

Determine the characteristic equation of the system by regarding the vectors x, u_{eq} and s as state variables

$$\det \begin{pmatrix} A - \lambda I_n & B & 0 \\ C & 0 & -I_m \\ 0 & 0 & \lambda I_m \end{pmatrix} = 0$$

or

$$\lambda^m \det \begin{pmatrix} A - \lambda I_n & B \\ C & 0 \end{pmatrix} = 0. \tag{11.14}$$

Since (11.13) has solution $s = \text{const}$ and in sliding mode $s = 0$, there is no particular solution corresponding to m zero eigenvalues of (11.14).

Comparison of (11.9) and (11.14) shows that the sliding mode motion is defined by the same eigenvalues as the slow motion in the high-gain linear system. Thus, in order to describe slow motions in a continuous linear system one may use (2.10) obtained by means of the equivalent control method. Notably, design procedures in discontinuous and continuous systems bear resemblance to each other. In the first place, motion on manifold $s = 0$ is designed, and then by means of (11.12) or similar invariant transformation (6.2) the desired convergence to this manifold is provided.

3 Equivalent Control Method for the Study of Non-Linear High-Gain Systems

Consider now non-linear systems where the right-hand side of differential equations is linearly dependent on control:

$$\dot{x} = f(x, t) + B(x, t)u. \tag{11.15}$$

Let control be continuous non-linear function of the state vector

$$u = ks(x) \tag{11.16}$$

with sufficiently high gain k. As was established in Sect. 1, the standard tools of the theory of singularly perturbed equations do not allow one to form the slow motion equations that basically define the properties of control. It will be demonstrated that this problem may be approached by the equivalent control method like in the case of linear time-invariant system [145].

To substantiate this assertion, let us consider the projection of motion on subspace s, then derive conditions for convergence of the trajectories to some vicinity of the origin, and complete by demonstrating that the motion on manifold $s = 0$ is described by the equations of the equivalent control method. The proof of these statements relies upon the following lemma.

Lemma. *Let in equations*

$$\frac{dy^*}{dt} = F(t)y^*, \tag{11.17}$$

$$\frac{dy}{dt} = (F(t) + \mu F_1(t))y + \mu f(t) \tag{11.18}$$

y^*, y and f be m-dimensional vectors, matrices $F(t)$ and $F_1(t)$ be bounded in norm by a positive number F_0, vector $f(t)$ be bounded by number f_0, and μ be a constant parameter.

If there exists an exponential function as an upper estimate for solution of (11.17) with initial condition $y(0)$

$$\|y^*(t)\| \leqq A^* \|y(0)\| e^{-\alpha^* t}, \quad t \geqq 0, \tag{11.19}$$

where A^* and α^* are positive numbers, $A^* \geqq 1$, then there will be positive numbers A, α and μ, such that for solution of (11.18) with the same initial condition $y(0)$ and $|\mu| < \mu_1$ the estimate

$$\|y(t)\| \leqq A \|y(0)\| e^{-\alpha t} + O(\mu), \quad \text{for } t \geqq 0 \tag{11.20}$$

is true, $\lim_{\mu\to 0} O(\mu) = 0$, $\lim_{\mu\to 0} |O(\mu)/\mu| \leqq H$ with H be a constant non-zero value.

Proof. System (11.18) is linear, therefore for any $t_1 > 0$ numbers M_0 and N_0 may be specified such that

$$\|y(t)\| \leqq \|y(0)\| M_0 + |\mu| + f_0 N_0, \quad t \in [0, t_1]. \tag{11.21}$$

Proceed now from differential to integral equations:

$$y^*(t) = y(0) + \int_0^t F(\gamma) y^*(\gamma) d\gamma, \tag{11.22}$$

$$y(t) = y(0) + \int_0^t (F(\gamma) y(\gamma) + \mu F_1(\gamma) y(\gamma) + \mu f(\gamma)) d\gamma. \tag{11.23}$$

Obtain from (11.21), (11.22), (11.23)

$$\|y(t) - y^*(t)\| \leqq F_0 \int_0^{t_1} \|y(\gamma) - y^*(\gamma)\| d\gamma + |\mu| F_0 \|y(0)\| M_0 t_1$$
$$+ \mu^2 F_0 f_0 N_0 t_1 + |\mu| f_0 t_1, \quad t \in [0, t_1].$$

Application of the Bellman-Gronwall lemma to this result leads to

$$\|y(t)\| \leqq \|y^*(t)\| + |\mu| M \|y(0)\| + |\mu| N, \tag{11.24}$$

where $M = F_0 M_0 t_1 e^{F_0 t_1}$, $N = (|\mu| F_0 f_0 N_0 t_1 + f_0 t_1) e^{F_0 t_1}$, and they are independent of the initial conditions. If t_1 and μ are chosen so that

$$A^* e^{-\alpha^* t_1} + |\mu| M = \xi^* < 1, \tag{11.25}$$

then according to (11.19) and (11.25)

$$\|y(t_1)\| \leqq \xi^* \|y(0)\| + |\mu| N. \tag{11.26}$$

Condition (11.25), obviously, may be met only for $|\mu| < \mu_1 = \xi^*/M$. For the initial condition

$$\|y(0)\| \geqq \frac{|\mu| N}{1 - \xi^*} \chi \quad \text{and any} \quad \chi > 1 \tag{11.27}$$

solution of (1.18) is estimated as

$$\|y(t_1)\| \leqq \xi y_0(0)\|, \quad \xi = \xi^* + \frac{1 - \xi^*}{\chi} < 1. \tag{11.28}$$

Inequality (11.28) means that for solution (11.18) with the initial condition from the domain (11.27) there exists an exponential estimate. If at some t_2

$$\| y(t_2) \| < \frac{|\mu| N}{1 - \xi^*} \chi \tag{11.29}$$

according to (11.26)

$$\| y(t_2 + t_1) \| < \frac{|\mu| N}{1 - \xi^*} \chi,$$

i.e. the state vector will be again in domain (11.29) within time t_1. The upper bound of vector $y(t)$ norm over interval $t \in [t_2, t_2 + t_1]$ may be estimated through (11.19), (11.21), (11.24) and (11.29):

$$\| y(t) \| \leqq A^* \frac{|\mu| N}{1 - \xi^*} \chi + |\mu| M \frac{|\mu| N}{1 - \xi^*} \chi + |\mu| N. \tag{11.30}$$

Until time t_2, solution of (11.18) has an exponential estimate, and after this instant it stays within domain (11.30). Consequently, estimate (11.20) holds for $t > 0$ and

$$\lim_{\mu \to 0} \left| \frac{O(\mu)}{\mu} \right| \leqq A^* \frac{N}{1 - \xi^*} + N = H. \quad \text{Q.E.D.}$$

Let us write the equation of motion projection of system (11.15) on subspace s in a stretched time scale:

$$\frac{ds}{d\tau} = \mu G f + G B s \tag{11.31}$$

where $\mu = 1/k, \tau = t/\mu, G\{\partial s/\partial x\}$. Equation (11.31) describes the fast motions. Let us formulate and prove the conditions under which fast motions converge to manifold $s(x) = 0$, and control u converges to equivalent control $u_{eq}(x, t)$.

Theorem. *If functions $f(x,t)$, $B(x,t)u_{eq}(x,t)$ and $B(x,t)$ meet the Lipschitz condition, the partial derivatives of u_{eq} with respect to all the arguments exist and are bounded on any limited domain, system*

$$\frac{ds}{dt} = G(x,t)B(x,t)s \tag{11.32}$$

is homogeneously exponentially stable with respect to any $x(t)$ and t, i.e. there exist positive numbers A and α such that

$$\| s(t) \| < A \| s(0) \| e^{-\alpha t}, \tag{11.33}$$

then for any positive numbers Δ, Δt_1 and $T(\Delta t_1 < T)$ one may find $\mu_0 > 0$

such that for (11.15) *with control* (11.16)

$$\| s(t) \| < \Delta, \tag{11.34}$$

$$\| u - u_{\mathrm{eq}} \| < \Delta \quad \textit{with} \quad 0 < \mu < \mu_0, \quad \Delta t_1 \leqq t \leqq T \tag{11.35}$$

As follows from the statement of the theorem, system behaviour is studied over a finite time interval. The further reasoning will be built around the assumption that over this interval solution of (11.15), (11.16) is bounded:

$$\| x(t) \| \leqq L_0, \quad t \in [0, T], \tag{11.36}$$

where L_0 is a positive number. For systems with infinitely increasing gain such an assumption may prove to be incorrect, therefore after the completion of proof we have to check its validity.

Introduce into consideration vector

$$v = u - u_{\mathrm{eq}} = \frac{1}{\mu} s - u_{\mathrm{eq}}(x, t)$$

characterizing the difference between the true and equivalent controls. Bearing in mind that $Gf + GBu_{\mathrm{eq}} = 0$, write the equation with respect to vector v in time scale τ:

$$\frac{dv}{d\tau} = (GB + \mu R_v)v + \mu R_s \tag{11.37}$$

with

$$R_v(x, \mu\tau) = -G_{\mathrm{eq}}B,$$

$$R_s(x, \mu\tau) = -G_{\mathrm{eqv}} - G_{\mathrm{eq}}Bu_{\mathrm{eq}} - \frac{\partial u_{\mathrm{eq}}}{\partial t},$$

$$G_{\mathrm{eq}}(x, \mu\tau) = \left\{ \frac{\partial u_{\mathrm{eq}}}{\partial x} \right\}.$$

What is special about the study of vector v behaviour is that with decrease of parameter μ its initial value

$$v(0) = \frac{1}{\mu} s(0) - u_{\mathrm{eq}}(0)$$

grows infinitely.

According to our assumption, (11.36), vector x and, consequently, matrices Gf, R_v and R_s are bounded in norm, therefore, according to the above lemma, there exist positive numbers A_s, α_s, A_v and α_v such that for solutions of (11.32) and (11.37) the following inequalities hold

$$\begin{aligned} &\| s(\tau) \| < A_s \| s(0) \| e^{-\alpha_s \tau} + O(\mu), \\ &\| v(\tau) \| < A_v \left\| \frac{s(0) - \mu u_{\mathrm{eq}}(0)}{\mu} \right\| + O(\mu), \quad O(\mu) > 0. \end{aligned} \tag{11.38}$$

If for $\tau = 0$

$$A_s \| s(0) \| + O(\mu) > \Delta,$$

$$A_v \left\| \frac{s(0) + \mu u_{eq}(0)}{\mu} \right\| + O(\mu) > \Delta$$

according to estimates (11.38)

$$\| s(\tau) \| < \Delta, \quad \| v(\tau) \| = \| u - u_{eq} \| < \Delta \quad \text{with} \quad \tau \geqq \tau_1 \tag{11.39}$$

with

$$\tau_1 = \max(\tau_1', \tau_1''), \qquad \tau_1' = \frac{1}{\alpha_s} \ln \frac{A_s \| s(0) \|}{\Delta - O(\mu)},$$

$$\tau_1'' = \frac{1}{\alpha_v} \ln \frac{A_v \| s(0) - \mu u_{eq}(0) \|}{\mu(\Delta - O(\mu))}.$$

In spite of the fact that τ_1 grows infinitely with decrease of μ, the interval of real-time fast motions $\Delta t_1 = \mu t_1$ diminishes with parameter μ. Over the rest of time interval $t \in [\Delta t_1, T]$ the trajectorties of the system state vector will run in Δ-vicinity of manifold $s = 0$, and control u also will differ in norm from u_{eq} by a value of the order of Δ. For any arbitrarily small Δ and Δt_1, $\mu_1 > 0$ may be always found such that for $0 < \mu < \mu_1$ conditions (11.39) will be satisfied over interval $[\Delta t_1, T]$. Such a motion in the boundary layer of manifold $s = 0$ was shown in Sect. 2.3 to be described by the equation of equivalent control method within $O(\Delta)$-accuracy[1]. This concludes the proof of the theorem.

It now remains only to make sure that our assumption that the state vector x is bounded in norm (11.36) holds for all $t \in [0, T]$. Rewrite (11.15) in time scale τ

$$\frac{dx}{d\tau} = \mu f(x, \mu\tau) + B(x, \mu\tau)s. \tag{11.40}$$

Consider separately two intervals $t \in [0, \mu\tau_1]$ and $t \in [\mu\tau_1, T]$. To estimate solution of (11.40) over the first interval, replace it by the integral equation

$$x(\tau) = x(0) + \int_0^\tau (\mu f(x, \mu\gamma) + B(x, \mu\gamma)s(x) - \mu f(0, \mu\gamma) - B(0, \mu\gamma)s(x))d\gamma$$
$$+ \int_0^\tau (\mu f(0, \mu\gamma) + B(0, \mu\gamma)s(x))d\gamma.$$

[1] In continuous systems control differs slightly from u_{eq} at motion in the vicinity of manifold $s = 0$, as distinct from discontinuous systems where control may arbitrarily vary within the boundaries u^+ and u^-.

Since, according to the theorem's conventions, there exists the Lipschitz constant for functions f and B, and $\mu\tau_1$ is bounded, therefore

$$\|x(\tau)\| \leqq x(0)\| + \int_0^{\tau}(\mu L + L_1\|s\|)\|x\|d\gamma$$

$$+ \int_0^{\tau_1}(\mu N + N_1\|s\|)d\gamma \quad \text{with} \quad \tau\in[0,\tau_1], \tag{11.41}$$

where L, L_1, N and N_1 are some positive numbers. By substituting into the last term the estimate of $\|s\|$ from (11.38), obtain

$$\int_0^{\tau_1}(\mu N + N_1\|s\|)d\gamma = \mu N\tau_1 + \frac{1}{\alpha_s}N_1 A_s\|s(0)\| + N_1\tau_1 O(\mu) \leqq N_0, \tag{11.42}$$

where $O < \mu \leqq \mu_2, \mu_2$ and N_0 are positive numbers. (Boundedness of integral (11.42) follows from the fact that according to (11.20) $O(\mu)$ is a first-order infinitesimal with respect to μ and lim $\mu\tau_1 = 0$ with $\mu \to \infty$).

With regard to (11.42), apply to (11.41) the Ballman-Gronwall lemma and obtain the following estimate

$$\|x(\tau)\| \leqq (\|x(0)\| + N_0)\exp\{\mu L\tau_1 + L_1\|s(0)\|A_s\frac{1}{\alpha_s} + \tau_1 O(\mu)\},$$

where $\tau\in[0,\tau_1]$. This inequality means that for $0 < \mu \leqq \mu_2$ vector $x(\tau)$ is bounded in norm over the interval of fast motions.

Over the second interval the motion in the system is described by the equation of the equivalent control method with an accuracy to $O(\Delta)$. According to the theorem's conventions, for the right-hand side of the equation there exists the Lipschitz constant, therefore, its solution will be bounded as well. As the result, we see that solution of the original system is bounded over all the interval under consideration $t\in[0, T]$ under a sufficiently small value of parameter μ. Remind that this assumption was used in the proof of the theorem. To summarize, all the assertions of the theorem hold true if $0 < \mu \leqq \mu_0$, $\mu_0 = \min(\mu_1, \mu_2)$.

Thus, the method of equivalent control allows one to directly form the equations of slow motions in non-linear high-gain systems that are linear with respect to control. Similar to the linear time-invariant systems, slow motions run on an $(n-m)$-dimensional manifold, and the conditions for convergence of fast motions to this manifold may be determined through the analysis of motion in the m-th order system.

Our study of motions in high-gain systems will be concluded by a discussion of the conditions of convergence of its fast component to manifold $s = 0$. This is due to the fact that, generally, it is difficult verify the existence of the exponential estimate (11.33) for solution of (11.32) and therefore determination of special cases where this problem is solvable would be of interest.

As shown in [122] system (11.32) is homogeneously exponentially stable if GB is Hadamard matrix: the magnitudes of all of its diagonal elements are negative, and the diagonal element of each column or row is greater than the sum of magnitudes of the rest of elements.

In cases where matrix B in (11.15) depends only on time t, its elements have bounded derivatives, rank $B(t)=m$ and det $GB \neq 0$, the conditions of fast motion convergence to manifold $s=0$ may be derived from the stability analysis of the time-invariant system. In order to make sure that it is the case, consider system behaviour in space (η, s), if

$$\eta = B_0(t)x, \quad s = s(x), \tag{11.43}$$

where $B_0(t)B(t)=0$ and $B_0(t)=n-m$ (i.e. the rows of matrix $B_0(t)$ and columns of $B(t)$ are orthogonal). According to the implicit function theorem, this transformation is non-singular because condition det $GB \neq 0$ means that rank $(B_0^{\mathrm{T}}, G^{\mathrm{T}})=n$. Hence, there is an inverse transformation $x=x(\eta, s, t)$. The behaviour of system (11.15), (11.16) in the space of new variables is described by

$$\frac{d\eta}{d\tau} = \mu R_1(\mu\tau, \eta, s) + \mu Q(\tau) R_2(\mu\tau, \eta, s), \tag{11.44}$$

$$\frac{ds}{d\tau} = \mu R_3(\mu\tau, \eta, s) + R_4(\mu\tau, \eta, s)s \tag{11.45}$$

with

$$R_1 = B_0(\mu\tau) f(x(\mu\tau, \eta, s)),$$

$$R_2 = x(\mu\tau, \eta, s),$$

$$R_3 = G(x(\mu\tau, \eta, s)) f(x(\mu\tau, \eta, s)),$$

$$R_4 = G(x(\mu\tau, \eta, s)) B(\mu\tau), \quad \text{rank } R_4 = m$$

and $Q(\mu\tau)$ is matrix equal to the time derivative of $B_0(t)$. Matrices R_1, R_2, R_3 and R_4 are assumed to meet the Lipschitz condition. In contrast to (11.40), the right-hand side of (11.44) at $\mu=0$ is zero. For this case, analysis of (11.44), (11.45) carried out according to the same scheme as for (11.31), (11.40) leads to the following result: if the equilibrium state $s=0$ of the system

$$\frac{ds}{dt} = R_4(\alpha, \beta, s)s$$

is homogeneously exponentially stable "in the large", for any numbers T, Δ and Δt_1 there will be μ_0 such that the motion is system (11.44), (11.45) will be described at $0 < \mu \leqq \mu_0$ by the equation of the equivalent control method accurate within Δ for time interval $t \in [\Delta t_1, T]$. In systems where control u (or vector $s(x)$) is linear, i.e. $G=\text{const}$, the same assertion may be made if matrix $GB(t)$ is the Hurwitz one for all t.

It will be noted that the method for investigation of fast and slow motions presented in this section is applicable also to the analysis of the regular case of singularly perturbed systems of the (5.1), (5.2) type. Rewrite (5.1), (5.2) as

$$\dot{x} = f(x, y, \mu) \tag{11.46}$$

$$\dot{y} = u \tag{11.47}$$

where $u = \frac{1}{\mu} s$, $s = g(x, y, t, \mu)$, $u \in \mathbb{R}^l$, $s \in \mathbb{R}^l$. As a result, we have obtained equations that may be regarded as describing a control system with high gain $k = 1/\mu$. Similar to (11.31), the equation of fast motions in the system (11.46), (11.47) is

$$\frac{ds}{d\tau} = \mu\left(G_x f + \frac{\partial s}{\partial t}\right) + G_y s$$

where

$$G_x = \left\{\frac{\partial s}{\partial x}\right\}, \quad G_y = \left\{\frac{\partial s}{\partial y}\right\}.$$

If system $ds/d\tau = G_y s$ is homogeneously exponentially stable, all the trajectories converge at μ tending to zero to manifold $s = 0$ and control u converges to

$$U_{\text{eq}} = -G_y^{-1}\left(G_x f + \frac{\partial s}{\partial t}\right)$$

(recall that u_{eq} is the solution of equation $\dot{s} = 0$ with respect to control). According to the equivalent control method, slow motion on manifold $s = 0$ at $k \to \infty$ or $\mu \to 0$ is described by the following equations

$$\dot{x} = f(x, y, t, 0), \quad \dot{y} = -G_y^{-1}\left(G_x f + \frac{\partial s}{\partial t}\right). \tag{11.48}$$

The order of the system may be reduced by l via making use of equation $s = 0$ and excluding vector y from (11.48). This procedure will naturally result in (5.5) derived by the methods of the theory of singularly perturbed equations.

4 Concluding Remarks

As we have found in this chapter, if a system is linear with respect to control, the equations of sliding modes in discontinuous systems and of slow motions in high-gain continuous ones coincide and have a smaller dimensionality as compared with the original system.

This similarity seems to have given rise to the conviction that these two types of systems are equivalent. Moreover, their "similarity" is due to the fact that the inputs of devices implementing discontinuous controls are in sliding mode close to zero and the averaged outputs differ from zero, i.e. that the averaged gains tend to infinity.

Nevertheless, substitution of an infinite-gain linear element for the discontinuous one with the aim of studying sliding modes might lead to qualitatively erroneous results because in these two types of systems the stability conditions for sliding modes in the discontinuous case and slow motions in the continuous one are not equivalent even for systems linear with respect to control.

By way of example [66] let us condier a system with two-dimensional control and assume that

$$\dot{s} = \begin{pmatrix} \dot{s}_1 \\ s_2 \end{pmatrix} = G(f + Bu) = -\begin{pmatrix} a_{11} & a_{12} \\ a_{21} & a_{22} \end{pmatrix}\begin{pmatrix} u_1 \\ u_2 \end{pmatrix}, \quad a_{ij} = \text{const.} \tag{11.49}$$

For the linear controls $u_1 = ks_1$ and $u_2 = ks_2$, write the stability conditions for (11.49) under arbitrary k (including $k \to \infty$)

$$a_{11} + a_{22} > 0, \quad a_{11}a_{22} - a_{12}a_{21} > 0. \tag{11.50}$$

This system, obviously, is stable at $a_{11} = 4$, $a_{12} = 9$, $a_{12} = -1$ and $a_{22} = -2$. But with the same coefficients and discontinuous controls $u_1 = M$ signs, and $u_2 = M$ sign s_2 $(m > 0)$ s_2 and $\dot{s}_2$ have the same signs, i.e. for $t \to \infty$ $\lim s_2 = \infty$ which is indicative of system instability. Hence, stability of an infinitely-high-gain system does not guarantee the existence of sliding modes if the inevitable limitations of control signals are taken into consideration.

Therefore, in this numerical example the infinite-gain system must be regarded as unstable if the constraints are taken into account. Indeed, the domain of initial conditions $\{s_1(0), s_2(0)\}$ such that for $t \in [0, \infty]$ the controls $u_1 = ks_1$ and $u_2 = ks_2$ do not reach the level of constraints $-M < u_1 < M$ and $-M < u_2 < M$ may be seen to contract to the point $\{0, 0\}$ with k tending to infinity.

Notably, at the inversion of signs of a_{ij} in the above example the linear system is unstable under any k, but the sliding mode existence conditions in discontinuous control system wil be fulfilled. Indeed, in this case s_2 and $\dot{s}_2$ have different signs, i.e. sliding mode will occur on the surface $s_2 = 0$. Substitution of $(M \operatorname{sign} s_2)_{\mathrm{eq}} = -(a_{21}/a_{22})u_1$ into the first equation of (11.49) demonstrates that s_1 and $\dot{s}_1$ have different signs as well, and sliding mode occurs on the intersection of both surfaces. The system with infinite gain becomes stable under the constraints.

What has been shown by this illustrative example? First of all, that the dynamic multi-variable systems with high gains or discontinuous controls are not equivalent. The properties of high-gain systems might change qualitatively under constraints on control even if the stability in the small is considered. In this sense, the substitution of linear elements with gains tending to infinity for discontinuous elements at the occurrence of sliding modes is not correct as well. It seems reasonable to analyze the multi-variable systems having sufficiently high gains and constraints in terms of sliding motions with emphasis on the allowance for levels of control constraints.

In addition to the convergence conditions, it might be well to point out to some other differences between the two approaches to inducing motions over some manifold in the state space. The first difference is that in the discontinuous systems motion on manifold $s=0$ is induced by finite controls while in the high-gain ones such a motion may be obtained only in the limit at infinitely large conrols.

Convergences to manifold $s=0$ are also of different nature. In the cotinuous systems, the state vector approaches it asymptotically; in the discontinuous systems, the Lyapunov functions, as it was shown Sect. 4.6, may be made finite, and as the result the sliding mode motion on the intersection of discontinuity surfaces $s=0$ begins after a finite time interval.

The last difference concerns the robustness of both types of systems with respect to small dynamic imperfections. Systems models always disregard small time constants caused, for instance, by the lags in actuators and instrumentation, small masses, inductances and capacities. When these small parameters are taken into consideration, the order of physical system grows in comparison with the ideal one. The behaviour of such a refined model can be described by equations of the (5.1), (5.2) type that are studied by the theory of singularly perturbed differential equations. As it was demonstrated in Sect. 5.3, solutions of the discontinuous systems depend continuously on the small time constants that were neglected by the ideal model. It is only within the framework of certain structures that the high gains do not lead to the loss of stability. For example, the single-loop time-invariant linear system may be made stable provided that the difference between the degrees of polynomials in the denominator and numerator of the transfer function does not exceed two [103]. Introduction of additional dynamic elements even with small time constants may force the system out the framework of "permissible" structures, and the unlimited increase in the gain will entail the loss of stability. Consequently, one may speak here only about choosing a gain that does not exceed some critical value defined by the small time constants disregarded in the ideal model.

Chapter 12
Control of Distributed-Parameter Plants

All the design methods that have been considered in this Part are oriented towards the control of dynamic plants described by oridinary differential equations. Theoretical generalizations to the infinite-dimensional cases involve basic difficulties due to the need of constructing a special mathematical apparatus for the study of discontinuous partial differential equations or, stated more generally, discontinuous equations in a Banach space [111]. The method of equivalent control developed in Part I for finite-dimensional systems was shown [110] to be applicable to the distributed systems described by second-order parabolic or hyperbolic equations with discontinuous distributed or lumped control. A wise range of processes (e.g. thermal and mechanical) is described by the equations of this kind.

This chapter discusses examples of designing systems with sliding modes for control of various distributed plants to which non-measurable disturbances may be applied.

1 Systems with Mobile Control

One of the promising fields of research in the theory of distributed-parameter systems is control by mobile action. The idea underlying this class of systems consists in providing the desired spatial distribution by means of a source of action that repeatedly follows a certain spatial trajectory [24–26]. Processes of this sort are exemplified by melting, thermal processing, zone refining of metals heated by moving electron, ion or laser beams, etc.

The basic difficulties involved into the control of mobile actions are due to the non-linearity of the plant equations with respect to the control defining the pattern of source motion. In the case of multiple-cycle motion with small cycle periods, an effective approach to these difficulties is offered by the use of time-averaged equations [87].

The use of programmed spatial source motion that may be determined through static-mode equations seems to be a natural approach of this problem. However, with incomplete information about the plant operator and external disturbances it may result in essential errors. If there is a possibility of obtaining information about the distribution of controlled parameter, the feedback system would considerably relax the dependence of solution on variations of unmeasurable disturbances and parameters in the process mathematical model. The interested reader is referred to [32] for the treatment of the basic engineering problems arising in the design of closed systems with mobile action, the approaches to them, control algorithms in the class of continuous functions, as well as of the particular devices used for their implementation.

This chapter demonstrates that sliding modes are effective also for plants with moving cyclic action in spatially distributed system under uncertainty conditions [22]. The process is modelled by an equation with time-averaged control which enables one to obtain, with rather limited *a priori* information about the nature of unmeasurable disturbances, a spatial distribution close to the given one. Discontinuous distributed control is designed by means of minimizing the rate of Lyapunov functional decrease.

Motion Equations

From now on consideration will be given to heating by a mobile source of the one-dimensional bar described by the following heat equation

$$\frac{\partial Q}{\partial t}-\frac{\partial^2 Q}{\partial x^2}+bQ=F(x,t), \quad 0\leqq x\leqq S, \quad t\geqq 0 \tag{12.1}$$

where $Q(x, t)$ is the temperature in point (x, t), b is a constant positive parameter, and S is the length of heated bar. Without the loss of generality one may confine oneself to the zero boundary conditions and initial distribution

$$Q(0,t)=Q(S,t)=0, \quad t\geqq 0 \tag{12.2}$$

$$Q(x,0)=0, \quad 0\leqq x\leqq S \tag{12.3}$$

Mobile cyclic control is implemented, for instance, by means of a laser beam repeatedly passing along the heated bar in both directions. If t_1, $t_2, \ldots, t_n, \ldots$ are instants when the beam comes alternately to the ends on the bar, and

$T_n = t_n - t_{n-1}$ is the duration of one cycle or the time of beam transit in one direction, function $F(x, t)$ representing this way of control is as follows

$$F(x,t) = \sum_{n=1}^{\infty} F_n(x,t) = p(t) \sum_{n=1}^{\infty} \delta(x - s_n(t))\chi_n(t), \tag{12.4}$$

where $p(t)$ is the source power, $s_n(t)$ is the profile of beam position in the n-th cycle, $0 \leqq s_n(t) \leqq S$, $t \in I_n = [t_{n-1}, t_n]$, $s_n(t)$ is a non-increasing or non-diminishing function, $\chi_n(t)$ is the characteristic function of the interval I_n:

$$\chi_n(t) = \begin{cases} 1 & \text{at} \quad t \in I_n, \\ 0 & \text{at} \quad t \notin I_n. \end{cases}$$

The source power and cycle lengths are assumed to be bounded as follows: $0 \leqq p(t) \leqq P_{\max}$, $T_n \leqq T_{\max}$, $n = 1, 2, \ldots$. In addition, the rate of source strength variation is bounded as well:

$$\frac{dp}{dt} = u_p, \quad |u_p| \leqq B, \tag{12.5}$$

where B is positive number, u_p is control input into the device that changes source power. The given state of the plant is defined by a function $Q^*(x)$ satisfying the boundary conditions (12.2). The aim of control is to choose $s_n(t)$ for $t \in T_n$, $n = 1, 2, \ldots$, and $p(t)$ such that beginning from some time t_p

$$Q(x,t) = Q^*(x), \quad x \in [0, S], \quad t > t_p \tag{12.6}$$

or $\lim Q(x, t) = Q^*(x)$ with $t \to \infty$.

There are some heuristic reasons that enable us to replace the mobile source problem by that of distributed control $U(x, t)$. Function $U(x, t)$ is the specific source power, i.e. $U(x, t)\,dx$ is the rate of heat received by element $[x, x + dx]$. Let us assume that the time during which the source traverses from one end of the bar to another is sufficiently small and we can regard $p(t)$ over the time interval as constant p_n. The relative amount of heat received by the element in n-th cycle is then $p_n(dx/v_n(x)T_n)$, and, correspondingly,

$$U_n(x) = \frac{p_n}{v_n(x)T_n}, \tag{12.7}$$

where $v_n(x) = \dot{s}_n(t)$ is the source speed distribution in n-th cycle, $v_n(x) = 1/(s_n^{-1}(x))'$, $s_n^{-1}(x)$ is a function inverse to $s_n(t)$,

$$T_n = s_n^{-1}(S) = \int_0^S \frac{dx}{v_n(x)}.$$

Under the above assumption, distributed control is representable as

$$U(x,t) = \sum_{n=1}^{\infty} \frac{p_n}{v_n(x) \int_0^S \frac{dx}{v_n(x)}} \chi_n(t). \tag{12.8}$$

Functions p_n and v_n are time-independent over each interval I_n. If the interval tends to zero, these piecewise constant functions approximate some continuous or piecewise continuous functions $p(t)$ and $v(x, t)$ at $t \geqq 0$, $x \in [0, S]$, and instead of (12.8) we have

$$U(x,t) = \frac{p(t)}{v(x,t)\int_0^S \frac{dx}{v(x,t)}} \tag{12.9}$$

which being substituted into (12.1) for $F(x, t)$ approximately describes the behaviour of mobile-source system. We have considered interval T_n for $\dot{s}_n > 0$, i.e. $v_n(x) > 0$, and the source moving from point $x = 0$ to point $x = S$. It will be readily seen that (12.9) holds for the reverse movement if speed in each point x is regarded as positive. This additional constraint

$$v(x, t) > 0 \tag{12.10}$$

should be taken into consideration when generating control action. (Constraint (12.10) is a natural one because the moving source heats the bar and cannot take off heat).

The problem, thus, has been reduced to the distributed control dependent on source power and source instantaneous velocity. The rightfulness of the passage to the averaged-control model (12.9) is justified in [22] by way of the N.N. Bogolyubov theorem [17] and is formulated as follows:

$$\lim_{T_{\max} \to 0} \int_0^S (Q_F(x,t) - Q_v(x,t))^2 \, dx = 0,$$

where Q_F and Q_v are the solutions of (12.1) with controls (12.4) and (12.9), respectively, under similar boundary and initial conditions.

In order to obtain the desired temperature field one may vary the source power under the constraint (12.5), and source speed in each point x. Both possibilities may be reasonably compared if we assume that the source power is upper-bounded by $P_{\max}$. Let us assume that the source motion law is fixed – for example, its speed is constant; then, $0 \leqq U(x, t) \leqq P_{\max}/S$. If it is the source power that is fixed, integration of (12.9) from 0 to S gives

$$0 \leqq \int_0^S U(x,t)\, dx \leqq P_{\max}$$

Comparing these relations one can conclude that the control of only source motion has a wider set of permissible functions $U(x, t)$ than the control of its power only. Consequently, if there is a possibility to control in each cycle both power and motion, the latter seems to be more advantageous because it provides a richer variety of controls and, hence, a richer set of possible controlled plant states.

Steady State. Equations in Deviations

Let now see how the desired distribution of temperature field Q^* may be provided by means of programmed control. Since the given distribution is chosen to be time-invariant, the control function that corresponds to the steady state also will be sought in the class of time-invariant functions:

$$U^* = \frac{p^*}{v^*(x) \int_0^S \frac{dx}{v^*(x)}}, \tag{12.11}$$

where p^* is the constant source power, $v^*(x)$ is the speed distribution in each cycle that depends only on the source coordinate. Let us determine these values.

Since in the static mode the temperature Q^* and control U^* do not vary in time, we obtain from (12.1) at $F(x, t) = U^*(x)$

$$-\frac{\partial^2 Q^*}{\partial x^2} + bQ^* = U^*(x). \tag{12.12}$$

By substituting the given function $Q^*(x)$ into the left side of (12.12) determine the desired value of control U^*. The constraint (12.10) or its equivalent constraint $U^* > 0$ defines the realizable class of permissible distributions $Q^*(x)$. In particular, if $b = 0$ (i.e. if no heat is dissipated from the surface of the heated bar) only convex functions may be taken as the desired temperature distributions. As mentioned above, the control $U(x, t)$ characterizes specific source power, therefore

$$p^* = \int_0^S U^*(x)\, dx. \tag{12.13}$$

which results from direct integration of (12.11) over the spatial variable.

Formally, the distribution of speed $v^*(x)$ may be determined from (12.11) with an accuracy to constant coefficient. This may be easily accounted for by the fact that under a constant source power proportional variation of speed does not change the distribution of heat coming to the heated body. But it will be recalled that control U (12.9) results from tending the duration of each cycle T_n to zero. Therefore, prior to choosing function $v^*(x)$ one needs to define $T_n = T^*$ for which approximation (12.9) enables the desired accuracy of process model having the distributed control instead of the mobile one. Since

$$T^* = \int_0^S \frac{dx}{v^*(x)}, \tag{12.14}$$

the desired distribution of speed $v^*(x)$ under chosen functions U^* (12.12) and p^* (12.13) may be determined through (12.11):

$$v^*(x) = \frac{p^*}{U^*(x) T^*} \tag{12.15}$$

(The validity of (12.14) is verified through integration of $1/v^*$ (12.15) with regard to (12.13)).

Since time t corresponding to source position in point x is determined in the n-the cycle from

$$t = t_{n-1} + \int_0^x \frac{d\gamma}{v^*(\gamma)},$$

the desired law of source motion $s_n^*(t)$ in the n-th cycle is the solution of

$$t = t_{n-1} + \int_0^{s^*} \frac{d\gamma}{v^*(\gamma)}. \tag{12.16}$$

This source motion law provides the desired temperature distribution in the static mode under power p^* (12.13).

Assume that the plant is subjected not only to control $U^*(x)$ but also to external disturbances $g(x, t)$ that are not measurable and about which it is only known that

$$|g(x,t)| \leqq A_1, \quad \left| \frac{d}{dt} \int_0^S g(x,t)\,dx \right| \leqq A_2, \tag{12.17}$$

where A_1 and A_2 are positive numbers. In order to offset the influence of disturbances and to provide convergence to the desired distribution $Q^*(x)$, it is intended to vary control $U^*(x)$ by varying the source power and speed distribution depending on the plant state. Let us form process equations with respect to the deviations from the desired mode and with allowance for the disturbing actions:

$$\frac{\partial \bar{Q}}{\partial t} - \frac{\partial^2 \bar{Q}}{\partial x^2} + b\bar{Q} = \bar{U}(x,t) + g(x,t), \quad 0 \leqq x \leqq S, \quad t \geqq 0 \tag{12.18}$$

with

$$\bar{Q}(x,t) = Q(x,t) - Q^*(x),$$
$$\bar{U}(x,t) = U(x,t) - U^*(x),$$
$$\bar{Q}(0,t) = \bar{Q}(S,t) = 0, \quad t \geqq 0,$$
$$\bar{Q}(x,0) = -Q^*(x), \quad 0 \leqq x \leqq S.$$

Since the deviation of control $U(x, t)$ from the programmed control $U^*(x)$ is attained by varying the source intensity and speed

$$p = p^* + \bar{p}(t), \quad v = v^*(x) + \bar{v}(x,t)$$

obtain

$$\bar{U}(x,t) = (p^* + \bar{p}(t)) \frac{1}{(v^* + \bar{v}) \int_0^S \frac{dx}{v^* + \bar{v}}} - p^* \frac{1}{v^* \int_0^S \frac{dx}{v^*}}.$$

Introduce notations

$$u^*(x)=\frac{1}{T^*v^*(x)},$$

$$\bar{u}(x,t)=\frac{1}{T^*}\frac{1}{v^*(x)+\bar{v}(x,t)}-u^*(x), \tag{12.19}$$

$$\mu(t)=\frac{1}{T^*}\int_0^S\frac{dx}{v^*(x)+\bar{v}(x,t)}-1.$$

It is assumed that only those speed $\bar{v}(x,t)$ variations may take place for which $v(x,t)=v^*(x)+\bar{v}(x,t)>0$. Rewrite the equation in deviations with the above notations:

$$\frac{\partial\bar{Q}}{\partial t}-\frac{\partial^2\bar{Q}}{\partial x^2}+b\bar{Q}=\bar{p}(t)\frac{u^*(x)+\bar{u}(x,t)}{1+\mu(t)}+p^*\frac{\bar{u}(x,t)-\mu(t)u^*(x)}{1+\mu(t)}+g(x,t). \tag{12.20}$$

$\bar{u}(x,t)$ and $\mu(t)$ in (12.20) characterize the deviations of control and cycle times due to variations in source speed.

Control System Structure

The task of control is to nullify the deviation from the desired temperature field $\bar{Q}(x,t)$ by varying $\bar{p}(t)$ and $\bar{u}(x,t)$. We assume that $\bar{u}(x,t)$ is a piecewise constant function bounded as follows

$$|u(x,t)|\leqq u_{\max},\quad 0\leqq u_{\max}<u^*(x) \tag{12.21}$$

that may take in the course of control only two extremal values, either $u_{\max}$ or $-u_{\max}$, its sign being dependent on the states $\bar{Q}(x,t)$ of plant (12.20). Now let us see which tasks should be executed by each of the two controls $\bar{p}(t)$ and $\bar{u}(x,t)$. If

$$\int_0^S g(x,t)\,dx\neq 0,$$

then an external disturbance changes the plant's heat content, and under fixed source power or $\bar{p}(t)=0$ its influence cannot be compensated.

This conclusion may be formally borne out in the following manner. Let us try to find a function $\bar{u}(x,t)$ such that the right-hand side of (12.20) is zero at $\bar{p}(t)=0$:

$$\bar{u}(x,t)=-\frac{1+\mu(t)}{p^*}g(x,t)+\mu(t)u^*(x). \tag{12.22}$$

Bearing in mind that

$$\int_0^S\bar{u}(x,t)\,dx=\mu(t)\quad\text{and}\quad\int_0^S u^*(x)\,dx=1$$

obtain after the integration of (12.22) from 0 to S that

$$\int_0^S g(x,t)\,dx \equiv 0.$$

Variation of source power $\bar{p}(t)$ will be chosen so that function

$$\bar{g}(x,t) = \bar{p}(t)\frac{u^*(x) + \bar{u}(x,t)}{1 + \mu(t)} + g(x,t) \tag{12.23}$$

in the right side of (12.20) satisfies the condition

$$\int_0^S \bar{g}(x,t)\,dx = 0 \quad \text{or} \quad \bar{p}(t) = -\int_0^S g(x,t)\,dx. \tag{12.24}$$

Then, it will be possible to select a control $\bar{u}$ for which the right side of (12.20) may be equal to zero. The specific character of this problem is that both this condition and (12.24) should be satisfied without measuring the disturbances.

Thus, the control system must include three subsystems: (i) programmed control providing the desired distribution $Q^*(x)$ in the steady state in the absence of disturbances; (ii) control of source power supporting equality (12.24); and (iii) control of source motion reducing error $\bar{Q}(x,t)$ to zero.

Source Power Control System

Let us integrate (12.20) with respect to x from 0 to S:

$$\frac{d}{dt}\int_0^S \bar{Q}(x,t)\,dx - \frac{\partial\bar{Q}}{\partial x}(S,t) + \frac{\partial\bar{Q}}{\partial x}(0,t) + b\int_0^S \bar{Q}(x,t)\,dx = \bar{p}(t) + \int_0^S g(x,t)\,dx. \tag{12.25}$$

Introduce notations

$$q(t) = \int_0^S \bar{Q}(x,t)\,dx, \tag{12.26}$$

$$\eta(t) = \frac{\partial\bar{Q}}{\partial x}(S,t) - \frac{\partial\bar{Q}}{\partial x}(0,t) - b\int_0^S \bar{Q}(x,t)\,dx, \tag{12.27}$$

$$\dot{q}_1 = \dot{q} - \eta \tag{12.28}$$

and rewrite (12.25) in compliance with (12.26–12.28)

$$\dot{q}_1 = \bar{p} + \int_0^S g(x,t)\,dx.$$

According to the previously formulated problem (12.24) that is solved by the source power control system, the right-hand side in this equality should be equal to zero. It suffices, therefore, to choose an algorithm of variation of $\bar{p}$ such that $\dot{q}_1$ is zero.

Taking into account (12.5), differentiate both sides of the obtained equality with respect to time

$$\ddot{q}_1 = u_p + R(t), \tag{12.29}$$

where

$$R(t) = \frac{d}{dt}\int_0^S g(x,t)\,dx.$$

Take control u_p in the form of discontinuous function

$$u_p = -B\,\mathrm{sign}\,s_p,$$

$$s_p = \dot{q}_1 = \frac{d}{dt}\int_0^S \bar{Q}(x,t)\,dx - \frac{\partial \bar{Q}}{\partial x}(S,t) + \frac{\partial \bar{Q}}{\partial x}(0,t) + b\int_0^S \bar{Q}(x,t)\,dx. \tag{12.30}$$

Since, according to (12.29), (12.30) and (12.17),

$$\dot{s}_p = -B\,\mathrm{sign}\,s_p + R(t),$$

there exists a finite number B such that s_p and $\dot{s}_p$ have opposite signs, i.e. $|s_p| = |\dot{q}_1|$ decreases with finite rate. Consequently, beginning from an instant of time sliding mode will be started in the system which enables solution of the posed problem $\left(\dot{q}_1 = \bar{p} + \int_0^S \bar{Q}(x,t)\,dx = 0\right)$ without measurement of the disturbance.

Source Motion Control System

Let us assume that the transient of the source power control system has completed and (12.24) is satisfied. Then (12.20) may be rewritten as

$$\frac{\partial \bar{Q}}{\partial t} - \frac{\partial^2 \bar{Q}}{\partial x^2} + b\bar{Q} = p^* \frac{\bar{u}(x,t) - \mu(t)u^*(x)}{1+\mu(t)} + \bar{g}(x,t) \tag{12.31}$$

with

$$\int_0^S \bar{g}(x,t)\,dx = 0 \tag{12.32}$$

Represent control $\bar{u}(x,t)$ as

$$\bar{u}(x,t) = \Delta u(x,t) + \delta u(x,t), \tag{12.33}$$

where function $\Delta u(x,t)$ satisfies equations

$$\Delta u(x,t) = -\frac{1+\mu(t)}{p^*}\bar{g}(x,t)$$

and $\int_0^S \Delta u(x,t)\,dx = 0.$ (12.34)

In compliance with conditions (12.17), (12.23) and (12.24), there exists a positive number λ such that $|\bar{g}(x,t)| \leqq \lambda A_1$. It will be assumed below that for the class of disturbances under consideration (12.34)

$$|\Delta u(x,t)| \leqq \frac{(1+\mu)\lambda}{p^*} A_1 \leqq u_{\max} \tag{12.35}$$

Keeping in mind that according to (12.14), (12.19) and (12.32)

$$\mu(t) = \int_0^S \bar{u}(x,t)\,dx \quad \text{and} \quad \int_0^S \Delta u(x,t)\,dx = 0$$

substitute (12.33) and (12.34) into (12.31):

$$\frac{\partial \bar{Q}}{\partial t} - \frac{\partial^2 \bar{Q}}{\partial x^2} + b\bar{Q} = p^* \frac{\delta u(x,t) - u^*(x)\int_0^S \delta u(x,t)\,dx}{1+\mu(t)}. \tag{12.36}$$

According to the previously established plan for solution of the control problem, function $\bar{u}(x,t)$ assumes either of the two extremal values: $u_{\max}$ or $-u_{\max}$. In conformity with (12.33) and (12.35), this condition means that the signs of $\bar{u}(x,t)$ and $\delta u(x,t)$ coincide.

As the result, the problem has been brought into the following formulation: deviation $\bar{Q}(x,t)$ may be reduced to zero if one can only choose the sign of $\delta u(x,t)$ but cannot affect its value. Take the Lyapunov functional

$$v(t) = \frac{1}{2}\int_0^S \bar{Q}^2(x,t)\,dx \tag{12.37}$$

characterizing the deviation from the desired temperature field. Determine the derivative of (12.37) on the solutions of (12.36):

$$\frac{dv}{dt} = \int_0^S \bar{Q}\left(\frac{\partial^2 \bar{Q}}{\partial x^2} - b\bar{Q} + p^* \frac{\delta u(x,t) - u^*(x)\int_0^S \delta u(x,t)\,dx}{1+\mu(t)}\right) dx$$

$$= \bar{Q}\frac{\partial \bar{Q}}{\partial x}\bigg|_0^S - \int_0^S \left(\frac{\partial \bar{Q}}{\partial x}\right)^2 dx - b\int_0^S \bar{Q}^2\,dx + \int_0^S p^*\bar{Q}\frac{\delta u(x,t) - u^*\int_0^S \delta u(x,t)\,dx}{1+\mu(t)}\,dx. \tag{12.38}$$

In the resulting expression, the first term is equal to zero, the second and third terms are always negative.

Denote

$$W(t) = p^*\int_0^S \bar{Q}(x,t)\frac{\delta u(x,t) - u^*(x)\int_0^S \delta u(x,t)\,dx}{1+\mu(t)}\,dx \tag{12.39}$$

The functional (12.39) is not linear with respect to $\delta u(x,t)$ since the time function $\mu(t) = \int_0^s \delta u(x,t)\,dx$ in the denominator depends on $\delta u(x,t)$. According to the assumptions pertaining to (12.19) the value $1+\mu(t)$ is positive, therefore the function $\delta u(x,t)$ will be designed such that the time derivative of the Lyapunov functional (12.39) be negative for any positive time function $1+\mu(t)$.

Changing integration order in (12.39) yields:

$$W(t) = \frac{p^*}{1+\mu(t)} \int_0^S \delta u(x,t)\Big(\bar{Q}(x,t) - \int_0^S \bar{Q}(x,t)u^*(x)\mathrm{d}x\Big)\,dx. \tag{12.40}$$

The functional (12.40) is negative if the functions $\delta u(x,t)$ and $\bar{Q}(x,t) - \int_0^S \bar{Q}(x,t)u^*(x)\,dx$ have opposite signs. According to (12.33) and (12.35) the control algorithm

$$\bar{u}(x,t) = -u_{\max}\,\mathrm{sign}\,s_u$$

$$s_u = \bar{Q}(x,t) - \int_0^S \bar{Q}(x,t)u^*(x)\,dx \tag{12.41}$$

allows one to satisfy condition $W(t) \leqq 0$ because $\delta u = \bar{u} - \Delta u$, and

$$W(t) = -\frac{p^*}{1+\mu(t)} \int_0^S (u_{\max} + \Delta u(x,t)\,\mathrm{sign}\,s_u)|s_u|\,dx. \tag{12.42}$$

The derivative of functional v is, thus, negative, i.e. function v is non-increasing.

Let us demonstrate that condition $dv/dt \leqq 0$ implies the pointwise convergence of deviation $\bar{Q}(x,t)$ to zero. This follows from the fact that the decrease of v may stop only if $\bar{Q}(x,t) \equiv 0$. Indeed, let at some time

$$\max_{x \in [0,S]} |\bar{Q}(x,t)| = \eta(t) > 0$$

and this maximum is attained in point x_0. Then

$$\left| \int_0^{x_0} \left(\frac{\partial \bar{Q}}{\partial x}\right) dx \right| = \eta(t).$$

By means of the Schwartz inequality estimate

$$\int_0^S \left(\frac{\partial \bar{Q}}{\partial x}\right)^2 dx$$

on which depends the time derivative (12.38) of functional (12.37):

$$\eta^2(t) = \left(\int_0^{x_0} \left(\frac{\partial \bar{Q}}{\partial x} \right) dx \right)^2 \leqq \left(\int_0^{x_0} \left(\frac{\partial \bar{Q}}{\partial x} \right)^2 dx \right) \left(\int_0^{x_0} 1 \cdot dx \right),$$

i.e.

$$\int_0^S \left(\frac{\partial \bar{Q}}{\partial x} \right)^2 dx \geqq \frac{\eta^2(t)}{S} \tag{12.43}$$

Thus, function $v \geqq 0$ diminishes with finite speed, and the process will be completed only at $\eta(t) = 0$[1].

The problem of source motion was solved by means of the auxiliary control $\bar{u}(x, t)$ (12.41). In order to determine the corresponding variations of source speed $v(x, t)$ with respect to the programmed motion v^*, make use of notations (12.19)

$$\bar{v}(x, t) = -v^*(x) \frac{\bar{u}(x, t)}{u^*(x) + \bar{u}(x, t)}.$$

As the result, deliberate introduction of sliding modes into the systems controlling source power and motion, has enabled us to obtain the desired temperature field with current process information and without measuring the external disturbances.

2 Design Based on the Lyapunov Method

In the area of control of lumped systems, discontinuous control inducing sliding mode has been designed with the Lyapunov functions that characterize the deviation of the state vector from the manifold with desirable trajectories. Importantly, in the class of discontinuous control systems this procedure does not require the measurement of external disturbances and plant parameters and is realizable with the knowledge of their ranges only.

Now we seek to apply this approach to distributed systems. In this case, the Lyapunov functional is constructed rather than function and used for determination of a control stabilizing the system [110]. Recall that just in this manner the source motion control (12.41) has been found by means of the functional (12.37).

---

[1] Notably, in studying by means of the Lyapunov functional [129] the stability of the solutions of the partial differential equations it is common practice to prove the convergence in the norm L_2, i.e.

$$\lim_{t \to \infty} \| \bar{Q}(x, t) \| = \lim_{t \to \infty} \left(\int_0^S \bar{Q}(x, t) dx \right)^{1/2} = 0,$$

rather than the pointwise convergence.

Distributed Control of Heat Process

Consider control of a one-dimensional heat process $Q(x,t)$ with heat-insulated ends that is described by the second-order nonlinear parabolic equation [161]

$$\rho(x)\dot{Q} = (k(x,Q)Q')' - q(x,Q)Q + u(x,t) + f(x,t) \tag{12.45}$$

$$0 \leqq x \leqq 1, \quad t \geqq 0, \quad Q(x,0) = Q_0(x), \quad Q'(0,t) = Q'(1,t) = 0 \tag{12.46}$$

with distributed control $u(x,t)$ and unmeasurable and bounded disturbance $|f(x,t)| \leqq C$. Natural constraints $\rho > 0, k > 0, q > 0$ are imposed on the unknown heat characteristics of the plant such at heat capacity, heat conductivity and heat exchange factor. Let the task of control be to create the temperature field $Q^*(x,t)$. Without the loss of generality, one may confine onself to the case

$$Q^*(x,t) = 0, \tag{12.47}$$

if suitable constraints are imposed on the smoothness of $Q^*(x,t)$, and then treat the equations in deviations. The deviation of heat field energy from the desired value will be used as the Lyapunov functional

$$v(t) = \frac{1}{2}\int_0^1 \rho(x)Q^2(x,t)\,dx. \tag{12.48}$$

Compute the time derivative of the functional $v(t)$ in virtue of (12.45) with boundary conditions (12.46)

$$\begin{aligned}\dot{v}(t) &= \int_0^1 Q(kQ')'\,dx - \int_0^1 qQ^2\,dx + \int_0^1 Q(u+f)\,dx \\ &= -\int_0^1 k(Q')^2\,dx - \int_0^1 qQ^2\,dx + \int_0^1 Q(u+f)\,dx.\end{aligned}$$

Taking $u = -M \operatorname{sign} Q(x,t)$, $M > c$, obtain $\dot{v}(t) < 0$ for $\int_0^1 Q^2\,dx \neq 0$, therefore $v(t) \to 0$ with $t \to \infty$. Hence, the discontinuous control $u = -M \operatorname{sign} Q$ provides convergence to zero of the heat field distribution in metric L_2. (It follows from (12.43) that in this case also the pointwise convergence stems from the convergence in metric L_2).

Distributed Control of Mechanical Processes

The Lyapunov method is applicable also to control of another commonly used class of distributed plants, the mechanical processes. For case of discussion let us consider an mechanical plant with fixed ends whose deviation from the equilibrium state is described by the function $\theta(x,t)$ that obeys the wave equation [161]

$$\ddot{\theta} = \theta'' - \gamma\theta + u(x,t) + \varphi(x,t), \quad 0 \leqq x \leqq 1, \quad t \geqq 0, \tag{12.49}$$

$$\theta(0,t) = \theta(1,t) = 0$$

(γ is elasticity coefficient).

For disturbance $\varphi(x,t)$ in the bounded range of variation $|\varphi(x,t)| \leqq c_\varphi$ and initial conditions $\theta(x,0) = \theta_0(x)$, $\dot{\theta}(x,0) = 0$, $0 \leqq x \leqq 1$, the design of control $u(x,t)$ is formulated in the following manner: obtain and maintain zero value of the system state norm: $\int_0^1 (\theta^2(x,t) + \dot{\theta}^2(x,t))\,dx = 0$ beginning from time t_0.

The difficulty of using the Lyapunov method lies usually in the selection of the functional. It is suggested here to take function

$$v(t) = \frac{1}{2}\int_0^1 (s^2 + (\theta')^2)\,dx$$

with $s = h\theta + \dot{\theta}$, $h > 0$, as such a functional. Then, in virtue of (12.49), its time derivative

$$\begin{aligned}
\dot{v}(t) &= \int_0^1 (s\dot{s} + \theta'\dot{\theta}')\,dx = \int_0^1 s(h\dot{\theta} + \ddot{\theta})\,dx + \int_0^1 \theta'\dot{\theta}'\,dx \\
&= \int_0^1 sh\dot{\theta}\,dx + \int_0^1 s(\theta'' - \gamma\theta + u + \varphi)\,dx + \int_0^1 \theta'\dot{\theta}'\,dx \\
&= \int_0^1 s(h\dot{\theta} - \gamma\theta + u + \varphi)\,dx \\
&\quad + \int_0^1 (h\theta + \dot{\theta})\theta''\,dx + \int_0^1 \theta'\dot{\theta}'\,dx \\
&= \int_0^1 s(h\dot{\theta} - \gamma\theta + u + \varphi)\,dx + (h\theta + \dot{\theta})\theta'\Big|_0^1 \\
&\quad - \int_0^1 h(\theta')^2\,dx - \int_0^1 \theta'\dot{\theta}'\,dx + \int_0^1 \theta'\dot{\theta}'\,dx \\
&= \int_0^1 s(h\dot{\theta} - \gamma\theta + u + \varphi)\,dx - \int_0^1 h(\theta')^2\,dx.
\end{aligned}$$

If now control $u(x,t) = -(F + R|\dot{\theta}| + N|\theta|)\operatorname{sign} s$, $F > c_\varphi$, $R > h$, $N > \gamma$ is chosen, then $\dot{v}(t) < 0$ for $v(t) \neq 0$ and, consequently, $v(t) \to 0$ or $\int_0^1 s^2(x,t)\,dx \to 0$ for $t \to \infty$. Since $\lim_{t\to\infty} s = \lim_{t\to\infty} (h\theta + \dot{\theta}) = 0$ and the exponent of power $-ht$ is the solution of $h\theta + \dot{\theta} = 0$, it follows that $\lim \theta = \lim \dot{\theta} = 0$ for $t \to \infty$, and the rate of convergence to zero of the norm of spacial distribution

$$\|\theta(x,t)\| = \int_0^1 (\theta^2(x,t) + \dot{\theta}^2(x,t)\,dx$$

is growing with coefficient h. Thus, a suitable choice of F, R, N and h allows one to obtain an approximate solution

$$\int_0^1 (\theta^2(x,t) + \dot{\theta}^2(x,t))\,dx < \varepsilon, \quad \varepsilon > 0$$

for $t \geqq t_0$ through discontinuous control having the form of $u(x,t) = -(F + N|\theta| + R|\dot{\theta}|)\,\mathrm{sign}(h\theta + \dot{\theta})$. It might be well to note that not only the plant state $\theta(x,t)$, but also its derivative $\dot{\theta}(x,t)$ are used in the control design. This resembles linear lumped second-order systems described by $\ddot{y}(t) = u$ where the stabilizing feedback should be designed in terms of the state vector $(y, \dot{y})$ (see the example of second-order relay system in Sect. 1.2).

System with Control by Maximal Deviation

Assume that control in (12.45) is mobile

$$u(x,t) = p(t)\delta(x - s(t)),$$

where $p(t)$ and $s(t)$ are, respectively, source power and position. In contrast to the case discussed in Sect. 1, control is non-cyclic and the values of $p(t)$ and $s(t)$ may be taken arbitrarily from some range $|p(t)| \leqq M$, $s(t) \in [0, 1]$ (recall that the rate of source power variation (12.5) and source speed (12.9), (12.10) were used as controlled parameters in (12.1)).

Consider again the problem of obtaining the zero distribution of the heat field (12.47) through the Lyapunov functional (12.48). Denote by $x_{\mathrm{ext}}(t) \in [0, 1]$ the point where the temperature deviation is maximal at time t

$$\max_{x \in [0,1]} |Q(x,t)| = |Q(x_{\mathrm{ext}}(t), t)|.$$

(If more than one point has maximal value, any of them is taken.) Control is constructed according to the following principle: source position $s(t)$ coincides with point. $x_{\mathrm{ext}}(t)$, and source power is maximal and has the sign opposite to that of deviation, i.e.

$$u(x,t) = -M(\mathrm{sign}\, Q(x_{\mathrm{ext}}(t), t))\delta(x - x_{\mathrm{ext}}(t)). \tag{12.50}$$

Compute the derivative of the Lyapunov functional (12.48) on the solutions of (12.45) with control (12.50):

$$\frac{dv}{dt} = -\int_0^1 k(Q')^2\,dx - \int_0^1 qQ^2\,dx - \int_0^1 Q\{M(\mathrm{sign}\, Q(x_{\mathrm{ext}}(t), t))\delta(x - x_{\mathrm{ext}}(t)) + f(x,t)\}\,dx$$

$$\leqq -\int_0^1 k(Q')^2\,dx - \int_0^1 qQ^2\,dx - M|Q(x_{\mathrm{ext}}(t), t)|\left(1 - \frac{C}{M}\right).$$

Obviously, for $M > C$ in the discontinuous control (12.50) $\dot{v} < 0$ if $\int_0^1 \rho Q^2\,dx \neq 0$ $(Q(x_{\mathrm{ext}}(t), t) \neq 0)$ and, consequently, $v(t) \to 0$ with $t \to \infty$ which is solution to the problem of stabilizing temperature field in the presence of unmeasurable disturbances and unknown plant parameters[1].

[1] From the physical viewpoint, source power $p(t)$ may turn out to be a value with permanent sign $(0 \leqq p \leqq M)$; in this case, its profile must be looked for also in the form of discontinuous function with values from the permissible domain, e.g. 0 and M.

3 Modal Control

The methods of heat process control design discussed in this section rest on the possibility of representing the partial differential equation in the form of infinite-dimensional system of ordinary first-order differential equations and, correspondingly, expanding the solution and control into a series (or into individual modes). Each control component is taken in the form of discontinuous state function of the corresponding mode, and the distributed control has the form of a sum of infinite series. Realization of this procedure allows one to make use of the sliding mode algorithms developed in the preceding chapters for control of finite-dimensional plants.

Control of the Plant Described by the Scalar Heat Equation

Let plant beahaviour be described by

$$\dot{Q} = Q'' + bu(x,t) + f(x,t), \quad 0 \leqq x \leqq 1, \quad t \geqq 0$$
$$Q(0,t) = Q(1,t) = 0, \quad Q(x,0) = Q_0(x), \tag{12.51}$$

where $Q(x,t)$ is plant heat field, $u(x,t)$ is control action, and $b \neq 0$ is a constant.

The unmeasurable disturbance $f(x,t)$ is assumed to satisfy the following conditions:

1) $f(0,t) = f(1,t) = 0,\ t \geqq 0$;
2) $f(x,t)$ is representable as a Fourier series

$$f(x,t) = \sum_{n=1}^{\infty} f_n(t) \sin \pi n x$$

with $|f_n(t)| \leqq M/n^2$ where M is a constant independent of n (this condition is known to be true if $|f''(x,t)| \leqq M$). The aim of control usually is to provide the desired temperature field $Q^*(x,t)$, but without the loss of generality it may be assumed that $Q^*(x,t) = 0$.

Represent (12.51) as series [135]

$$Q(x,t) = \sum_{n=1}^{\infty} q_n(t) \sin \pi n x,$$

where $q_n(t)$ satisfy equations $\dot{q}_n = -(\pi n)^2 q_n + bu_n(t) + f_n(t)$, $q_n(0) = a_n$; a_n, $f_n(t)$ and $u_n(t)$ are Fourier coefficients of, respectively, initial distribution $Q_0(x)$, disturbance $f(x,t)$ and control $u(x,t)$. Assuming that

$$u_n = -\frac{M_1}{n^2} \operatorname{sign} bq_n, \quad n = 1, 2, \ldots, M_1 > \frac{M}{|b|},$$

write the solution of n-th mode

$$q_n(t) = a_n e^{-(\pi n)^2 t} + \int_0^t e^{-(\pi n)^2 (t-\tau)} (bu_n(\tau) + f_n(\tau)) d\tau.$$

Whence $q_n(t)=0$ at some $t_n^0 < t_n = \frac{1}{(\pi n)^2}\ln(1+\pi^2 a_n n^4/(b_1 M_1 - M))$ and since the values of q_n and $\dot{q}_n$ have opposite signs for $q_n \neq 0$, sliding mode with respect to the coordinate q_n will occur in the system at least at $t > t_n$, i.e. $q_n(t)=0$ for $t > t_n$. Most important is the existence of the upper bound T for all t_n, $n=1,2,\ldots$, and $\lim T = 0$ with $M_1 \to \infty$. It follows from the above said that for the class of disturbances, that are not accessible to measurement, the required temperature field will be attained in the system in finite time. Solution of this problem by means of the finite distributed control has been enabled by the fact that the series $\sum_{n=1}^{\infty} u_n(t)\sin \pi n x$ converges absolutely.

Modal Control of Multi-Variable Thermal Plant

Let us consider stabilization of a set of r interconnected heat plants by means of m distributed controls

$$\dot{Q} = \Lambda Q'' + RQ + Bu + Df, \quad 0 \leqq x \leqq 1, \quad t \geqq 0,$$
$$Q(0,t) = Q(1,t) = 0, \quad Q(x,0) = Q_0(x), \tag{12.52}$$

where p-dimensional disturbance $f(x,t)$ obeys the same constraints as above; Λ is positive definite diagonal matrix; $D \in B$, i.e. D is representable as $D = BA$ (similar to (10.5) for finite-dimensional systems), $D \in B$ is the necessary invariance condition).

Expanding (12.52) into modes, obtain

$$\dot{q}_n = -(\pi n)^2 \Lambda q_n + Rq_n + Bu_n + Df_n, \quad n = 1,\ldots,N, \tag{12.53}$$

$$\dot{q}_k = -(\pi k)^2 \Lambda q_k + Rq_k + Bu_k + Df_k, \quad k = N+1,\ldots \tag{12.54}$$

(as before, $u_n(t)$ and $f_n(t)$, $n=1,2,\ldots$, are the Fourier coefficients of control and disturbance, respectively). Number N is chosen so as to provide stability of any of the subsystems of (12.54) with $u_k = f_k = 0$. Its existence may be easily verified by forming the Lyapunov function as $v = q_k^{\mathrm{T}} q_k$. For any α_0, there will be N such that the convergence rate of any subsystem of (12.54) with $u_k = f_k = 0$, $k > N$, will be defined by $\dot{v} \leqq -\alpha_0 v$. For the class of disturbances under study $\| f_k(t) \| \leqq M/k^2$ relation $\dot{v} \leqq -\alpha_0 v(t) + |B^{\mathrm{T}} q_k| \left(-\frac{M_1}{k^2} + \frac{CM}{k^2}\right)$ holds for $u_k = -\frac{M_1}{k^2}\operatorname{sign}(B^{\mathrm{T}} q_k)$ where the positive constant C is defined by matrix A. If $M_1 > CM$, obviously, $\dot{v} \leqq -\alpha_0 v$ which means that in each subsystem of (12.54) the desired convergence rates are provided.

The system (12.53) is finite-dimensional, and if it is controllable, it also may be made stable with the desired transients, and invariant to disturbances $f_n(t)$, $n = 1,2,\ldots,N$ using discontinuous control (see Chap. 10).

Thus, discontinuous controls and deliberately induced sliding modes allow one to make the original system (12.52) invariant to the disturbances having the desired transients. Similar to the one-dimensional case, these features are provided by the discontinuous finite distributed control $u(x, t)$.

4 Design of Distributed Control of Multi-Variable Heat Processes

To solve the zero distribution problem for a set of interconnected heat fields let us take advantage of the Lyapunov functional-based design method discussed in Sect. 2 as applied to the plants described by the scalar heat equation.

Assume that the matrix R in the equation of the multivariable heat plant which characterizes heat exchange between the system elements and the environment is a Hurwitz one and, therefore, the Lyapunov equation $R^{\mathrm{T}}W + WR = -S(S > 0)$ has positive definite solution $W > 0$. Assume also that matrix $W\Lambda$ is positive definite as well (for instance, all the plant elements are of one type and $\Lambda = \lambda I_r$, scalar $\lambda > 0$, or $-(R + R^{\mathrm{T}}) > 0$ and $W = I_r$, or R is a Hadamard matrix and $W = I_r$).

Choosing control in the form of discontinuous system state function $u = -M\,\mathrm{sign}(B^{\mathrm{T}}WQ)$, $M > 0$, compute the time derivative of the positive definite functional $v = \frac{1}{2}\int_0^1 Q^{\mathrm{T}}WQ\,dx$ on the trajectories of (12.52)

$$\dot{v}(t) = -\int_0^1 (Q')^{\mathrm{T}}W\Lambda Q'dx - \int_0^1 Q^{\mathrm{T}}SQ\,dx - M\int_0^1 |Q^{\mathrm{T}}WB|\left(1 - \frac{Q^{\mathrm{T}}WBAf}{M|Q^{\mathrm{T}}WB|}\right)dx.$$

Since disturbance $f(x, t)$ is assumed to be bounded (and, thus, the components of vector $Af(x, t)$ are bounded), there will be a number M such that $\dot{v} < 0$. This means that the distribution $Q(x, t)$ converges to zero both in L_2 and pointwise (the justification is similar to (12.43) of Sect. 1). If pair $\{R, B\}$ is controllable (in this case matrix R does not need to be Hurwitz one), one can provide in (12.52) the invariance to disturbances and the desired transients by taking control in the form of $u = FQ - M\,\mathrm{sign}\,B^{\mathrm{T}}WQ$ and making a suitable choice of the eigenvalues of matrix $R + BF$.

In all the cases considered in this chapter it was assumed that the plant is stable and the corresponding Lyapunov functional is known. Discontinuous control was chosen so as to suppress the effect of disturbances and to preserve the sign of Lyapunov functional time derivative. After the completion of the transient, the system is for sure sliding mode because the argument of discontinuous control function is zero and corresponds to the discontinuity points. However, sliding mode does not necessarily occur over all the discontinuity manifold; as a result, the possibility of sliding mode-based decomposition is not used.

This chapter will be concluded by a design procedure emphasizing the major point of sliding mode-based algorithms for finite-dimensional systems, namely the choice of the manifold with desirable motion and use of discontinuous controls in order to induce sliding modes.

Assume that all the heat conductivities making up matrix Λ in (12.52) are similar, i.e. $\Lambda = \lambda I_r$, scalar $\lambda > 0$, pair $\{R, B\}$ is controllable. As it was already noted in Sect. 7.4, one may assume without loss of generality that rank $B = m$. Enumerate the elements of Q so as to satisfy condition $\det B_2 \neq 0$ for matrix $B^T = (B_1^T, B_2^T)$ in the Eq. (12.52) with respect to $Q^T = (Q_1^T, Q_2^T)$, $Q_1 \in \mathbb{R}^{r-m}$, $Q_2 \in \mathbb{R}^m$.

Let us decide that the manifold on which sliding mode motion must be induced is linear

$$s = CQ_1 + B_2^{-1} Q_2, \quad s \in \mathbb{R}^m, \quad C - \text{const} \tag{12.55}$$

and consider motion equation with respect to (Q_1, s) if similar to (12.52) $D = BA$

$$\dot{Q}_1 = \lambda Q_1'' + R_{11} Q_1 + R_{12} s + B_1 u + B_1 A f \tag{12.56}$$

$$\dot{s} = \lambda s'' + R_{21} Q_1 + R_{22} s + u + Af, \tag{12.57}$$

where $R_{ij} (i, j = 1, 2)$ are some constant matrices.

Assume that we have succeeded in inducing sliding mode motion on the manifold $s = 0$ by means of discontinuous control u. The sliding equation may be obtained through substitution of $s = 0$ and $u_{eq} = -R_{11} Q_1 - Af$ (solution of (12.57) with respect to u with $s'' = 0$ and $\dot{s} = 0$) into (12.56):

$$\dot{Q}_1 = \lambda Q_1'' + R_0 Q_1, \quad R_0 = R_{11} - B_1 R_{21}. \tag{12.58}$$

The applicability of the equivalent control method to derivation of the sliding equation for infinite-dimensional systems is substantiated in [111].

Prior to analyzing sliding mode system behaviour, note that for the finite-dimensional system $\dot{Q} = RQ + Bu$ the equation of sliding on manifold $s = CQ_1 + B_2^{-1} Q_2 = 0$ has the form of $\dot{Q}_1 = R_0 Q_1$ according to the equivalent control method. As it has been shown in Sect. 7.4, for a controllable system the desired spectrum of matrix R_0 in the sliding equation may be obtained by a suitable choice of matrix C. Proceeding from this fact we can demonstrate that in (12.58) the desired rates of convergence to the zero distribution of heat field $Q \equiv 0$ may be provided.

Form a Lyapunov functional

$$v(t) = \tfrac{1}{2} \int_0^1 Q_1^T W Q_1 \, dx,$$

where W is positive definite solution of the Lyapunov equation $R_0^T W + W R_0 = -I_{r-m}$. Let us determine the time derivative of functional $v(t)$ on the trajectories of (12.58):

$$\frac{dv}{dt} = -\lambda \int_0^1 (Q_1')^T W Q_1 \, dx - \int_0^1 Q_1^T Q_1 \, dx.$$

Remembering that $Q_1^T W Q_1 \leqq \lambda_{max} Q_1^T Q_1$ (λ_{max} is the maximal eigenvalue of matrix W), obtain

$$\frac{dv}{dt} \leqq -\frac{1}{\lambda_{max}} v(t). \tag{12.59}$$

Since $W = \int_0^\infty e^{R_0^T t} e^{R_0 t} dt$ [6], it is possible to make the norm of matrix W and, therefore, λ_{max} arbitrarily small by choosing the eigenvalues for R_0 with sufficiently great negative real parts. According to (12.59), this implies that by means of matrix C one can provide the desired rates of convergence to zero of the Lyapunov functional or vector Q_1 in metric L_2.

At the final stage, a discontinuous control is chosen that provides occurrence of sliding mode or convergence of the trajectories to the manifold $s = 0$. Compute the derivative of Lyapunov functional $v_s(t) = \frac{1}{2}\int_0^1 s^T s dx$ on the trajectories of (12.57) with control $u = -M \operatorname{sign} s, M > 0$:

$$\frac{dv_s}{dt} = -\int_0^1 (s')^T s' dx - M \int_0^1 |s| \left(1 - \frac{s^T(R_{21}Q_1 + R_{22}s + Af)}{M|s|} \right) dx. \tag{12.60}$$

Keeping in mind that $s = CQ_1 + B_2^{-1}Q_2$ and disturbances $f(x,t)$ are bounded, one can select numbers α and M_0 such that the second term in (12.60) will be negative for $M = d|Q| + M_0$. In consequence, obtain that $\dot{v}_s < 0$ for $v_s \neq 0$ which means that the trajectories converge to manifold $s = 0$ in metric L_2, whence the pointwise convergence follows according to the concluding remarks in Sect. 12.1. Since the sliding equation is independent of disturbances, the discontinuous control provides invariance to disturbances along with the desired dynamic properties.

Chapter 13

Control Under Uncertainty Conditions

1 Design of Adaptive Systems with Reference Model

Direct application of the methods of the linear optimal system theory to control of multi-variable plants meets most commonly with two practical problems. First, it is difficult to formulate the objectives in terms of the criterion to be minimized. Second, the wide range of unpredictable variations of the plant operator parameters prevents the optimal algorithms to be realized in all the modes.

A possible approach to overcoming these difficulties is the use of a reference model in the system; in this case, the objective of control is to nullify the mismatch between the model and plant state vectors. Studies were carried out oriented to using this approach for the design of multi-variable linear systems with linear models through the hyper-stability method [94] and the second Lyapunov method [114, 164]. It should be noted that for systems with scalar inputs and outputs the possibility of coordinating the model and plant state vectors by means of sliding modes is discussed in the works [35, 167]. This section summarizes the solution to the problem of designing multi-variable model reference systems described in [168].

Let the behaviour of controlled plant be described by the linear equation

$$\dot{x} = A(t)x + Bu, \quad x \in \mathbb{R}^n, u \in \mathbb{R}^m, \tag{13.1}$$

the parameters of matrix $A(t)$ vary arbitrarily over a limited range, matrix B

be time invariant. The reference model equation will be taken linear as well:

$$\dot{x}_M = A_M x_M + B_M u_M, \tag{13.2}$$

where x_M is the state vector of model $x_M \in \mathbb{R}^n, u_M$ is model input vector (e.g. the set of reference inputs), constant matrices A_M and B_M being chosen so as to provide the desired dynamic processes in the model.

The task of control is posed as follows: through the information about the state vectors of plant x and model x_M and the model input vector u_M it is required to design a control function u such that the mismatch $e = x_M - x$ will be nullified.

In order to establish the conditions enabling solution of this problem and design the proper control function, let us write the motion equation with respect to the error vector:

$$\dot{e} = A_M e + (A_M - A)x + B_M u_M - Bu. \tag{13.3}$$

Since mismatch e is to be nullfied, it would be natural to regard the vectors $(A_M - A)x$ and $B_M u_M$ as disturbances. According to the invariance conditions of Ch. 10, their contribution to the processes in (13.3) might be suppressed by sliding modes if the vectors belong to the subspace formed by the base vectors of matrix B or if

$$\operatorname{rank} B = \operatorname{rank}\{B, B_M\} = \operatorname{rank}\{B, A_M - A\}. \tag{13.4}$$

(Notably, (13.4) are the "perfect model matching" conditions of [29]). The formulated above problem will be solved if (13.4) is satisfied and if on an $(n-m)$-dimensional manifold

$$s = Ce = 0 \tag{13.5}$$

there are stable "in the large" sliding modes and solutions of the linear differential equation describing sliding mode motion are asymptotically stable.

As it was demonstrated in Sect. 7.4, the eigenvalues of the characteristic equation of sliding mode motion may be arbitrarily allocated if the pair $\{A_M, B\}$ is controllable and the matrix C in the discontinuity surface Eq. (13.5) satisfies condition

$$\det CB \neq 0. \tag{13.6}$$

In order to determine the sliding equation, one has first to determine the equivalent control from the algebraic system $\dot{s} = 0$ (u_{eq} exists if (13.6) is satisfied):

$$u_{\text{eq}} = (CB)^{-1} C(A_M e + (A_M - A)x + B_M u_M) \tag{13.7}$$

and to substitute it into (13.3)

$$\dot{e} = A_M e - B(CB)^{-1} C(A_M e + (A_M - A)x + B_M u_M),$$
$$s = 0.$$

If (13.4) is satisfied, motion is independent of both the variable matrix $A(t)$ and the inputs of model u_M:

$$\dot{e} = (I_n - B(CB)^{-1} C) A_M e. \tag{13.8}$$

The control design procedure of Sect. 10.3 for combined control systems is completely applicable to the case in question. Indeed, the equivalent control (13.7) is a linear combination of the components of vectors e, x and u_M with bounded coefficients, therefore there exists an upper estimate for it of the following form

$$E(e, x, u_M) = \alpha|e| + \beta|x| + \gamma|u_M|,$$

where α, β, γ are positive numbers. In order to provide stable "in the large" sliding modes on the manifold $s = 0$, control algorithms similar to (10.11) or (10.13) may be then used

$$u = k(-\alpha|e| - \beta|x| - \gamma|u_M|)\operatorname{sign} s^*, s^* = Ds, k > 1$$
$$u = -\Psi e - \Psi^x x - \Psi^M u_M,$$

where matrix D and vector s^* are defined by the form of invariant transformation of s, the elements of piecewise-constant matrices Ψ, Ψ^x and Ψ^M have discontinuities on the planes $s_i^* = 0$ $(s^{*\mathrm{T}} = (s_1^*, \ldots, s_m^*))$ and at the changes of the signs of components of e, x and u_M. (We omit the relations for calculation of the numerical values of elements of matrices C, Ψ, Ψ^x and Ψ^M because all the design methods described in Chaps. 7, 9 and in Sect. 10.3 are applicable to this case).

Thus, if sliding modes are invariant to the vectors $(A_M - A)x$ and $B_M u_M$ (13.4) and the pair $\{A_M, B\}$ is controllable, the control design procedure consists of the following stages: selection of a model with the desired dynamic properties, determination of the discontinuity surface equations with respect to some criterion (e.g. eigenvalue allocation or minimum of quadratic functional) and, finally, determination of control u in the form of piecewise-linear function of vectors e, x and u_M providing stability "in the large" of sliding modes.

How model is used in order to provide adaptation is exemplified by the control system of longitudinal motion of the Convair C-131B airplane. For linearized equation of this motion, the authors of [94] choose a linear model and propose a linear control algorithm enabling coordination of model and airplane state vectors. For the case under consideration, parameters in (13.1) and (13.2) have the following numerical values:

$$A_M = \begin{bmatrix} 0 & 1.0 & 0 & 0 \\ 5.31 \cdot 10^{-7} & -0.4179 & -12.02 & 2.319 \cdot 10^{-3} \\ -4.619 \cdot 10^{-9} & 1.0 & -0.7523 & -2.387 \cdot 10^{-2} \\ -0.5614 & 0 & 0.3002 & -1.743 \cdot 10^{-2} \end{bmatrix},$$

$$B_M = \begin{bmatrix} 0 & 0 \\ -0.1717 & 7.451 \cdot 10^{-6} \\ -0.0238 & -7.738 \cdot 10^{-5} \\ 0 & 3.685 \cdot 10^{-3} \end{bmatrix},$$

$$A=\begin{bmatrix} 0 & 1.0 & 0 & 0 \\ 1.401\cdot 10^{-4} & \sigma & -1.9513 & 0.0133 \\ -2.505\cdot 10^{-4} & 1.0 & -1.3239 & -0.0238 \\ -0.561 & 0 & 0.358 & -0.0279 \end{bmatrix},$$

$$B=\begin{bmatrix} 0 & 0 & 0 \\ -5.3307 & 6.447\cdot 10^{-3} & -0.2669 \\ -0.16 & -1.155\cdot 10^{-2} & -0.2511 \\ 0 & 0.106 & 0.862 \end{bmatrix}.$$

The components u_1, u_2 and u_3 of three-dimensional control vector are, respectively, the commands to elevator, flaps and engine; the components u_M^1 and u_M^2 of two-dimensional vector u_M are the reference inputs of elevator and engine. State vector $x^{\mathrm{T}}=(x_1, x_2, x_3, x_4)$ consists of the values of pitch, rate of pitch, angle of attack and flight speed. In matrix A that characterizes the airplane dynamics, the only varying parameters is σ which lies within the range $-0.558 \leqq \sigma(t) \leqq -3.558$.

Model choice is followed by determination of the set of discontinuity surfaces. To this end, let us use the procedure of Sect. 7.4. In the case under discussion, (13.3) has the form of (7.45), (7.46) because the first rows of matrices B and B_M are zero. In the notations of (7.45), $A_{11}=0$, $A_{12}=(1,0,0)$, $e_1=e^1$, $e_2^{\mathrm{T}}=(e^2, e^3, e^4)$ if $e=(e^1, e^2, e^3, e^4)$ and the equations of discontinuity surfaces and of sliding mode on their intersection (7.47) and (7.48) are, correspondingly,

$$\begin{aligned} &s=c_1 e_1+e_2=0, \\ &\dot{e}_1=(A_{11}-A_{12}c_{\mathrm{T}})e_1, \quad \dot{e}_1=-c_{11}e_1 \end{aligned} \tag{13.9}$$

with $s \in \mathbb{R}^3, c_1^{\mathrm{T}}=(c_{11}, c_{12}, c_{13})$.

Notably, (13.4) is satisfied for the "plant-model" system, and, therefore, the sliding equations are independent of vectors $(A-A_M)x$ and $B_M u_M$. Let us require that in sliding mode $e^3=0$, and $e^4=0$ (i.e. $c_{12}=c_{13}=0$), and take coefficient c_{11} defining the behaviour of the coordinate e^1 in sliding mode equal to 10. As a result, obtain three discontinuity planes of the components u_1, u_2, u_3 of the vector u:

$$s_1=e^3=0, s_2=e^4=0, s_3=10e^1+e^2=0.$$

In order to induce sliding mode on the manifold $s=0$, let us use the method of scalar hierarchy of control described in Sect. 4.5. According to this method, control u_1 must provide sliding mode motion on plane $s_1=0$ independently of the values of controls u_2 and u_3, u_2 – on the intersection of $s_1=0$ and $s_2=0$ independently of u_3, and, finally, u_3 – on manifold $s=0$.

Control is designed in the reverse order. By assuming that there is motion on the intersection of planes $s_1=0$ and $s_2=0$, obtain from equations $\dot{s}_1=0$ and $\dot{s}_2=0$ the equivalent controls $u_{1\mathrm{eq}}$ and $u_{2\mathrm{eq}}$ dependent on u_3 and substitute them into the motion equations. Next, reasoning from the conditions

$\dot{s}_3 > 0$ for $s_3 < 0$

$\dot{s}_3 <$ for $s_3 > 0$

obtain control u_3. Then, determine control u_2 for the chosen control u_3 under the assumption that sliding mode occurs on the plane $s_1 = 0$ and that $u_{1\text{eq}}$ is substituted for u_1. At the last step, determine the component u_1 inducing sliding mode on plane $s_1 = 0$ for known functions u_2 and u_3.

Now we present the outcome of the above design procedure:

$$u_i = ((k_e^i)^{\mathrm{T}}|e| + (k^i)^{\mathrm{T}}|x| + (k_{\mathrm{M}}^i)^{\mathrm{T}}|u_{\mathrm{M}}|)\,\mathrm{sign}\, s, \quad i = 1, 2, 3,$$

$$(|e|)^{\mathrm{T}} = (|e^1|, |e^2|, |e^3|, |e^4|), \quad (|x|)^{\mathrm{T}} = (|x_1|, |x_2|, |x_3|, |x_4|),$$

$$(|u_{\mathrm{M}}|)^{\mathrm{T}} = (|u_{\mathrm{M}}^1|, |u_{\mathrm{M}}^2|),$$

$$\begin{aligned}
(k_e^1)^{\mathrm{T}} &\geqq (2.08 \quad 12.82 \quad 11.49 \quad 1.80),\\
(k_e^2)^{\mathrm{T}} &\geqq (6.11 \quad 3.26 \quad 6.10 \quad 0.89),\\
(k_e^3)^{\mathrm{T}} &\geqq (0.27 \quad 3.05 \quad 3.07 \quad 0.12),\\
(k^1)^{\mathrm{T}} &\geqq (4 \cdot 10^{-3} \quad 1.65 \quad 8.50 \quad 0.23),\\
(k^2)^{\mathrm{T}} &\geqq (4.6 \cdot 10^{-3} \quad 0.82 \quad 2.99 \quad 0.19),\\
(k^3)^{\mathrm{T}} &\geqq (10^{-3} \quad 0.41 \quad 2.19 \quad 0.07),\\
(k_{\mathrm{M}}^1)^{\mathrm{T}} &\geqq (0.32 \quad 0.03),\\
(k_{\mathrm{M}}^2)^{\mathrm{T}} &\geqq (0.09 \quad 0.05),\\
(k_{\mathrm{M}}^3)^{\mathrm{T}} &\geqq (0.08 \quad 1.5 \cdot 10^{-3}),
\end{aligned}$$

where all the inequalities are to be satisfied componentwise.

The results of system simulation for control

$$\begin{pmatrix} u_1 \\ u_2 \\ u_3 \end{pmatrix} = \mathrm{diag}(\mathrm{sign}\, s_1, \mathrm{sign}\, s_2, \mathrm{sign}\, s_3) \left\{ \begin{bmatrix} 13 & 13 & 12 & 2 \\ 7 & 4 & 7 & 1 \\ 1 & 4 & 4 & 1 \end{bmatrix} |e| \right.$$

$$\left. + \begin{bmatrix} 0.01 & 2 & 9 & 1 \\ 0.01 & 1 & 3 & 1 \\ 0.001 & 1 & 3 & 0.1 \end{bmatrix} |x| + \begin{bmatrix} 1.0 & 0.1 \\ 0.1 & 0.1 \\ 0.1 & 0.01 \end{bmatrix} |u_{\mathrm{M}}| \right\}$$

are described in [168].

It follows from the experiments that under the initial conditions $e^1(0) = 1$, $e^2(0) = e^3(0) = e^4(0) = 0$ and time varying reference inputs u_{M} and parameter σ, sliding mode motion begins in the system in time $t \leqq 0.3$ s. During the motion $e^3 = e^4 = 0$ and the component e^1 exponentially decays with time constant 0.1 s according to (13.9).

2 Identification with Piecewise-Continuous Dynamic Models

The above approach to the design of adaptive systems makes the state vectors of plant and model equal, and the dynamic properties of control do not vary no matter what values are assumed by the plant parameters. But it should be noted that the form of optimal, in a sense, processes, generally, must vary with variations of parameters. In this connection, it is of interest to consider the use of adjustable model (or variation of controller parameters so as to adjust them to other versions of control algorithms for time-varying plants) if plant characteristics begin to vary in the course of operation.

It will be demonstrated here that the discontinuous control algorithms enable identification of plant operator. In particular, for the adaptive model reference system of Sect. 1 the equivalent control (13.7) depends on matrix A or on the plant parameters; therefore, one could attempt to determine their current values by defining u_{eq} with the aid of the first-order filters (see Sect. 2.4).

From here on consideration will be given to the identification of a dynamic plant under the assumption that the right-hand side of its differential equation may be expanded into finite-dimensional series with respect to a system of known functions $\varphi_i(x, t)$ $(i = 1, \ldots, N)$:

$$\dot{x} = A\varphi(x, t), \tag{13.10}$$

where $x \in \mathbb{R}^n$, $\varphi^{\mathrm{T}}(x, t) = (\varphi_1(x, t), \ldots, \varphi_N(x, t))$ and A is constant $n \times N$ matrix whose elements a_{ij} are expansion coefficients. Functions $\varphi_i(x(t), t)$ are assumed to be linearly independent on the solutions $x(t)$ everywhere except for the points of zero measure.

With such a formulation, to identify plant operator is to determine parameters a_{ij} if the state vector $x(t)$ is known. It will be assumed below that the range of possible values of the coefficients a_{ij} is bounded and known.

To solve this problem with respect to (13.10), design a model described by equations of the (13.10) type whose coefficients are discontinuous functions. Let us attempt to vary model structure so as to make the plant and model considered as an interconnected system to enter always sliding modes during which the processes in the plant and model will be identical. In virtue of the fact that the motion equations of plant and model have similar structures and their outputs coincide, one may judge about the unknown plant parameters by the average values of model discontinuity coefficients in sliding mode. The average value of control is known to coincide in sliding mode with the equivalent control, this value being readily measurable by first-order filter with sufficiently small time constant.

Thus, along with (13.10) consider the model equation

$$\dot{y} = u, \tag{13.11}$$

where y and u are n-dimensional vectors of model state and control, respectively, with components $(y_1, \ldots, y_n)$ and $(u_1, \ldots, u_n)$. Let us design the control vector in the form of the piecewise-linear function of the components of φ:

$$u_i = \sum_{j=1}^{N} b_{ij}\varphi_j(x,t) + \left(\sum_{j=1}^{N} d_{ij}|\varphi_j(x,t)|\right)u_i^*, \; b_{ij} \quad \text{and} \quad d_{ij} - \text{const}, \tag{13.12}$$

$$u_i^* = \begin{cases} 1 & \text{for} \quad s_i > 0 \\ -1 & \text{for} \quad s_i < 0, \quad i = 1, \ldots, n \end{cases}$$

and functions s_i defining the discontinuity surfaces are equal to the mismatches between the corresponding coordinates of plant and model:

$$s_i = x_i - y_i. \tag{13.13}$$

By computing from (13.10) through (13.13)

$$\dot{s}_i = \sum_{j=1}^{N} ((a_{ij} - b_{ij})\varphi_j - d_{ij}|\varphi_j|u_j^*)$$

we see that s_i and $\dot{s}_i$ will have opposite signs, i.e. the conditions of sliding mode stability "in the large" will be satisfied on each surface $s_i = 0$ if

$$d_{ij} > \max_{a_{ij}} |a_{ij} - b_{ij}|. \tag{13.14}$$

This implies that at some instant of time in the dynamic "plant-model" system under study sliding mode motion will be started on the intersection of all the discontinuity surfaces, all the mismatches between the plant and model outputs being identical to zero.

According to the equivalent control method, in order to determine the sliding equations one must substitute for u_i in the model equation functions u_{ieq} that are solutions of $\dot{s}_i = 0$ $(i = 1, \ldots, n)$ with respect to controls. It is clear from this that for $u_i^* = u_{ieq}^*$ all $\dot{s}_i$ are identical to zero, i.e.

$$\sum_{j=1}^{N} l_{ij}\varphi_j(x,t) = \varepsilon_i^0, \quad (i = 1, \ldots, n), \tag{13.15}$$

where

$$l_{ij} = a_{ij} - b_{ij}, \quad \varepsilon_i^0 = \sum_{j=1}^{N} d_{ij}u_{ieq}^*|\varphi_j(x,t)|. \tag{13.16}$$

The components u_{ieq}^* coincide with the average values of u_i^* and may be obtained at the outputs of filters having small time constants if the real controls are fed into the inputs of filters. (The choice of parameters and errors of measurement were discussed in Sect. 2.4). If it turns out that all the u_{ieq} do not vary in time over the intervals of constant sign of $\varphi_j(x,t)$, then obtain the desired plant parameters from (13.15) in virtue of linear independence of functions $\varphi_j(x,t)$:

$$a_{ij} = b_{ij} + d_{ij}u_{ieq}^* \operatorname{sign} \varphi_j(x,t). \tag{13.17}$$

In the general case, however, the functions u_{ieq}^* that are solutions of (13.15) and (13.16)

$$u_{ieq}^* = \frac{\sum_{j=1}^{N} (a_{ij} - b_{ij})\varphi_j(x, t)}{\sum_{j=1}^{N} (d_{ij} \operatorname{sign} \varphi_j(x, t))\varphi_j(x, t)} \tag{13.18}$$

are time-varying. If one chooses the model parameters so that $d_{ij} \operatorname{sign} \varphi_j$ are proportional to $a_{ij} - b_{ij}$ with the same factor of proportionality, and then considers that part of the process where $\varphi_j(x, t)$ does not change its sign, it is obvious that over this interval $u_{ieq}^* = \text{const}$. It goes without saying that one cannot directly choose the parameters b_{ij} and d_{ij} since the coefficients a_{ij} are unknown. But this reasoning demonstrates that there exist the coefficients b_{ij} and d_{ij} under which $u_{ieq}^* = \text{const}$, and the problem boils down to their determination.

We describe now one of possible procedures for searching model parameters for which all the u_{ieq}^* are constant and, therefore, the desired plant parameters may be found through (13.17). Let us complete (13.16) by similar equations whose left-hand sides instead of the time functions $\varphi_j(x, t)$ contain time functions $\varphi_j^k(x(t), t) = L^k\{\varphi_j(x(t), t)\}$, that has been obtained by the k-fold application of the linear operator L to $\varphi_j(x(t), t)$. In this case, the right-hand sides are, generally, time functions.

Introduce notations

$$\sum_{j=1}^{N} l_{ij}\varphi_j^k(x, t) = \varepsilon_i^k(t), \quad k = 0, \ldots, N-1. \tag{13.19}$$

(Attention is drawn to the fact that functions ε_i^k are not equal to $L^k\{\varepsilon_i^0\}$ because the coefficients l_{ij} will vary in time during the search of model parameters, and the left-hand side of (13.19) is not then the outcome of the k-fold application of the operator L to the linear combination $\sum_{j=1}^{N} l_{ij}\varphi_j$ which is equal to ε_i^0 according to (13.15)).

Now we demonstrate that all the functions ε_i^k are directly measurable in spite of the fact that the plant parameters a_{ij} on which the coefficients l_{ij} in (13.19) depend are unknown. It follows from the conditions $\dot{s}_i = 0$ for $u_i^* = u_{ieq}^*$ that

$$\sum_{j=1}^{N} a_{ij}\varphi_j(x, t) = \sum_{j=1}^{N} b_{ij}\varphi_j(x, t) + \sum_{j=1}^{N} d_{ij}|\varphi_j(x, t)|u_{ieq}^*. \tag{13.20}$$

By applying the linear operator L^k to both sides of (13.20), obtain

$$\sum_{j=1}^{N} a_{ij}\varphi_j^k(x, t) = L^k\left\{\sum_{j=1}^{N} b_{ij}\varphi_j(x, t) + \left(\sum_{j=1}^{N} d_{ij}|\varphi_j(x, t)|u_{ieq}^*\right)\right\} \tag{13.21}$$

which allows one to rewrite (13.19) as follows:

$$L^k\left\{\sum_{j=1}^{N} b_{ij}\varphi_j(x,t)+\sum_{j=1}^{N} d_{ij}|\varphi_j(x,t)|u^*_{i\text{eq}}\right\}-\sum_{j=1}^{N} b_{ij}\varphi_j^k(x,t)=\varepsilon_i^k. \tag{13.22}$$

It follows from (13.22) that all the functions in the left-hand side are known and ε_i^k can be, therefore, computed.

The linear operator L will be chosen so as to fulfill the conditions

$$\det\|\varphi_j^k(x,t)\|\neq 0 \quad (j=1,\dots,N;\ k=0,\dots,N-1) \tag{13.23}$$

for functions $\varphi_j^k(x,t)$.

Let us study the behaviour of auxiliary functions $v_i=\frac{1}{2}\sum_{j=1}^{N} l_{ij}^2$, if the model coefficients b_{ij} vary as follows:

$$\dot{b}_{ij}=c\sum_{k=0}^{N-1}\varepsilon_i^k\varphi_j^k(x,t),\quad c=\text{const},\quad c>0. \tag{13.24}$$

Write the time derivatives of v_i:

$$\dot{v}_i=-\sum_{j=1}^{N} l_{ij}\dot{b}_{ij},$$

or, taking into consideration (13.24),

$$\dot{v}_i=-c\sum_{j=1}^{N} l_{ij}\sum_{k=0}^{N-1}\varepsilon_i^k\varphi_j^k(x,t). \tag{13.25}$$

According to (13.19), Eq. (13.25) is reducible to

$$\dot{v}_i=-c\sum_{k=0}^{N-1}(\varepsilon_i^k)^2,$$

i.e. the functions v_i are lower-bounded ($v_i\geqq 0$) and monotonously decreasing for at least one of ε_i^k being non-zero. The process will be over when all the ε_i^k are zero.

It follows from (13.19) and (13.23)[1] that in this case $\lim_{t\to\infty} l_{ij}=0$ or $\lim_{t\to\infty} b_{ij}=a_{ij}$. Note that $u^*_{i\text{eq}}$ tends to zero according to (13.18). The search procedure (13.24), therefore, enables identification of all the plant parameters for linear operators of the (13.23) type.

The pure-delays, first-order filters or differentiating elements were shown [20, 83] to be suited for identification of linear plants through the above search procedure because their operators satisfy the condition (13.23).

Let us demonstrate now by means of the linear time-invariant system

$$\dot{x}=Ax+Bu,\quad x\in\mathbb{R}^n,\quad u\in\mathbb{R}^m, \tag{13.26}$$

---

[1] The model parameters adjustment is assumed to be much faster than the plant processes, and if $\lim_{t\to\infty}\det\|\varphi_j^k\|=0$, then $\lim_{t\to\infty}(\varepsilon^k/\det\|\varphi_j^k\|)=0$.

that the problem of identification of the parameters of matrices A and B can be solved without using the additional functions resulting from k-fold application of the operator L if the components of vector $z^{\mathrm{T}} = (x^{\mathrm{T}}, u^{\mathrm{T}})$ are linearly independent and input u makes it possible to satisfy the condition [93]

$$\int_0^\infty \| w \|^2 dt = \infty \tag{13.27}$$

for any linear combination

$$w = kz, \quad k = \text{const} \neq 0. \tag{13.28}$$

The elements of matrices A and B can be determined by designing an adjustable-parameter model which coincides structually with the plant (13.26). As a rule, with such an approach the parameter variation algorithm is chosen so as to provide the decrease of the Lyapunov function that depends on the mismatches between the parameters of model and plant and their state vectors [94]. Sliding mode algorithm enables one to independently coordinate the state vectors and provide convergence of the model parameters to those of the plant. The identification algorithm will be as follows [146]:

$$\frac{dy}{dt} = \bar{A}x + \bar{B}u + v$$

$$v = \psi^A x + \psi^B u, \quad \psi^A = \| \psi^A_{ij} \|, \quad \psi^B = \| \psi^B_{ij} \|; \quad y, v \in \mathbb{R}^n, \tag{13.29}$$

$$\psi^A_{ij} = \begin{cases} \alpha^A_{ij} & \text{for} \quad x_j s_i > 0 \\ \beta^B_{ij} & \text{for} \quad x_j s_i < 0, \end{cases} \qquad \psi^B_{ij} = \begin{cases} \alpha^B_{ij} & \text{for} \quad u_j s_i > 0 \\ \beta^B_{ij} & \text{for} \quad u_j s_i < 0, \end{cases}$$

$$s_i = x_i - y_i, \quad \beta^A_{ij} < a_{ij} < \alpha^A_{ij}, \quad \beta^B_{ij} < b_{ij} < \alpha^B_{ij};$$

$$\frac{d\bar{A}}{dt} = -\lambda v x^{\mathrm{T}}, \quad \frac{d\bar{B}}{dt} = -\lambda v u^{\mathrm{T}}, \quad \lambda > 0. \tag{13.30}$$

It follows from (13.26) and (13.29) that functions

$$\dot{s}_i = \sum_{j=1}^{n} (a_{ij} - \psi^A_{ij}) x_j + \sum_{j=1}^{m} (b_{ij} - \psi^B_{ij}) u_j$$

have signs opposite to s_i. Consequently, in the dynamic system (13.26), (13.29), (13.30) sliding mode will be started on the manifold $s = 0 (s = x - y)$, and in order to derive its mathematical description one has to substitute into (13.30) the equivalent control $v_{\mathrm{eq}} = \Delta A x - \Delta B u$ which is the solution of the equation $\dot{s} = 0$ with respect to $v (\Delta A = A - \bar{A}, \Delta B = B - \bar{B})$. Denoting $\Delta C = (\Delta A, \Delta B)$, construct the Lyapunov function as

$$V = \tfrac{1}{2} \mathrm{tr}(\Delta C \Delta C^{\mathrm{T}}) \geqq 0 (V = 0 \text{ if } \Delta A = 0, \ \Delta B = 0).$$

Compute the time derivative of V on the trajectories of the systems (13.30) or $\Delta \dot{C} = -\lambda v z^{\mathrm{T}}$ after the origination of sliding mode (i.e. for $v = v_{\mathrm{eq}} = \Delta C z$):

$$\dot{V} = -\lambda \, \mathrm{tr}(\Delta C Z Z^{\mathrm{T}} \Delta C^{\mathrm{T}}) = -\lambda \| w \|^2, \quad w = \Delta C z.$$

Since the components of z are linearly independent, $\dot{V} \leqq 0$ and $\dot{V} \not\equiv 0$ and therefore the function V is monotoneously decreasing. If the condition (13.27), (13.28) is satisfied, integral $\int_0^\infty \|w\|^2 dt$ diverges and, therefore,

$$\lim_{t\to\infty} V = 0, \quad \lim_{t\to\infty} \Delta C = 0,$$

or

$$\lim_{t\to\infty} \Delta A = 0, \quad \lim_{t\to\infty} \Delta B = 0.$$

The elements of $\bar{A}$ and $\bar{B}$, thus, tend to the values of the parameters in A and B, and the identification problem is solved. In contrast to the methods emphasizing reference models, no assumption is made about the asymptotic stability, and the Lyapunov function, correspondingly, is independent of the mismatch between the model and plant state vectors [94, 114], the plant itself being in this case the reference model. When sliding mode starts, the dynamic system (13.29), (13.30) becomes similar to the identification systems discussed in [21, 93], but the proposed algorithm may be realized without direct measurement of the state vector derivative or approximate differentiation by physically realizable elements.

3 Method of Self-Optimization

There is a wide range of applied problems where the control system must select a set of plant input parameters for which the scalar output becomes extremal. Among examples one may quote realization of the desired temperature field in heating furnaces where the sum of temperature deviations from the reference values is minimized for a finite set of points, efficiency maximization, minimization of electric motor heat losses through current minimization under fixed torque or power, etc.

Optimization of this sort is usually desired in the environment of significant uncertainty about the plant operator and very limited information about the current plant state that handicap the *a priori* determination of the desirable operation modes. Therefore, they are to be determined by the control system itself. The insufficiency of information largely defines the design of discrete and continuous algorithms providing convergence to the optimal mode. For the sake of definiteness, it will be assumed that the optimal mode is defined by the minimum point of a function.

As a rule, the search problem is solved in two stages: first, the gradient of the function to be minimized (or its projections, their signs, etc) is determined, and, second, in the input parameter space motion is arranged towards the optimal point [85], although sometimes (e.g. in [51]) this decomposition is not explicit. The information about the gradient of minimized function is most commonly obtained by feeding probe signals into the plant and analysing its response.

The self-optimization method discussed in this section relies upon the following approach. Since the plant output must decrease, let it be "modelled" by a monotone decreasing time function, and it is the task of the control system to track this reference input. We are going to demonstrate that the problem stated in this manner is solvable by discontinuous controls and sliding mode motions.

To begin with, consider the scalar case where the minimized value depends on a single variable

$$y = f(x). \tag{13.31}$$

Assume that $f(x)$ is differentiable and $df/dx \neq 0$ everywhere but in the minimum point x^*. In accordance with our approach outlined before consider analogue-type procedures that are equivalent to a system of differential equations where x is the state coordinate and the right-hand side depending on x and y is defined by the search algorithm. Figure 19 depicts a system for determination of the optimal x^* that is described by

$$\begin{aligned} &\dot{x} = u, \quad u = u_0 \operatorname{sign}(s_1 s_2), \quad s_1 = \varepsilon, \\ &s_2 = \varepsilon + \delta, \quad \varepsilon = g - y, \quad \dot{g} = \rho - Mv(s_1, s_2), \end{aligned} \tag{13.32}$$

where $g(t)$ is reference input; u_0, δ, M are positive constants; ρ is positive value (that is constant or varying depending on the particular search technique); the function $v(s_1, s_2)$ is implemented by a three-level relay element. The functions u and v are plotted in Fig. 20. The hysteresis width 2Δ should not exceed δ, and inequality

$$M > u_0 \left| \frac{df}{dx} \right| + \rho \tag{13.33}$$

should be satisfied for M. For the initial conditions $(s_1 - \Delta)(s_2 + \Delta) > 0$ the time derivatives $\dot{s}_1$ and $\dot{s}_2$

$$\dot{s}_1 = \dot{s}_2 = -\frac{df}{dx} u_0 - \rho + Mv$$

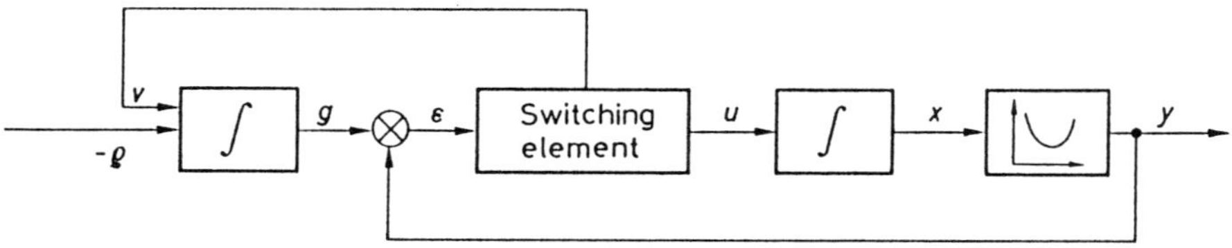

Fig. 19

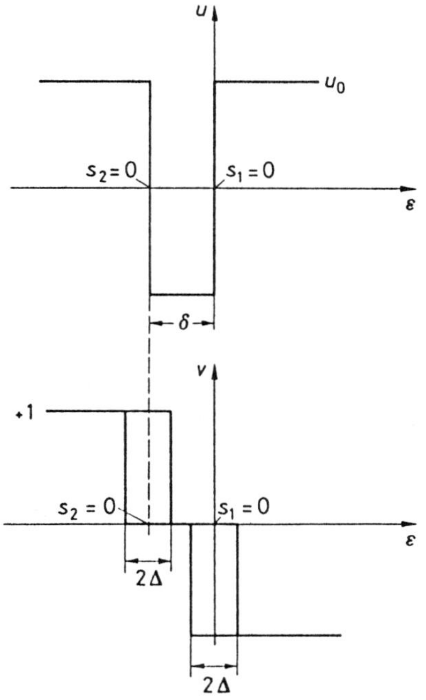

Fig. 20

according to (13.48) have the signs opposite to s_1 and s_2 which implies that at some instant of time the system state will be defined by the inequality $(s_1 + \Delta)(s_2 - \Delta)$, and function v will vanish. For further motion in the system, functions s_1 and s_2 will have opposite signs, and their derivatives

$$\dot{s}_1 = \dot{s}_2 = -\frac{df}{dx} u_0 \operatorname{sign}(s_1 s_2) - \rho. \tag{13.34}$$

If inequality

$$\left|\frac{df}{dx}\right| u_0 > \rho, \tag{13.35}$$

is obeyed, then, depending on the sign of df/dx, that of s_1 or s_2 will be opposite to the sign of rate and, consequently, one of these variables will be at some time equal to zero.

Since any change of the sign of s_1 and s_2 results in the change of the sign of their derivative (13.34), sliding mode will occur in the system. During this mode, either of s_1 and s_2 is identical to zero provided that the element realizing u is ideal. In physical system, the frequency of switching will be finite and, consequently, the amplitude of oscillations of s_1 or s_2 will be non-zero as well. In order to nullify v in sliding mode, the hysteresis width Δ must be taken greater than this amplitude. For $v = 0$ the reference input (13.32) decreases monotonically with rate $-\rho$. In sliding mode either $s_1 = g - y$, or $s_2 = g - y + \delta$ is close to zero, and, therefore, the plant output y which is to be minimized will be monotonously decreasing as well. This process will go on until the vicinity of the minimum point is reached where (13.35) is not satisfied. The inequality (13.35), therefore, my serve as an estimate of the accuracy of self-optimization[1].

In order to enhance the convergence rate of the algorithm described here, one has to increase the rate of decrease of reference input ρ. This, however, would deteriorate search accuracy because the domain (13.35) where the sliding mode existence conditions are violated grows with ρ. It is desirable, in this connection, to design a system where the rate of reference input would decrease with df/dx.

Now we describe algorithms with varying search rate [79]. Information about the derivative df/dx may be obtained through the measurement of equivalent control which is the solution of $\dot{s}_1 = \dot{s}_2 = 0$:

$$u_{\text{eq}} = -\frac{\rho}{df/dx}. \tag{13.36}$$

By determining u_{eq} by means of the first-order filter (Sect. 2.4) and keeping in mind that $|u_{\text{eq}}| \leqq u_0$, let us generate a reference input in the form of decreasing function

$$\dot{g} = -\rho = -\rho_0\left(1 - \frac{u}{u_0}\operatorname{sign} u_{\text{eq}}\right), \tag{13.37}$$

where ρ_0 is a positive number. As may be seen from (13.36) and (13.37), the decrease rate of g will automatically diminish with the approach to the extremum, i.e. with the decrease of the magnitude of df/dx. Importantly, for the reference input (13.37) the sliding mode existence condition is always satisfied. Indeed, if sliding mode terminates, control becomes equal to u_0 or $-u_0$, and u_{eq} also will be equal to this value. In this case, $\rho = 0$, and (13.35) will be obeyed everywhere but for the extremum point, i.e. sliding mode motion will occur again[2].

Thus, in the system with varying rate of search (13.37) sliding mode motion always occurs, its equation can be obtained by substitution of u_{eq} for u:

$$\dot{g} = -\rho = -\rho_0\left(1 - \frac{|u_{\text{eq}}|}{u}\right). \tag{13.38}$$

In (13.38) u_{eq} is a continuous time function lying between $-u_0$ and u_0, therefore for any ρ_0 and $df/dx \neq 0$ the output y will approach its minimum, tracking the reference input $g(t)$.

---

[1] As it was demonstrated in [79] in the vicinity of the minimum point where (13.35) is not satisfied, an oscillatory component appears in the output y, but its average value, nevertheless, remains a decreasing time function. After the transient is over the plant output does not exceed $f(x^*) + \Delta$.

[2] This fact is substantiated in [79] also for the case where a non-ideal sliding mode occurs in the system and the time constant of filter used to compute u_{eq} is finite.

An important feature of the system with varying rate of search is that with the increase of parameter ρ_0 the rate of motion towards the extremum grows and tends to the maximal possible value u_0. Since the maximal possible rate of y is $|df/dx|u_0$ and $y = g(t)$, the rate of g is also bounded by this value. According to (13.38), this constraint means that $|u_{eq}|$ must tend to u_0 with increase of ρ_0. During sliding mode $\dot{x} = u_{eq}$ and, consequently, the input varies with maximal possible rate under sufficiently great ρ_0.

Let us try now to generalize the one-dimensional optimization algorithm to the case where the parameter to be minimized is the function of n inputs, i.e. $x^T = (x_1, \ldots, x_n)$, and the problem again consists in finding a vector x^* such that $\min f(x) = f(x^*)$. It will be also assumed that $f(x)$ is differentiable and its gradient is zero only in the point of minimum. In order to solve the multi-dimensional problem of optimization one may use any of the one-dimensional methods by varying somehow the direction of search in the n-dimensional space of input parameters [163] either periodically, or depending on system state, or randomly, etc. The motion of this system is described by

$$\dot{x} = ku, \tag{13.39}$$

where k is piecewise-constant vector with components $k_1, \ldots, k_n$, u is scalar control defined by the chosen method of one-dimensional optimization.

Consider now the system behaviour for the case where control u is chosen according to (13.32) for varying search rate (13.37). Since $\dot{y} = (\operatorname{graf} f, k)u$, for the constant vector k a conditional extremum along this direction will be found defined by the point where $(\operatorname{grad} f, k) = 0$. After the attainment of the minimum, the vector k defining the direction of search must be changed. We now demonstrate that with the algorithm (13.32), (13.37) one can fix the instants when conditional extrema are attained by measuring equivalent control. By solving $\dot{s}_1 = \dot{s}_2 = 0$ with respect to u obtain

$$|u_{eq}| = \frac{\rho_0}{\rho_0 - u_0(\operatorname{grad} f, k) \operatorname{sign} u_{eq}} u_0, \tag{13.40}$$

$$u_{eq} = -\frac{\rho_0\left(1 - \dfrac{|u_{eq}|}{u_0}\right)}{(\operatorname{grad} f, k)}. \tag{13.41}$$

Like in the case of one-dimensional optimization, $|u_{eq}|$ does not exceed u_0; therefore, it follows from (13.41) that

$$\operatorname{sign} u_{eq} = -\operatorname{sign}(\operatorname{grad} f, k). \tag{13.42}$$

Rewrite (13.40) with allowance for (13.42):

$$|u_{eq}| = \frac{\rho_0}{\rho_0 + u_0|(\operatorname{grad} f, k)|} u_0.$$

Obviously,

$$\lim_{(\operatorname{grad} f, k) \to 0} |u_{eq}| = u_0,$$

therefore, the closeness of $|u_{\text{eq}}|$ to u_0 testifies to approaching the conditional extremum and the direction of search must be changed. The decrease of input y terminates if the scalar product $(\operatorname{grad} f, k) = 0$ along any direction defined by the vector k, i.e. $\operatorname{grad} f = 0$, and the minimum point is found.

In conclusion it may be said that the above principles of designing optimization systems are applicable to the cases where input parameters are bounded by equalities or inequalities:

$$h_i(x) = 0, \quad i = 1, \dots, m, \tag{13.43}$$

$$h_i(x) \leqq 0, \quad i = m + 1, \dots, m + l. \tag{13.44}$$

The design is built around the introduction of an additional vector control with discontinuities on the boundaries of the permissible domain into a system performing unconditional minimization, i.e. minimization of $f(x)$ without constraints. For the (13.43)-type constraints, the additional control induces sliding modes on the intersection of surfaces $h_i(x) = 0$ $(i = 1, \dots, m)$. If the minimum search procedure results in the trajectories in space x that intersect a surface $h_i(x) = 0$ $(i = m + 1, \dots, m + l)$ sliding mode motion occurs also on this surface. As the result, the constraints (13.43), (13.44) are satisfied automatically, and the problem reduces to application of an extremum search procedure (like that described in this section) in no-constraint environment. In the presence of the constraints of only (13.43)-type the search algorithm is, for example, as follows

$$\dot{x} = ku + B(x)U, \tag{13.45}$$

where U is m-dimensional vector with components

$$U_i = \begin{cases} U_i^+(x) & \text{for} \quad h_i(x) > 0 \\ U_i^-(x) & \text{for} \quad h_i(x) < 0 \end{cases} \quad i = 1, \dots, m,$$

$U_i^+(x)$ and $U_i^-(x)$ are continuous functions of vector x; and matrix $B(x)$ should be selected so as to induce sliding mode on the manifold $h(x) = 0$, $h^{\text{T}} = (h_1, \dots, h_m)$. Assume that $m < n$, the gradient matrix $G = \{\partial h / \partial x\}$ has the maximal rank, and $\det GB \neq 0$. At the occurrence of sliding mode, the equation of motion on the manifold $h(x) = 0$ as obtained through the equivalent control method has then the form of (2.10):

$$\dot{x} = (I_n - B(GB)^{-1}G)ku.$$

In compliance with this equation, control U "directs" the rate vector $\dot{x}$ along the intersection of surfaces $h_i(x) = 0$ $(i = 1, \dots, m)$ for any vector $k(\dot{h} = G(I_n - B(GB)^{-1}G)ku = 0)$. The search procedure developed for the unconditional optimization can ensure also in this case the local decrease of $f(x)$ if search direction k is changed at attainment of the conditional extremum.

The problem with inequality-type constraints may be reduced to that with equality-type constraints by introducing new variables $z_1, \dots, z_l$ and increasing problem dimensionality

$$\dot{x} = k^x u + B_x U, \quad \dot{z} = k^z u + B_z U + U_z \tag{13.46}$$

with $z^T = (z_1, \dots, z_l)$, U is $(m + l)$-dimensional discontinuous control inducing sliding mode motion on the surfaces

$$h_i(x) = 0, \quad i = 1, \dots, m, \tag{13.47}$$

$$h_i^*(x, z) = h_{m+i}(x) + z_i = 0, \quad i = 1, \dots, l, \tag{13.48}$$

k^x and k^z are piecewise-constant vectors, and B_x and B_z are matrices chosen so as to induce sliding mode motion on the manifold (13.47), (13.48).

If the l-dimensional vector $U_z = 0$, the $(n + l)$-dimensional optimization problem with constraints of the type of (13.47), (13.48) may be solved by the algorithm (13.46) from analogy with (13.45). Since the minimized function $f(x)$ is independent of the vector z which can assume arbitrary values, the constraint (13.48) is insignificant. However, it follows from the condition $h_i^*(x, z) = 0$ that $h_{m+i}(x) > 0$ for $z_i > 0$, and the constraint (13.44) will be violated. Let us choose control $U_z^T = (U_{z1}, \dots, U_{zl})$ so as to prevent the component of z from taking negative values:

$$U_{zi} = \begin{cases} M_i & \text{for} \quad z_i < 0, \\ 0 & \text{for} \quad z_i > 0, \end{cases} \quad i = 1, \dots, l, \tag{13.49}$$

M_i exceeds the magnitude of the i-th component of vector $k^z u + B_z U$. Control (13.49) provides the motion towards the permissible domain if any of the constraints of inequality (13.44)-type is violated in the initial point. As soon as all the constraints (13.44) are satisfied, we deal with the above optimization with equality – type constraints.

We have discussed here only the principles of designing a self-optimization system under constraints on the permissible domain of the input parameter vector. References [20, 81, 82, 143] describe different versions of search algorithm design depending on the available information about the functions that define the constraints and about their gradients.

Chapter 14

State Observation and Filtering

1 The Luenberger Observer

In the previous chapters, the discussion of problems (in particular, eigenvalue allocation in Chap. 7) was based on the assumption that the system state vector is known. In practice, however, only a part of its components or some of their functions may be measured directly. This gives rise to the *problem of determination* or *observation* of the state vector through the information on the measured variables. Below, consideration will be given to the problem formulated in this way for the linear time-invariant system

$$\dot{x} = Ax + Bu, \quad x \in \mathbb{R}^n, \quad u \in \mathbb{R}^m, \quad A, B = \text{const.} \tag{14.1}$$

and it will be assumed that one can measure the vector y that is a linear combination of the system state vector components:

$$y = Kx, \quad y \in \mathbb{R}^l, \quad 1 \leqq l < n, \quad K - \text{const.} \tag{14.2}$$

Without loss of generality we assume that rank $K = l$. It is assumed also that the pair $\{K, A\}$ is observable and matrices A, B, K are known. In this case one can design a linear observer with input vector y whose state vector asymptotically approaches the control system state vector [6, 97]. We shall discuss the design methods for these observers, find out their properties and describe an observer operating in sliding mode.

Let us design an n-th order linear observer in the form of dynamic system described by

$$\frac{d\bar{x}}{dt} = A\bar{x} + Bu - L(y - K\bar{x}), \tag{14.3}$$

where the state vector $\bar{x}$ is directly measurable. Vector $y - K\bar{x}$ characterizes the mismatch between the measured vector y and its estimate $\bar{y} = K\bar{x}$ made up of the components of $\bar{x}$. We demonstrate that it is always possible to find a matrix L such that $\lim \bar{x} = x$ for $t \to \infty$ and the mismatch between the system state vector and its estimate $\hat{x} = x - \bar{x}$ is the solution of a homogeneous differential equation with any desirable allocation of the eigenvalues. To this end, write according to (14.1)–(14.3) the equation describing the behaviour of $\hat{x}$

$$\frac{d\hat{x}}{dt} = A\hat{x} + LK\hat{x}. \tag{14.4}$$

Since the eigenvalues of matrices $A + LK$ and $A^{\mathrm{T}} + K^{\mathrm{T}}L^{\mathrm{T}}$ coincide, any of their allocations may be obtained (see Sect. 7.3) through a suitable choice of L^{T} if the pair $\{A^{\mathrm{T}}, K^{\mathrm{T}}\}$ is controllable. Direct comparison of the necessary and sufficient conditions of controllability (7.3) for $\{A^{\mathrm{T}}, K^{\mathrm{T}}\}$ and observability (9.9) for $\{K, A\}$ reveals their equivalence. The system (14.1), (14.2) is assumed to be observable; therefore, the problem of desired eigenvalue allocation for the differential equation (14.4) is solvable. A suitable choice of matrix L in the observer equation (14.4), thus, can bring about the desired decay of mismatch $\hat{x}$ or convergence of the estimate $\bar{x}$ to the system state vector x.

2 Observer with Discontinuous Parameters

Let us discuss now the possibility of decomposing the observer motion through the deliberate introduction of sliding mode. Since rank $K = l$, the observed vector y may be represented as

$$y = K_1x_1 + K_2x_2, \quad x^{\mathrm{T}} = (x_1^{\mathrm{T}}, x_2^{\mathrm{T}}), \quad x_2 \in \mathbb{R}^l, \quad \det K_2 \neq 0.$$

Write the equation of system (14.1) motion in space x_1, y:

$$\dot{x}_1 = A_{11}x_1 + A_{12}y + B_1u, \tag{14.5}$$

$$\dot{y} = A_{21}x_1 + A_{22}y + B_2u, \tag{14.6}$$

where

$$TAT^{-1} = \begin{bmatrix} A_{11} & A_{12} \\ A_{21} & A_{22} \end{bmatrix}, \quad TB = \begin{bmatrix} B_1 \\ B_2 \end{bmatrix}, \quad T = \begin{bmatrix} I_{n-l} & 0 \\ K_1 & K_2 \end{bmatrix}$$

(this coordinate transformation is, obviously, non-singular).

Prove the following ancillary statement: *if the pair* $\{K, A\}$ *is observable, the system* $\dot{x}_1 = A_{11}x_1$ *with the output vector* $z = A_{21}x_1$ *or pair* $\{A_{21}, A_{11}\}$ *is observable as well.*

Assume that this condition is not obeyed. In this case, (14.5) where y and u are regarded as inputs is representable together with z in the canonical observability form (9.19), (9.20):

$$\dot{x}'_1 = A'_1x'_1 + A'_{12}y + B'_1u, \tag{14.7}$$

$$\dot{x}''_1 = A''_1x'_1 + A_2x''_1 + A''_{22}y + B''_1u, \tag{14.8}$$

$$z = A'_{21}x'_1, \quad x_1^{\mathrm{T}} = [(x'_1)^{\mathrm{T}}, (x''_1)^{\mathrm{T}}].$$

In (14.7), (14.8) the pair $[A'_{21}, A'_1]$ is observable, and x''_1 is the subspace of non-observable states. In compliance with the above transformation, (14.6) becomes

$$\dot{y} = A'_{21}x'_1 + A_{22}y + B_2u \tag{14.9}$$

The subspace x''_1 will be non-observable, obviously, not only with respect to z but also to y which is the output of the original system (14.1) or its equivalent (14.7)–(14.9). This conclusion contradicts the assumption about observability of the pair $\{K, A\}$, hence, the pair $\{A_{21}, A_{11}\}$ is observable as well, and the statement is proved.

Let us proceed now to the design of state observer with sliding mode where the motion preceding this mode and motion along the intersection of discontinuity surfaces in the state space may be considered independently. This observer is described by the following differential equations

$$\frac{d\bar{x}_1}{dt} = A_{11}\bar{x}_1 + A_{12}\bar{y} + B_1u - L_1v, \tag{14.10}$$

$$\frac{d\bar{y}}{dt} = A_{21}\bar{x}_1 + A_{22}\bar{y} + B_2u + v, \tag{14.10}$$

where $\bar{x}_1$ and $\bar{y}$ are the estimates of the system state,

$$v = M \operatorname{sign} \hat{y}, \quad M > 0, \quad M - \text{const.}, \tag{14.11}$$

$\hat{y} = y - \bar{y}$, $(\operatorname{sign} \hat{y})^{\mathrm{T}} = (\operatorname{sign} \hat{y}_1, \ldots, \operatorname{sign} \hat{y}_l)$, the vectors $\bar{x}_1$ and $\bar{y}$ and, therefore, $\hat{y}$ are measurable.

The discontinuous vector function v is chosen so as to induce sliding mode motion on the manifold $\hat{y} = 0$, and, thus, $y = \bar{y}$ holds. Now, a matrix L_1 must be found such that stability of this motion and the desired eigenvalue allocation in the equation with respect to the mismatch $\hat{x}_1 = x_1 - \bar{x}_1$ is provided. Equation with respect to $\hat{x}_1$ and $\hat{y}$ may be obtained from (14.5), (14.6), (14.10):

$$\frac{d\hat{x}_1}{dt} = A_{11}\hat{x}_1 + A_{12}\hat{y} + L_1v,$$

$$\frac{d\hat{y}}{dt} = A_{21}\hat{x}_1 + A_{22}\hat{y} - v, \tag{14.12}$$

or according to (14.11)

$$\frac{d\hat{x}_1}{d\tau} = \mu(A_{11}\hat{x}_1 + A_{12}\hat{y}) + L_1 \operatorname{sign}\hat{y},$$

$$\frac{d\hat{y}}{d\tau} = \mu(A_{21}\hat{x}_1 + A_{22}\hat{y}) - \operatorname{sign}\hat{y},$$

where $\tau = 1/\mu$, $\mu = M^{-1}$.

Since for $\mu = 0$ in a system with a stretched time scale the sign of each component of $\hat{y}$ is opposite to that of its derivative, the state vector in finite time will reach the manifold $\hat{y} = 0$ whereupon sliding mode motion starts. According to the theorem on continuous dependence of solution on the parameter, sliding mode occurs also at sufficiently small but non-zero value of μ (or at sufficiently great M). Hence, if one assumes that the initial condition domain in the space $(\hat{x}_1, \hat{y})$ or spaces x and $\hat{x}$ is bounded, there exists a value of parameter M such that sliding mode motion in (14.12) will always occur. In order to derive the sliding equation through the equivalent control method it is necessary to solve $d\hat{y}/dt = 0$ with respect to v, substitute this solution $v = v_{eq} = A_{21}\hat{x}_1$ into the first equation of (14.12) and take y be equal to zero:

$$\frac{d\hat{x}_1}{dt} = A_{11}\hat{x}_1 + L_1 A_{21}\hat{x}_1. \tag{14.13}$$

Since the pair $\{A_{21}, A_{11}\}$ is observable, it is possible to choose matrix L_1 like in (14.4) so that $\lim \hat{x}_1 = 0$ for $t \to \infty$ and the eigenvalues of $A_{11} + L_1 A_{21}$ have the desired values.

As the result, the observer (14.10) enables determination of the vector x_1 to the accuracy of the decaying transient component. The components of x_2 that are also included into x are computed through

$$x_2 = -K_2^{-1}(y - K_1 x_1). \tag{14.14}$$

Notably, this method for determination of x is basically a sliding-mode realization of $(n-l)$-dimensional asymptotic Luenberger observer [6, 97]. The possibility of reducing the observer's order is due to the fact that it suffices to determine $n - l$ components of x, the rest being computable from relations of the (14.14) type. The design procedure of this observer rests upon the coordinate transformation

$$x' = x_1 + L_1 y, \tag{14.15}$$

and system behaviour is considered in the space (x', y), the coordinate transformation, being obviously, non-singular for any matrix L_1. The equation with respect to x' may be obtained from (14.5), (14.15):

$$\dot{x}' = (A_{11} + L_1 A_{21})x' + A'_{12} y + (B_1 + L_1 B_2)u,$$

$$A'_{12} = A_{12} + L_1 A_{22} - (A_{11} + L_1 A_{21})L_1.$$

The observer is chosen in the form of a dynamic $(n-l)$-th order system described by

$$\frac{d\bar{x}'}{dt} = (A_{11} + L_1 A_{21})\bar{x}' + A'_{12}y + (B_1 + L_1 B_2)u \tag{14.16}$$

where $\bar{x}'$ is the state vector, and y and u are regarded as inputs. The mismatch $\hat{x}' = x' - \bar{x}'$ is governed by the following differential equation

$$\frac{d\hat{x}'}{dt} = (A_{11} + L_1 A_{21})\hat{x}' \tag{14.17}$$

which coincides with (14.13). Stated differently, a suitable choice of L_1 in this case also enables one to obtain the desired properties of solution with respect to $\hat{x}'$. As the result, the vector x' may be determined, and the rest of the state vector components may be calculated through (14.15) and (14.14).

Thus, the discontinuous parameter observer in sliding mode is equivalent to the reduced-order Luenberger one. However, in the case where the plant and observed signal are affected by noise, the non-linear observer may happen to be preferable due to the properties of filtering because its structure coincides with that of Kalman filter. These issues will be discussed below in Sect. 14.4.

3 Sliding Modes in Systems with Asymptotic Observers

The preceding chapters, have demonstrated the possibilities of discontinuous control systems that are due to the deliberate induction of sliding modes. Like with any other principle of control, their practical application meets with the problem of correspondence between the ideal models of sliding mode motion and physical processes. The need to investigate it is twofold: first, the models of plant, instrumentation and actuator usually neglect the small time constants; second, in practical applications one cannot directly measure all the coordinates of the state vector, and, therefore, they are restored through the observable variables by means of auxiliary dynamic systems that also give rise to motions that differ from those in the system having ideal information about all the state vector components.

If there is a discrepancy between the plant and model operators (which is always the case) and the data about system state are distorted, the use of sliding mode control algorithms can significantly deteriorate the system efficiency. This situation has been met in applications that used the algorithms of variable-structure systems where the state vector components were restored by differentiating filters (including variable-structure ones). Sliding modes had a high-frequency component, and even small time constants of the differentiators resulted in a significant reduction in the switching frequency of control and, as a result, in system accuracy, occurrence of intolerable self-oscillations, higher heat losses in electrical power circuits, and higher wear of moving mechanical parts. Owing to the above reasons, at the first development stage of variable-structure systems where the scalar control was generated as function of the error coordinate and its derivatives determined by physical differentiators, practical implementation of systems of this class often met with insuperable barriers.

Below, consideration will be given to the design of sliding modes in the cases of [18]:

—discrepancy between the model and physical control system which is due to the lags of instrumentation or controlled plant neglected by the model when there is complete information about the model state vector;

—an asymptotic observer is used in the system for state vector restoration from the observed variables (see Sect. 1), and when there is complete correspondence between the model and the plant; and

—a combination both cases.

As it will be shown below, in the case of dynamic discrepancy between the model and physical plant where only a part of the state vector components is used for generation of control actions, the asymptotic observer enables an ideal sliding mode and, thus, preservation of all the advantages that systems have when in this kind of motion.

Case 1. Let us discuss at first which situations might occur in discontinuous control system if the state vector is assumed to be known but the model of measuring devices or plant disregards the small time constants.

Let state vector x in time-invariant system

$$\dot{x} = Ax + Bu, \quad x \in \mathbb{R}^n, \quad u \in \mathbb{R}^m, \quad \operatorname{rank} B = m \tag{14.18}$$

be measured by inertial measuring device

$$\mu \dot{z} = Dz + Hx, \quad z \in \mathbb{R}^q, \tag{14.19}$$ [1]

where u is control, z is the state vector of an inertial measuring device, μ is the parameter defining the rate of motion in the measuring device.

The matrix D is assumed to be the Hurwitz one (i.e. the matrix D^{-1} exists), and the pair $\{A, B\}$ is assumed to be controllable. The components of the discontinuous control being designed have discontinuities on some surfaces $s_i(x, z) = 0$, $i = 1, \ldots, m$ in the state space of system (14.18), (14.19):

$$u_i(x, z) = \begin{cases} u_i^+(x, z) & \text{with} \quad s_i(x, z) > 0, \\ u_i^-(x, z) & \text{with} \quad s_i(x, z) < 0, \end{cases} \tag{14.20}$$

[1] Differentiating filters can be such a dynamic measuring element.

where $u_i(x, z)$ are the components of $u; u_i^+$, u_i^- and s_i are continuous functions of x and z; $u_i^+(x, z) \neq u_i^-(x, z)$.

Let the vector $s^T = (s_1, \ldots, s_m)$ be composed of the observable components of x and z:

$$s = R_s z + K_s x; \quad R_s, K_s - \text{const.} \tag{14.21}$$

It is assumed that for the ideal model ($\mu = 0$) one can always choose matrices R_s and K_s such that the desired sliding mode occurs on the manifold $s_{\mu=0} = Cx = 0, C = -R_s D^{-1} H + K_s$ (i.e. $\det CB \neq 0$). Determine the time derivative of s on the trajectories of the system (14.18), (14.19):

$$\dot{s} = \frac{1}{\mu} R_s D z + \frac{1}{\mu} R_s H x + K_s A x + K_s B u. \tag{14.22}$$

According to the above design procedures, for sliding mode to occur along the manifold $s = 0$ the matrix $K_s B$ in (14.22) must have full rank. Practically, this is not the case for arbitrary system (14.19) e.g. in the case of $K_s = 0$.

As it was illustrated in [143] by the scalar case ($m = 1$), instead of ideal sliding mode there will be selfoscillations in the vicinityof manifold $s = 0$ of the order μ with finite switching frequency as smaller as greater μ.

Assume now that the model of controlled plant with inertialess measuring devices disregards the dynamic elements with rapidly decaying eigenmotions and all the state vector components may be used for control generation. In this case, the plant is described by the following system of equations:

$$\dot{x} = Ax + B\tilde{u}, \tag{14.23}$$

$$\tau \dot{z} = Dz + B_u u, \quad \tilde{u} \in \mathbb{R}^m, \quad z \in \mathbb{R}^q. \tag{14.24}$$

It is assumed that a part of vector $\tilde{u}$ components is formed without dynamic distortions, and the rest of them is subjected to transformation by (14.24) with small time constants of the order of τ that define the neglected dynamics in the controlled plant:

$$\tilde{u} = K_z z + R_u u, \quad K_z, R_u - \text{const.} \tag{14.25}$$

In the absence of dynamic discrepancy between the plant and model, motion equations (14.18) and (14.23) should coincide, and from (14.24) obtain for $\tau = 0$ that

$$z_0 = -D^{-1} B_u u \tag{14.26}$$

and the equality condition for u and $\tilde{u}$ is determined from (14.25) and (14.26):

$$-K_z D^{-1} B_u + R_u = I_m, \tag{14.27}$$

where I_m is an identity matrix.

Compose a function s from the components of vector x:

$$s = Cx, \quad C - \text{const.} \tag{14.28}$$

and determine its derivative in virtue of (14.23) and (14.25):

$$\dot{s} = CAx + CBK_z + CBR_u u. \tag{14.29}$$

Like in the systems with neglected small parameters in measuring elements, for the existence of sliding mode along the manifold $s=0$ the matrix CBR_u should have full rank which is not the case in practice. For instance, the actuator having control at its input is inertial and, as a result, $R_u = 0$. It is clear that in this case for $\tau \neq 0$ (like for $\mu \neq 0$) modes might occur having finite switching frequency that may be intolerably small and significantly deteriorate system's dynamic performance and accuracy.

Case 2. Assume now that the model in question (14.18) completely corresponds to the physical plant, but only a vector of output variables is directly measurable.

Let output vector

$$y = Kx, \quad y \in \mathbb{R}^l \tag{14.30}$$

be observed in a system with controlled plant described by (14.18) and inertialess measuring device. The pair $\{K, A\}$ is assumed to be observable. In compliance with the procedure of Sect. 1, the state observer is designed for the system (14.18), (14.30):

$$\frac{d\bar{x}}{dt} = A\bar{x} + Bu + L(K\bar{x} - y), \quad \bar{x} \in \mathbb{R}^n. \tag{14.31}$$

The motion equations with respect to the mismatch $\varepsilon = \bar{x} - x$ will be obtained from (14.18) and (14.31)

$$\dot{\varepsilon} = (A + LK)\varepsilon. \tag{14.32}$$

For the observable pair $\{K, A\}$, the matrix L can be always found such that the matrix $A + LK$ will have the desired eigenvalues and the mismatch ε will tend to zero with given decay rate or $\lim_{t \to \infty} \bar{x} = x$. Form now the switching function as

$$s_0 = C\bar{x} \tag{14.33}$$

and determine its derivative along the trajectories of (14.31):

$$\dot{s}_0 = CA\bar{x} + CBu + CLK\varepsilon. \tag{14.34}$$

Let the components of u have discontinuities on surface $s_0 = 0$ similar to (14.20). In the system (14.18), (14.30), (14.31), (14.33), sliding mode may be induced if the matrix CB has full rank. This condition has no relation to the observer parameters and may be always satisfied together with providing the desired dynamics of sliding modes in systems with complete information about the state vector.

In the system (14.18), (14.30), (14.31), (14.33) with an additional dynamic subsystem (observer), sliding mode, thus, is not eliminated and occurs for sufficiently small ε exactly as in the system with complete information about the state vector. The sliding equations may be obtained through the equivalent control method. To this end, the equivalent control is determined from (14.30) (the solution of $\dot{s}_0 = 0$ with respect to u) and substituted into the original system. Since in sliding mode $s_0 = 0$, $n - m$ components of the vector $\bar{x}$ may be expressed in terms of the rest of its m components. Substitution of $x + \varepsilon$ for $\bar{x}$ leads to an $(2n - m)$-th order system describing sliding mode motion for the second case

$$\dot{x}_1 = A_1 x_1 + Q\varepsilon, \quad x_1 \in \mathbb{R}^{n-m}, \qquad (14.35)$$
$$\dot{\varepsilon} = (A + LK)\varepsilon,$$

where the matrices A_1 and Q depend on those of the initial system. One may provide the desired dynamic properties in (14.35) by independent assignment of the eigenvalues of matrices A_1 and $A + LK$ through a suitable choice of the matrices C and L. If motion rate in the observer is one order of magnitude higher than that with respect to the state vector x, sliding mode motion will be not exactly on manifold $s = Cx = 0$ as in the complete-information system, but in its small vicinity defined by $s_0 = Cx + C\varepsilon = 0$, where ε rapidly decays in virtue of the autonomous second equation in (14.35).

Case 3. Consider now the case where in the system with an asymptotic observer inertial measuring devices are used or the model does not correspond to the physical plant. Let in the system neglecting the dynamics of measuring device whose motion is described by (14.18), (14.19) an observer of the form of (14.31) be used for design of discontinuity surfaces.

Vector

$$y = R_z z + K_x x, \quad y \in \mathbb{R}^l \qquad (14.36)$$

is observed in the system. If there is no dynamic discrepancy between the model and plant ($\mu = 0$ in (14.19))

$$z_0 = D^{-1} H x \qquad (14.37)$$

and

$$y_0 = (-R_z D^{-1} H + K_x) x = Kx, \qquad (14.38)$$

i.e. in the case of inertialess measurement the vectors y_0 and y coincide. Introduce new variables

$$\lambda = z - z_0 = z + D^{-1} H x. \qquad (14.39)$$

According to (14.18), (14.39):

$$\mu\dot{\lambda} = D\lambda + \mu D^{-1} H(Ax + Bu) \qquad (14.40)$$

Taking into account (14.36), (14.38), (14.39), y may be rewritten as

$$y = R_z \lambda + Kx \qquad (14.41)$$

The equations of motion with respect to x, ε and λ will be for this system as follows

$$\dot{x} = Ax + Bu,$$
$$\dot{\varepsilon} = (A + LK)\varepsilon - LR_z\lambda, \tag{14.42}$$
$$\mu\dot{\lambda} = D\lambda + \mu D^{-1}H(Ax + Bu).$$

Take the equation of function that defines the discontinuity surfaces in the form of (14.33) and compute its derivative

$$\dot{s}_0 = CA\bar{x} + CBu + C(A + LK)\varepsilon - CLR_z\lambda. \tag{14.43}$$

Also in this case the ideal sliding mode may be, obviously, provided because the condition $\det CB \neq 0$ may be satisfied independently of the parameters of observer and neglected dynamic elements. The motion equations of $(2n - m + q)$-th order for sliding mode may be obtained similar to (14.35):

$$\dot{x}_1 = A_1x_1 + Q\varepsilon + V_\mu\lambda,$$
$$\dot{\varepsilon} = (A + LK)\varepsilon + V_{\mu\lambda}\lambda,$$
$$\mu\dot{\lambda} = D\lambda + \mu(Q_{x\mu}x_1 + Q_{\varepsilon\mu}\varepsilon + Q_{\lambda\mu}\lambda), \tag{14.44}$$

where the matrices Q, V_μ, $V_{\mu\lambda}$, $Q_{x\mu}$ and $Q_{\lambda\mu}$ depend on those of the original system (14.42).

Consider now sliding mode design in a system described by the equations (14.23)–(14.25) if the switching function is formed of the components of vector $\bar{x}$ of the observer (14.30), (14.31), (14.33). Obviously, sliding mode in this case exists because the equation for s_0 is similar to (14.34), and the matrix CB in it has full rank. Let us determine the sliding mode equations. Taking into consideration (14.25), write the equations (14.23), (14.24) with respect to x, z and ε:

$$\dot{x} = Ax + BK_zz + BR_uu,$$
$$\dot{\varepsilon} = (A + LK)\varepsilon + B(I_m - R_u)u - BK_zz,$$
$$\tau\dot{z} = Dz + B_uu \tag{14.45}$$

From $\dot{s}_0 = C\dot{x} + C\dot{\varepsilon} = 0$ determined with regard to (14.45), compute the equivalent control

$$u_{\mathrm{eq}} = -(CB)^{-1}(CAx + C(A - LK)\varepsilon). \tag{14.46}$$

After substitution of u_{eq} into (14.45) obtain

$$\dot{x} = Ax + BK_zz - BR_u(CB)^{-1}(CAx + C(A + LK)\varepsilon),$$
$$\dot{\varepsilon} = (A + LK)\varepsilon - B(I_m - R_u)(CB)^{-1}(CAx + C(A + LK)\varepsilon) - BK_zz,$$
$$\tau\dot{z} = Dz - B_u(CB)^{-1}(CAx + C(A + LK)\varepsilon). \tag{14.47}$$

Denote

$$z_{01} = D^{-1}B_u(CB)^{-1}(CAx + C(A + LK)\varepsilon)$$

and

$$\lambda = z - z_{01}.$$

Then, given (14.27) one can rewrite (14.47) with respect to the coordinates x, ε and λ in the form of

$$\begin{aligned}
\dot{x} &= (I_n - B(CB)^{-1}C)Ax - B(CB)^{-1}C(A + LK)\varepsilon + BK_z\lambda,\\
\dot{\varepsilon} &= (A + LK)\varepsilon - BK_z\lambda,\\
\tau\dot{\lambda} &= D\lambda - \tau D^{-1}B_u(CB)^{-1}C(A(I_n - B(CB)^{-1}C)Ax\\
&\quad + (A + LK - AB(CB)^{-1}C)(A + LK)\varepsilon - LKBK_z\lambda).
\end{aligned} \tag{14.48}$$

It must be noted that its order may be reduced to $2n - m + q$ like with the systems (14.35) and (14.44):

$$\begin{aligned}
\dot{x}_1 &= A_1x_1 + Q\varepsilon + V_\tau\lambda,\\
\dot{\varepsilon} &= (A + LK)\varepsilon + V_{\tau\lambda}\lambda,\\
\tau\dot{\lambda} &= D\lambda + \tau(Q_{x\tau}x_1 + Q_{\varepsilon\tau}\varepsilon + Q_{\lambda\tau}\lambda),
\end{aligned} \tag{14.49}$$

where the matrices V_τ, $V_{\tau\lambda}$, $Q_{x\tau}$, $Q_{\varepsilon\tau}$ and $Q_{\lambda\tau}$ depend on those of (14.48).

Thus, sliding mode in systems with asymptotic observer is preserved even in the presence of lags in the controlled plant and measuring device that were neglected in the ideal model. It follows from the comparison of (14.44) and (14.49) that the sliding mode equations for both cases are equivalent.

The study of motion in the systems (14.44), (14.49) with small parameters μ and τ is a standard problem of the theory of singularly perturbed equations. The methods of this theory enable decomposition of the overall motion in the system into a set of partial motions of lower dimensionality and different rates, which is important for the engineering analysis and design of control systems.

Denote the eigenvalues of (14.44), (14.49) by p_i. The parameter ξ is equal to μ for the neglected lags in measuring device and to τ in the plant. The rates of motion in these systems are defined by three groups of eigenvalues:

$$\begin{aligned}
p_i &= \frac{1}{\xi}(p_i\{D\} + O(\xi)), \quad i = 1, \ldots, q,\\
p_i &= p_i\begin{Bmatrix} A_1 & Q \\ 0 & A + LK \end{Bmatrix} + O(\xi), \quad i = q + 1, \ldots, 2n - m + q,
\end{aligned} \tag{14.50}$$

where $O(\xi)$ is infinitesimal of the order of ξ. Since the observer equations are autonomous with respect to the vector x_1, the latter group of eigenvalues in (14.50) may be rewritten as

$$p_i = p_i\{A + LK\} + O(\xi), \quad i = q + 1, \ldots, n + q$$

$$p_i = p_i\{A_1\} + O(\xi), \quad i = n + q + 1, \ldots, 2n - m + q.$$

It is relevant to note here that for $\mu = 0$ in (14.44) and for $\tau = 0$ in (14.49) both systems coincide with (14.35) that describes sliding modes in the case where the model exactly corresponds to the plant.

Thus, we have discussed the three cases of deviation of system motion from the ideal sliding mode. In all the cases, motion takes place in the vicinity of switching manifold $s = 0$. The vicinity is defined in the first case by the parameters μ and τ that characterize the dynamic elements with fast motions that are neglected in the model, in the second case by autonomous observer dynamics, and in the third case by a combination of both. However, in the systems with an observer (cases 2 and 3) sliding mode is ideal, and without an observer (case 1) control switching frequency is finite. In the average, all these systems are equivalent, but systems without observer as a rule are inpracticable because of the low-frequency motion component and self-oscillations. Observers in discontinuous control systems enable elimination of these phenomena and effective use of sliding modes in applications.

4 Quasi-Optimal Adaptive Filtering

Let us see how the state of a linear dynamic system may be estimated through measurements of its output in the presence of random disturbances that are close to white noises. This problem will be treated by means of a filter with discontinuous parameters in the right side of its equation. The system will be analyzed by means of the averaging theory.

For the conventional problem of time-varying linear filtration, the system equation and output vector are

$$\dot{z} = A(t)z + B(t)u + \zeta_0, \quad y = C(t)z + \eta_0, \tag{14.51}$$

where $z = \mathbb{R}^n$, $y \in \mathbb{R}^m$, $u \in \mathbb{R}^l$, A, B and C are measurable matrix time functions bounded over any finite interval, and ζ_0 and η_0 are assumed to be independent white noises. It is desired to find the estimate $\bar{z}(t)$ of the state $z(t)$ of (14.51) through observations of its output vector y over the interval $[0, t]$. For non-singular η_0, the linear estimate optimal with respect to the r.m.s. criterion $E \| \bar{z}(t) - z(t) \|^2$ (where E is expectation) is defined by the Kalman–Bucy filter

$$\dot{\bar{z}} = A(t)\bar{z} + B(t)u + L(t)(y - C(t)\bar{z}), \tag{14.52}$$

where $L(t) = P(t)C^{\mathrm{T}}(t)(RR^{\mathrm{T}})^{-1}$, $P(t)$ is the unique positive definite solution of the Riccati equation

$$\dot{P}(t) = A(t)P(t) + P(t)A^{\mathrm{T}}(t) - P(t)C^{\mathrm{T}}(t)(RR^{\mathrm{T}})^{-1}C(t)P(t) + QQ^{\mathrm{T}}, \tag{14.53}$$

and RR^{T} and QQ^{T} are the intensities of noises $\eta_0(t)$ and ζ_0 respectively [92].

Implementation of (14.52) requires exact knowledge of RR^{T} and QQ^{T}, but in practice this information is usually unavailable and only the intervals are known where the elements of these matrices may occur. Moreover the noise characteristics sometimes can vary if, for instance, control system mode changes. In this case they may be regarded as quasi-stationary, i.e. having constant intensities over sufficiently large time intervals, and the filter must be changed depending on the changing noise characteristics. Solution of this problem is usually sought in the class of adaptive systems, noise intensities being estimated by one or another technique. These systems, howeve, may be difficult to implement, and when during long time intervals no change is required, the "adaptive part" of the filter is actually idling. Significant complication is, therefore, unwarranted.

Of practical interest is design of a filter that would, on the one hand, enable close-to-optimal estimation error and, on the other hand, have low sensitivity to the inaccurate knowledge of noise intensities.

The approach used below relies upon the well known property of discontinuous non-linearities to change the equivalent gain with the amplitude of the output high-frequency signal. From this viewpoint, it seems reasonable to introduced into the filter equation sign $(y - C\bar{z})$ instead of the mismatch between system output and its estimate $y - C\bar{z}$. As the result, obtain instead of (14.52) the following filter

$$\dot{\bar{z}} = A(t)\bar{z} + B(t)u + D(t)\operatorname{sign}(y - C(t)\bar{z}), \tag{14.54}$$

where $D(t)$ is a matrix to be chosen below. This filter could be expected to be adaptive with respect to the variations of noise intensity in the observation vector.

Let us write (14.54) with respect to the estimation error $x = z - \bar{z}$ as

$$\dot{x} = A(t)x - D(t)\operatorname{sign}(C(t)x + \eta(t)) + \zeta(t). \tag{14.55}$$

with the right side being non-linear with respect to the process $(\eta^{\mathrm{T}}(t), \zeta^{\mathrm{T}}(t))$ which is regarded as being close to the multidimensional white noise and enabling solution in the conventional sense.

Following the ideas of the regularization methods, let us consider the sequence or, more precisely, the family of random processes $(\eta^{\mathrm{T}}(\varepsilon, t), \xi^{\mathrm{T}}(\varepsilon, t))$ that almost everywhere have continuous trajectories and at $\varepsilon \to 0$ tend, in a sense, to the white noise. Let us consider the asymptotic behavior of the system

$$\dot{x} = A(t)x + D(\varepsilon, t)\operatorname{sign}(C(t)x + \eta(\varepsilon, t)) + \zeta(\varepsilon, t) \tag{14.56}$$

at $\varepsilon \to 0$ under the assumption that the process

$$\left(\left(\int_0^t \eta(\varepsilon, \tau)\,d\tau\right)^{\mathrm{T}}, \left(\int_0^t \zeta(\varepsilon, \tau)\,d\tau\right)^{\mathrm{T}}\right)$$

weakly converges in the sense of [14] to the process $((Rw_1(t))^{\mathrm{T}}, (Qw_2(t))^{\mathrm{T}})$ where $w_1(t)$, $w_2(t)$ are, respectively,, m- and n-dimensional independent Wiener processes.

The system (14.56) has discontinuous right-hand side under fixed sampled trajectories of $\eta(\varepsilon, t)$ and $\zeta(\varepsilon, t)$. That is why like in the above cases it does not always have solution in the conventional sense so far as sliding modes can occur in it.

Let us take $\eta(\varepsilon, t) = \varepsilon^{-1}\xi_1(t\varepsilon^{-2})$ and $\zeta(\varepsilon, t) = \varepsilon^{-1}\xi_2(t\varepsilon^{-2})$ as the family of processes $(\eta^{\mathrm{T}}(\varepsilon, t), \zeta^{\mathrm{T}}(\varepsilon, t))$, where $\xi_1(t)$ and $\xi_2(t)$ are stationary processes having zero expectations $E\xi_1(t) = 0$ and $E\xi_2(t) = 0$ and satisfying the strong mixing condition (Sect. 2.5). Under these assumptions, the assertion that the processes

$$\int_0^t \eta(\varepsilon,\tau)d\tau = \varepsilon^{-1}\int_0^t \xi_1(\tau\varepsilon^{-2})d\tau$$

and

$$\int_0^\tau \zeta(\varepsilon,\tau)d\tau = \varepsilon^{-1}\int_0^\tau \xi_2(\tau\varepsilon^{-2})d\tau$$

weakly converge to $Rw_1(t)$ and $Qw_2(t)$ for $\varepsilon \to 0$ is simply the central limit theorem for random processes (e.g. see [160]), matrices R and Q being defined by

$$RR^{\mathrm{T}} = 2\int_0^\infty E\xi_1(0)\xi_1^{\mathrm{T}}(\tau)d\tau, \quad QQ^{\mathrm{T}} = 2\int_0^\infty E\xi_2(0)\xi_2(\tau)d\tau.$$

The choice of process family converging to the white noise enables one to apply the methods of the averaging theory as developed in [75, 76] to the systems of (14.56) type. In these references, however, all the proofs are based on the assumption that the right-hand sides satisfy the Lipschitz condition or even are smooth, whereas the right-hand side of (14.56) is discontinuous.

The following theorem will be used as the major tool in the study of this system.

Theorem. *Let $D(\varepsilon,t) = D(t)\varepsilon^{-1}$ where $D(t)$ is continuous norm-bounded functional matrix, $\xi_1(t)$ and $\xi_2(t)$ are independent time-invariant random processes such that*

(1) *the property of uniformly strong mixing is inherent into the process $\xi(t) = (\xi^{\mathrm{T}},(t), \xi_2^{\mathrm{T}}(t))^{\mathrm{T}}$,*

(2) $E\xi(t) = 0$,

(3) $E\,\mathrm{sign}\,\xi_1(t) = 0$,

(4) $\int_0^\infty \mathrm{sign}\,\xi_1(0)\,(\mathrm{sign}\,\xi_1(\tau))^{\mathrm{T}} < \infty$,

(5) *the density of $\xi_1(t)$ obeys the Lipschitz condition.*

For $\varepsilon \to 0$, the solution $x(\varepsilon,t)$ of (14.56) with the initial condition $x(0) = x_0$ weakly converges to the process $x^0(t)$ which is the solution of a system of stochastic differential equations of the (14.51) type:

$$\dot{x}^0 = (A(t) + D(t)FC(t))x^0 + D(t)H\eta_0 + Q\zeta_0 \tag{14.57}$$

with the initial condition $x^0(0) = x_0$. Matrices F, H and Q are defined as follows:

$$F = \mathrm{diag}(f_1,\ldots,f_m) \quad \text{with} \quad f_i = \frac{d}{d\gamma}E\,\mathrm{sign}(\gamma + \xi_{1i})|_{\gamma=0}, \tag{14.58}$$

$$HH' = 2\int_0^\infty E\,\mathrm{sign}\,\xi_1(0)(\mathrm{sign}\,\xi_1(\tau))^{\mathrm{T}}d\tau, QQ^{\mathrm{T}} = 2\int_0^\infty E\xi_2(0)\xi_2^{\mathrm{T}}(\tau)d\tau.$$

In addition, $\lim_{\varepsilon\to 0} E\,\|x(t)\|^2 = E\,\|x^0(t)\|^2$.

The reader is referred to [39] for the proof.

One can easily see that (14.57) is similar to the equation of filtration error $x = z - \bar{z}$ of the linear Kalman–Bucy filter (14.52):

$$\dot{x} = (A(t) + L(t)C(t))x + L(t)R\eta_0 + Q\zeta_0. \tag{14.59}$$

Let $\det F \neq 0$, then denote $\tilde{L} = DF$, $\tilde{R} = F^{-1}H$ and rewrite (14.57) as

$$\dot{x}^0 = (A(t) + \tilde{L}(t)C(t))x^0 + \tilde{L}(t)\tilde{R}\eta_0 + Q\zeta_0. \tag{14.60}$$

If $\tilde{P}$ is a positive definite solution of the Riccati equation

$$\dot{\tilde{P}}(t) = A(t)\tilde{P}(t) + \tilde{P}(t)A^{\mathrm{T}}(t) - \tilde{P}(t)C^{\mathrm{T}}(t)(\tilde{R}\tilde{R}^{\mathrm{T}})^{-1}C(t)\tilde{P}(t) + QQ^{\mathrm{T}}, \tag{14.61}$$

by taking $D(t) = F^{-1}\tilde{P}(t)C^{\mathrm{T}}(t)(\tilde{R}\tilde{R}^{\mathrm{T}})^{-1}$ one obtains for the filter (14.54) the r.m.s. – criterion-minimal error with $\varepsilon \to 0$.

In fact (14.59) and (14.60) differ only in noise intensities RR^{T} and $\tilde{R}\tilde{R}^{\mathrm{T}}$; therefore, it is evident that the difference between the criteria for (14.52) and (14.54) is defined by the relation between R and $\tilde{R}$. But if noise intensity differs from the estimated value or varies in time, the behaviours of (14.52) and (14.54) are basically different. Indeed, assume that the observation noise has changed by a factor of k and is now $\eta_1 = k\eta$; its intensity, then, will be $R_1R_1^{\mathrm{T}} = k^2RR^{\mathrm{T}}$, and since $\tilde{R} = F^{-1}H$, one can easily see that $\tilde{R}_1\tilde{R}_1^{\mathrm{T}} = k^2\tilde{R}\tilde{R}^{\mathrm{T}}$, i.e. RR^{T} and $\tilde{R}\tilde{R}^{\mathrm{T}}$ vary proportionally by a factor of k^2. With this variation of noise, the error variance in the Kalman–Bucy filter varies as well, and the filter (14.52, 14.53) will not give the optimal estimate any more because $P(t)$ does not satisfy the Riccati equation depending on unknown noise parameters and $L = PC(RR^{\mathrm{T}})^{-1}$ for any k whereas the similar variable $\tilde{L}_1(t)$ for (14.54) depends on noise characteristics and will change so that $\tilde{L}_1 = k^{-1}FD = k^{-1}\tilde{L}$. It is this fact that allows one to assert that the filter (14.54) is partially adaptive.

Below, a first-order system is used in order to illustrate the comparison of criterion values in the steady state if $\xi_1(t)$ is a Gaussian process, and the characteristics of the filters (14.52) and (14.54) under deviations of noise intensity from the initial one are analyzed.

Let us estimate random process $z(t)$ satisfying

$$\dot{z} = az + \zeta(t)$$

($a < 0$) through observations of the process $y = z + \eta(t)$, where $\eta(t) = \varepsilon^{-1}\xi_1(t\varepsilon^{-2})$, $\zeta(t) = \varepsilon^{-1}\xi_2(t\varepsilon^{-2})$ and $\xi_1(t)$ and $\xi_2(t)$ obey the theorem conditions. The process $\bar{z}(t)$ defined by

$$\dot{\bar{z}} = a\bar{z} + d\,\mathrm{sign}(y - \bar{z}) \tag{14.62}$$

will be used as the estimate. Let $d = d_0\varepsilon^{-1}$, then the estimation error $x = z - \bar{z}$ will be described by equation

$$\dot{x} = ax - d_0\varepsilon^{-1}\,\mathrm{sign}(x + \varepsilon^{-1}\xi_1(t\varepsilon^{-2})) + \varepsilon^{-1}\xi_2(t\varepsilon^{-2}).$$

In virtue of the theorem, $x(t)$ weakly converges over any finite interval to the diffusion process $x^0(t)$:

$$\dot{x}^0 = (a - d_0 f)x^0 + d_0 h\eta_0 + q\zeta_0,$$

$$h^2 = 2\int_0^\infty E\,\mathrm{sign}\,\xi_1(0)\xi_1(\tau)d\tau, \quad q^2 = 2\int_0^\infty E\xi_2(0)\xi_2(\tau)d\tau,$$

$f = 2P_\xi(0)$, where $P_\xi(x)$ is the density of ξ_1.

For the problem of filtering consider now the Kalman–Bucy filter which for $\varepsilon \to 0$ is as follows:

$$\dot{\bar{z}} = a\bar{z} + l(y - \bar{z}).$$

The filtering error $\bar{x} = z - \bar{z}$ satisfies

$$\dot{\bar{x}} = (a - l)\bar{x} + lr\eta_0 + q\zeta_0,$$

where r is intensity of noise in observations

$$r = 2\int_0^\infty E\xi_1(0)\xi_1(\tau)d\tau.$$

Over large time intervals, p, which is equal to the variance of error $E|\bar{x}|^2$, and l are defined as

$$p = r^2(a + \sqrt{a^2 + q^2/r^2}), \quad l = a + \sqrt{a^2 + q^2/r^2} \tag{14.63}$$

The analogous value for (14.62) has the form of

$$\tilde{p} = \tilde{r}^2(a + \sqrt{a^2 + q^2/\tilde{r}^2}), \tag{14.64}$$

where $\tilde{r} = hf^{-1}$.

Thus, comparison of p and $\tilde{p}$ requires preliminary computation of r and $\tilde{r}$. Let us make this comparison for a Gaussian process $\xi_1(t)$ with correlation function $k(\tau) = E\xi_1(0)\xi(\tau)$ and variance $\sigma^2 = k(0)$. Direct calculations lead to

$$E\,\mathrm{sign}\,\xi_1(0)\xi_1(\tau) = \frac{2}{\pi}\arcsin\frac{k(\tau)}{\sigma^2}, \quad f = \sqrt{\frac{2}{\pi}}\frac{1}{\sigma}.$$

The obvious inequality

$$x \leqq \arcsin x \leqq \frac{\pi}{2}x$$

leads to

$$r \leqq \tilde{r} \leqq \sqrt{\frac{\pi}{2}}r$$

or, according to (14.62), (14.63), to

$$p \leqq \tilde{p} \leqq \frac{\pi}{2}p,$$

the right inequality may be usually made more accurate if a is known, for example, if $a = 0$

$$p \leqq \tilde{p} \leqq \sqrt{\frac{\pi}{2}}p.$$

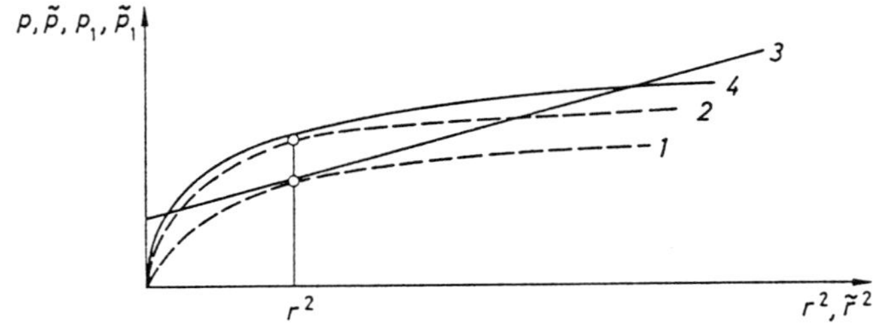

Fig. 21

The filter (14.62), thus gives an error that is close to the optimal one for $\varepsilon \to 0$.

Let us analyze now the filtering error under deviations of noise intensities from the estimates.

If the process $\xi_1(t)$ varies so that its variance becomes σ_1^2 while h^2 and

$$s^2 = 2 \int_0^\infty \frac{k(\tau)}{\sigma^2} d\tau$$

do not change which is equivalent to the multiplication of $\xi_1(t)$ by the constant σ_1/σ and if the stability condition ($a < 0$) is fulfilled, the variance of steady-state errors (that now are not optimal) over a considerable time interval are as follows:

$$p_1 = \frac{l^2 r^2 + q^2}{2(l-a)},$$
$$\tilde{p}_1 = \frac{d_0^2 h^2 + q^2}{2\left(\frac{d_0 s}{\tilde{r}} \sqrt{\frac{2}{\pi}} - a\right)}. \tag{14.65}$$

Approximate profiles ($1 - p(r^2)$, $2 - \tilde{p}(\tilde{r}^2)$, $3 - p_1(r^2)$, $4 - \tilde{p}_1(\tilde{r}^2)$) are plotted in Fig. 21. The line $p(r^2)$ is the lowermost one because it is the minimal error variance provided by the Kalman–Bucy filter for accurate measurement of noise intensity; $\tilde{p}(\tilde{r}^2)$ for (14.62) is similar to $p(r^2)$ and is greater than p but does not exceed $(\pi/2)p$. If one fixes l corresponding to the Kalman–Bucy filter for observation noise intensity r^2, it follows from (14.65) that at another intensity the dependece of variance p_1 on r^2 is a straight line tangent to that of p in the initial point. For the filter (14.62) with $\varepsilon \to 0$, $\tilde{p}_1$ that is similar to p_1 for the stable system ($a < 0$), will be less slanting owing to the adaptation of filter to noise characteristics and for significant deviations of r^2 from the initial value the inequality $\tilde{p}_1 < p_1$ will hold, i.e. under changing noise characteristics the increment of error variance of (14.62) will be smaller than that of the Kalman–Busy filter.

Notably, since the partial adaptivity of (14.62) is obtained automatically without a recourse to special tools used for estimation of noise characteristics, its implementation may be expected to be simpler than those of the existing adaptive filters.

Chapter 15

Sliding Modes in Problems of Mathematical Programming

1 Problem Statement

This chapter addresses the methods for solution of mathematical programming problems that enable one to find the extremum point by generating sliding modes on the boundary of the domain of permissible values of the arguments of the function to be minimized [80–82].

Consideration is given to the common problem of convex programming: determine the minimum point of a scalar function $f(x)$ of vector argument with components $x_1, \dots, x_n$ provided that

$$\begin{aligned} &h_i(x) = 0 \quad (i = 1, \dots, m), \\ &h_i(x) \leqq 0 \quad (i = m+1, \dots, m+l). \end{aligned} \tag{15.1}$$

The minimized function $f(x)$ as well as all the functions $h_i(x)$ defining the permissible domain of vector x are continuously differentiable, $h_i(x)$ $(i = 1, \dots, m)$ are linear, $h_i(x)$ $(i = m+1, \dots, m+1)$ are convex. Then any local extremum x^* is global [60]. Denote by X^* the set of all the solutions x^*; it is desired to determine a point of this set and the value of $f(x^*)$.

This problem may be reduced to determination of the free extremum of the Lagrange function

$$L(x, \lambda) = f(x) + \sum_{i=1}^{m} \lambda_i h_i(x) + \sum_{i=m+1}^{m+l} \lambda_i h_i(x)$$

with respect to $n+m+l$ variables $x, \lambda_1, \ldots, \lambda_{m+l}$ under an additional condition [47] $\lambda_i \geqq 0$ $(i=m+1, \ldots, m+l)$. This approach may prove to be ineffective owing to the high dimensionality of the problem in question under multiple constraints.

Another possibility is to determine the free minimum of an auxiliary function of n (rather than $n+m+l$) variables

$$F(x)=f(x)+H(x), \tag{15.2}$$

where $H(x)=\lambda H_0(x)$ is the penalty function which is zero in the permissible domain and positive in the forbidden one, $\lambda>0$. If function $H_0(x)$ characterizing the penalty for violation of the constraints is continuously differentiable, then according to [47] for the minimum point $x^0(\lambda)$ of $F(x)$

$$\min_x F(x, \lambda)=F^*(\lambda)$$

the following relations hold

$$\lim_{\lambda \to \infty} x^0(\lambda)=x^*, \quad \lim_{\lambda \to \infty} F^*(\lambda)=f(x^*).$$

This implies that under a sufficiently great penalty coefficient one may solve approximately the original problem by determining the minimum of $F(x)$ without any constraint on the permissible values of arguments. If the penalty function is convex, its minimum, for instance, may be found by means of a gradient procedure (it will be recalled that $F(x)$ is continuously differentiable and, therefore, vector $\operatorname{grad} F$ is defined everywhere): $\dot{x}=-\operatorname{grad} F$; then $\dot{y}=-(\operatorname{grad} F)^{\mathrm{T}}\dot{x}=-\|\operatorname{grad} F\|^2$, and function y diminishes to the point $\operatorname{grad} F=0$, i.e. to the minimum point. The requirement that the penalty function be continuously differentiable enables the minimization of $F(x)$ by means of the standard gradient procedure, but at the same time the exact solution of the original problem becomes determinable only asymptotically with coefficient λ tending to infinity. Function $H_0(x)$ over the permissible domain reaches the minimum and is continuously differentiable, therefore $\operatorname{grad} H_0(x^*)=0$ and if point x^* does not coincide with the minimum point of $f(x)$, (i.e. $\operatorname{grad} f(x^*) \neq 0$), then

$$\dot{x}=-\operatorname{grad} f(x^*)-\lambda \operatorname{grad} H_0(x^*) \neq 0.$$

Consequently, the minimum of auxiliary function $F(x)$, generally, is outside the permissible domain and approaches it with the growth of penalty coefficient λ.

In this connection it would be of interest to abandon the continuously differentiable penalty function in favour of a peicewise smooth function with discontinuities of the gradient on the boundaries of permissible domain. If outside the domain $\|\operatorname{grad} H_0\| \geqq \varepsilon_0$ where ε_0 is a positive number, the equilibrium point in the gradient procedure basically cannot be in the forbidden domain under a sufficiently great finite penalty coefficient. The question of what motions occur on the discontinuity boundaries of the gradient of $H_0(x)$ needs, of course, further investigation. Let us take a convex piecewise smooth function in the form of

$$H(x) = h^{\mathrm{T}}(x)u \tag{15.3}$$

where with

$$h^{\mathrm{T}}(x) = (h_1(x), \ldots, h_{m+l}(x)),$$

$$u^{\mathrm{T}} = (u_1, \ldots, u_{m+l}),$$

$$u_i = \begin{cases} \lambda_i & \text{at} \quad h_i > 0, \\ -\lambda_i & \text{at} \quad h_i < 0, \end{cases} \quad i = 1, \ldots, m,$$

$$u_i = \begin{cases} \lambda_i & \text{at} \quad h_i > 0, \\ 0 & \text{at} \quad h_i < 0, \quad \lambda_i > 0, \quad \lambda_i = \text{const.} \end{cases} \quad i = m+1, \ldots, m+l.$$

As before, function $H(x)$ chosen in this fashion characterizes the penalty for violation of the constraints because, according to (15.1), it is equal to zero in the permissible domain and positive in the forbidden one. As it was demonstrated in [170] for penalty functions of the (15.3) type, there exists a positive number λ_0 such that for all $\lambda_i \geqq \lambda_0$ the minimum of $F(x)$ coincides with solution of the original problem. This property allows one to reduce the problem of programming to that without constraints and to determine the exact solution under finite penalty function coefficients without increasing problem's dimensionality (as distinct from the Lagrange function-based algorithms). It may be in order to draw attention here to the fact that $F(x)$ is convex because it suffices to determine at least one local minimum.

Obviously, such a change of problems requires, first, estimation of λ_0 and, second, organization of a procedure for determination of the minimum of piecewise smooth function $F(x)$. The solution of the first problem enables one to choose the coefficients λ_i of penalty function (15.3). Issues specific to the determination of the minimum of $F(x)$ may be easily revealed if the gradient procedure is formally written for all the points where the gradient of this function is defined (i.e. outside surfaces $h_i(x) = 0$):

$$\dot{x} = -\operatorname{grad} f - G^{\mathrm{T}}u, \tag{15.4}$$

where x and $\operatorname{grad} f$ are n-dimensional column vectors, $G^{\mathrm{T}} = (g^1, \ldots, g^{m+l})$, columns $g^i = \operatorname{grad} h_i$. The gradient of $F(x)$ will have discontinuities on surfaces $h_i(x) = 0$; that is why they will be referred to in what follows as *discontinuity surfaces*. Outside the discontinuity surfaces of space $x_1, \ldots, x_n$ motion goes on according to the antigradient algorithm (15.4), and as a result $F(x)$ is locally diminishing. If the state trajectory did not intersect the discontinuity surfaces or if the set of intersection points had the zero measure, this motion would end, in virtue of the convexity of $F(x)$, at some stationary point that is the minimum point of F, which solves the problem under investigation. The points where the right side of (15.4) vanishes will be referred to below as *stationary points*.

In dynamic systems described by differential equations with discontinuous right-hand sides (such as (15.4)), sliding modes may occur during which the trajectories lie on the intersection of discontinuity surfaces. In particular, if for some surface $h_i(x)=0$ conditions (1.9) are satisfied

$$\lim_{h_i \to 0} h_i(x) < 0 \quad \text{and} \quad \lim_{h_i \to -0} (x) > 0$$

sliding mode occurs on this surface, i.e. the set of points where the penalty function gradient undergoes discontinuities is not a zero measure set (with respect to t). That is why the behaviour of (15.4) in sliding moded deserves further consideration. This is the specificity of the problem requiring substantiation of the convergence to extremum for the gradient procedure[1].

2 Motion Equations and Necessary Existence Conditions for Sliding Mode

Let us write relations that will be required later in the behaviour studies of dynamic systems of the (15.4) type. Assume that sliding mode occurs at the intersection of (not necessarily all) discontinuity surfaces $h_i(x)=0$, and let functions $h_i(x)$ that define them constitute vector h' and functions having discontinuities on them constitute vector u'. The rest of the components of vector h are non-zero and constitute vector h'', and, correspondingly, the rest of vector u components assume either of the two possible values and constitute vector u''.

Let us describe a procedure that enables us to form sliding mode motion equation according to the equivalent control method of Sect. 2.3. The time derivative of h' on the trajectories of (15.4) should be equated to zero:

$$\dot{h}' = -G' \operatorname{grad} f - G'(G')^{\mathrm{T}}u' - G'(G'')^{\mathrm{T}}u'' = 0, \tag{15.5}$$

where G' and G'' are matrices whose rows are the gradients of functions constituting vectors h' and h″. The equation (15.5) must be solved with respect to u', and its solution which is the equivalent control u'_{eq} must be substituted into (15.4) for u'. As a result, obtain equation of sliding mode motion. In particular, if matrix G' has maximal rank or $\det(G'(G')^{\mathrm{T}}) \neq 0$, equivalent control and sliding mode equation are as follows

$$\begin{aligned} u'_{\mathrm{eq}} &= -(G'(G')^{\mathrm{T}})^{-1}(G' \operatorname{grad} f + G'(G'')^{\mathrm{T}}u''), \\ \dot{x} &= -P(\operatorname{grad} f + (G'')^{\mathrm{T}}u''), \\ P &= I_r - (G')^{\mathrm{T}}(G'(G')^{\mathrm{T}})^{-1}G'. \end{aligned}$$

---

[1] It should be noted that for piecewise smooth functions the authors of [45, 159] studied the convergence of gradient minimization procedures by postulating motion equations while the physical substantiation of this description of the dynamic system behaviour on the discontinuity boundary was let open.

If $\det(G'(G')^{\mathrm{T}})=0$, (15.5) has multiple solutions with respect to u'. This situation corresponds to the third type of singular cases discussed in Sect. 3.3 where sliding equations may be unambiguously determined. (15.4) is representable as (3.16) if $B'=(\bar{G}')^{\mathrm{T}}$, $C'=\bar{G}'$, and $(\bar{G}')^{\mathrm{T}}$ is matrix composed of the base column vectors of matrix $(G')^{\mathrm{T}}$. Although u'_{eq} is not unique, sliding equations are determined unambiguously

$$\dot{x}=-P(\operatorname{grad} f+(G'')^{\mathrm{T}}u''), \tag{15.6}$$

where

$$P=I_r-(\bar{G}')^{\mathrm{T}}(\bar{G}'(\bar{G}')^{\mathrm{T}})^{-1}\bar{G}'.$$

In addition to the sliding mode equation, we need its existence condition. As applied to (15.4), this condition (see Sect. 4.1) is formulated as follows: if sliding mode arises on manifold $h'(x)=0$, there will be at least one equivalent control u'_{eq} such that all of its components satisfy conditions

$$\begin{aligned} -\lambda_i \leqq u_{i\mathrm{eq}} \leqq \lambda_i \quad &\text{for} \quad i\leqq m,\\ 0\leqq u_{i\mathrm{eq}} \leqq \lambda_i \quad &\text{for} \quad i>m. \end{aligned} \tag{15.7}$$

3 Gradient Procedures for Piecewise Smooth Function

Let us consider the behaviour of dynamic system (15.4). Since outside the surfaces $h_i(x)=0$ the right-hand side of (15.4) is the antigradient of $F(x)$ (15.2), (15.3), this function is decreasing in time everywhere with the exception of discontinuity surfaces. The equation (15.6) enables also investigation of the behaviour of (15.4) when the state vector moves in the space $(x_1,\ldots,x_n)$ over various intersections of discontinuity surfaces in sliding mode. This section is addressed to the properties of the gradient procedure (15.4) taking into account the possible occurrence of such motions.

Property 1. *$F(x)$ is increasing time function*
Let sliding mode motion occur along some intersection of discontinuity surfaces, and let the rest of vector $h(x)$ components that constitute vector $h''(x)$ be non-zero. According to (15.2) and (15.3), $F(x)$ has during this motion the following form

$$F(x)=f+(h'')^{\mathrm{T}}u''.$$

Since $f(x)$ and $h''(x)$ are continuously differentiable and $u''=\text{const}$,

$$\dot{F}=(\operatorname{grad} f+(G'')^{\mathrm{T}}u'')^{\mathrm{T}}\dot{x}. \tag{15.8}$$

Matrix P in (15.6) is symmetrical, and $P = P^2 = P^{\mathrm{T}}P$ [1]. By substituting matrix $P^{\mathrm{T}}P$ for P in (15.6) and then derivative $\dot{x}$ into (15.8), obtain

$$\dot{F} = - \| P(\operatorname{grad} f + (G'')^{\mathrm{T}}u'' \|^2 \leqq 0, \tag{15.9}$$

i.e. $F(x)$ is a decreasing function.

Property 2. *Time derivative of $F(x)$ is zero only in stationary points* (it will be recalled that in these points $\dot{x} = 0$). This statement directly follows from the comparison of (15.6) and (15.9).

Property 3. *The set of stationary points coincides with the set X_F^* of minimum points x_F^* of function $F(x)$.*

Let x^0 be a stationary point of sliding mode motion on manifold $h'(x) = 0$. In compliance with the procedure for derivation of sliding mode equation (Sect. 2), u'_{eq} should be substituted into (15.4) for u', and, therefore, in the stationary point x^0 the following relation holds

$$-\operatorname{grad} f(x^0) - (G'(x^0))^{\mathrm{T}}u_{\mathrm{eq}}(x^0) - (G''(x_0))^{\mathrm{T}}u''(x^0). \tag{15.10}$$

The Eq. (15.5) from which u'_{eq} is found may have multiple solutions, but in spite of this fact, the sliding mode motion Eq. (15.6) is determined unambiguously. Hence, the study of this motion may be based on any function u'_{eq}. It was also noted in Sect. 2 that among all the functions u'_{eq} there will be at least one whose components satisfy (15.7). Later on we shall make use of namely this equivalent control. Since the components of h'' are non-zero, vector $u''(x^0)$ is unambiguously determined from (15.3).

Consider function

$$\bar{F}(x) = f(x) + (h')^{\mathrm{T}}u'_{\mathrm{eq}}(x^0) + (h'')^{\mathrm{T}}u''.$$

The components of vector $u'_{\mathrm{eq}}(x^0)$ related to the constraints of inequality type are positive in virtue of (15.7), and the functions defining constraints of the equality type are linear. Consequently, in spite of the fact that some components of the vector $u'_{\mathrm{eq}}(x^0)$ may be negative, $\bar{F}(x)$ is convex and has a unique point of minimum. Since in the point x^0 the gradient of $\bar{F}(x)$ is continuous and equal to zero, $\min_x \bar{F}(x) = \bar{F}(x^0)$. It follows from (15.7) that $F(x) \geqq \bar{F}(x)$; since $F(x^0) = \bar{F}(x^0)$, $F(x)$ also reaches its minimum in x^0, and any stationary point of procedure (15.4) is the minimum point of a piecewise smooth function $F(x)$. The opposite statement is proved in an obvious manner: if $\min_x F(x) = F(x_F^*)$ and x_F^* is not the stationary point of (15.4), in virtue of Property 2 this function will decrease in point x_F^* and assume values smaller than $F(x_F^*)$.

---

[1] It is relevant to note here that matrix P is obviously singular because $P(G')^{\mathrm{T}} = 0$.

Property 4. *If domain X_F^* of all the minimum points x_F^* of $F(x)$ is bounded, the gradient procedure (14.4) provides convergence to this domain from any initial position $x(t_0)$.*

Convergence will be understood in the following sense:

$$\lim_{t\to\infty} r^*(x) = 0, \quad r^*(x) = \min_{X_F^*} \| x - x_F^* \| \tag{15.11}$$

and, as a consequence, $\lim_{t\to\infty} F(x) = F^* = F(x_F^*)$. It follows from the continuity of $F(x)$ that the bounded domain X_F^* is closed, and, therefore, function $r^*(x)$ always exists.

This property is proved by the same reasoning as the Barbashin–Krasovsky theorem on symptotic stability in the whole [9]. But at first let us prove that if $F_0 = F(x(t_0))$, the closed set $X_0\{x: F(x) \leqq F_0\}$ is bounded. To this end, consider set of points X_1 on a sphere with the center in point x_F^* and radius R such that $X_1\{x: \| x_F^* - x \| = R\}$, $\min_{X_1} F(x) - F^* = \Delta F > 0$. Existence of this sphere follows from the boundedness of domain X_F^*. The following inequality holds for convex functions

$$\frac{F_0 - F^*}{\| x(t_0) - x_F^* \|} R > F\left(x_F^* + \frac{x(t_0) - x_F^*}{\| x(t_0) - x_F^* \|} R \right) - F^* \geqq \Delta F > 0$$

at $\| x(t_0) - x_F^* \| \geqq R$. Had X_0 been unbounded, there would have been in it a sequence of points $x(t_0), x_{01}, x_{02}, \ldots$ tending to infinity with value of function $F(x_{0i}) < F_0$. The left side of the inequality, then, should tend to zero and at the same time be greater than positive number ΔF. This contradicts our assumption that domain X_0 is unbounded.

In addition to the bounded closed domain X_0, introduce the following domains: S_ε is ε-hull of X_F^* and for each of its points $\tau^*(x) = \varepsilon$; X^ε is the set of points x for which $r^*(x) < \varepsilon$; X_F^ε is bounded open set of points for which $F(x) < F_\varepsilon = \min_{S_\varepsilon} F(x)$. Let us assume that ε was chosen sufficiently small and $X_F^\varepsilon \subset X_0$. Consider the behaviour of the dynamic system (15.4) in the bounded closed domain $X_0 \backslash X_F^\varepsilon$. The right-hand sides of (15.4) and, consequently, of (15.8) and (15.9) that define $\dot{F}$ are bounded at any x for functions f, h_i and H under study. Indeed, the components of vector u assume either of the two possible values if there is no sliding mode on the corresponding discontinuity surface and, according to (15.7), an intermediate values should sliding mode occur. (It will be recalled that the sliding equation (15.6) is independent of the equivalent control chosen, and at least one of the controls satisfies (15.7)). On the other hand, there is no minimum or stationary points in the domain $X_0 \backslash X_F^\varepsilon$ (Property 3), and outside the stationary points $F(x)$ is always decreasing (Properties 1 and 2). Thus, function $\dot{F}$ in the closed bounded domain $X_0 \backslash X_F^\varepsilon$ is always negative and reaches its upper bound, i.e. there exists a positive number m_0 such that in this domain

$$\dot{F}(x) \leqq -m_0. \tag{15.12}$$

It follows from (15.12) that a trajectory going out of point $x(t_0)$ will not leave domain X_0 and within a finite time will reach domain X_F^ε and stay there (otherwise, $F(x)$ becomes less than F^* within finite time). Since $S_F^\varepsilon \subset X^\varepsilon$ and ε may be chosen arbitrarily small, $\lim_{t\to\infty} r^*(x) = 0$ by definition of the limit. It follows from the continuity of $F(x)$ that

$$\lim_{r^*(x)\to 0} F(x) = F^* \quad \text{or} \quad \lim_{t\to\infty} F(x) = F^*.$$

Property 4 is proved.

Attention is drawn to the fact that proved is not only the convergence to set X_F^*, but also its asymptotic stability in the sense that for any its ε-vicinity (domain X^ε may be such a vicinity) there will be δ-vicinity (X_F^ε in our case) such that any motion beginning from δ-vicinity will not leave ε-vicinity and, moreover, $\lim_{t\to\infty} r^*(x) = 0$.

4 Conditions for Penalty Function Existence. Convergence of Gradient Procedure

In order to substantiate the possibility of solving the problem of convex programming by means of piecewise smooth penalty functions, let us make use of the properties of gradient procedures as described in the last section. As a preliminary, introduce a set of $(m+l)$-dimensional column vectors U with components U_i:

$$-\lambda_i \leqq U_i \leqq \lambda_i, \quad i = 1, \ldots, m, \qquad (15.13)$$

$$0 \leqq U_i \leqq \lambda_i$$

for essential constraints, i.e. for $i > m$ and $h_i(x^*) = 0$ ($U_i = 0$ for unessential constraints, i.e. for $i > m$ and $h_i(x^*) < 0$).

Theorem 1. *For the minimum of function $F(x)$ (14.2), (14.3) to coincide with the value of $f(x^*)$, it is necessary and sufficient that condition*

$$\operatorname{grad} f(x^*) = -G^{\mathrm{T}} U^0 \qquad (15.14)$$

be met at least in one point x^* where U^0 is a vector from the set (15.13).

Sufficiency. Condition (15.14) implies that the gradient of a convex function of the $\bar{F}(x)$ type is zero in point x^*, and, consequently, x^* is the minimum point of $\bar{F}(x)$. Since set X^* of points x^* lies within the permissible domain where penalty function $H(x)$ is zero, and $F = \bar{F} = f$,

$$\min_x F(x) = f(x^*) \qquad \text{Q.E.D.}$$

Necessity. Independently of whether sliding mode occurs or not, all the components of vector u in dynamic system (15.4) satisfy (15.7). This suggests that if vector grad f is not representable in the form of (15.14) in any point of X^*, the right-hand side of (15.4) is non-zero on X^*, and there is no stationary point in this set. According to Property 2, $\dot{F}(x^*)$ is strictly negative in any point x^*. Since $F(x^*) = f(x^*)$ and $\dot{F}(x^*) < 0$, function $F(x)$ becomes smaller than $f(x^*)$.

Q.E.D.

Corollary. *If (15.14) is satisfied at least in one point x^* of domain X^*, it is satisfied in any other point of this set.*

Assume that in point x^* vector grad f is representable in the form of (15.14), and in point $x^{**} \in X^*$ it is not representable. Then $F(x)$ reaches in x^* minimum equal to $f(x^*)$, and in non-stationary point x^{**} decreases and becomes smaller than $F(x^{**}) = f(x^{**})$. But $f(x) = f(x^*) = \min_x F(x)$, therefore, the assumption that (15.14) is violated in some point of the domain X^* is not true.

Theorem 2. *Procedure (15.4) provides convergence to set X^* in the sense of (15.11) if this set is bounded, condition (15.14) is satisfied at least in one of the points x^*, and vector U_0 is inner point of set U. (Dimensionality of this set is equal to the sum of the number of equality-type constraints and the number of essential inequality-type constraints.)*

Condition (15.14) is satisfied, obviously, if the conditions of Theorem 2 are satisfied, i.e. according to Theorem 1

$$\min_x F(x) = F(x^*) = f(x^*).$$

Represent $F(x)$ as

$$F(x) = F_0(x) + \Delta F(x) \tag{15.15}$$

where $F_0 = f + h^T U^0$, $\Delta F = h^T(u - U^0)$. The components of vector U^0 satisfy conditions (15.13), and function $F_0(x)$ is, therefore, convex like $F(x)$, but unlike it is continuously differentiable. The gradient of $F_0(x)$ in point x^* is zero according to the condition of Theorem 2 and, therefore,

$$\min_x F_0(x) = F_0(x^*). \tag{15.16}$$

If the conditions of Theorem 2 are satisfied, function ΔF is strictly positive in the forbidden domain and zero in the permissible one, therefore, $F_0(x^*) = F(x^*)$ and, according to (15.15), (15.16), for the points of forbidden domain the following inequality holds

$$F(x) > F_0(x) \geqq F(x^*) = f(x^*). \tag{15.17}$$

This means that domain X_F^* of the minimum points of $F(x)$ concides with domain X^*. Since in the cases under study domain X^* is bounded, the validity of Theorem 2 follows from Property 4.

Remark. Since the minimum points of $F(x)$ are stationary for (15.14),

$$\operatorname{grad} f + (G')^{\mathrm{T}} u'_{\mathrm{eq}} + (G'')^{\mathrm{T}} u'' = 0 \tag{15.18}$$

in these points.

Like in Sect. 2, matrix $(G')^{\mathrm{T}}$ consists of the gradients to the surfaces on whose intersection sliding mode exists in the point under consideration. It follows from (15.18) that the components of vectors u'_{eq} and u'' are the values of dual variables in the optimal point. If the components of vector u'' are directly measurable, one may use for determination of u'_{eq} according to Sect. 2.4 the method of averaging by means of a filter described by equation $\tau \dot{u}_{\mathrm{av}} + u_{\mathrm{av}} = u'$, where τ is scalar parameter, and output vector u_{av} is related to u'_{eq} by

$$\lim_{\substack{\tau \to 0, \Delta \to 0 \\ \Delta/\tau \to 0}} u_{\mathrm{av}} = u'_{\mathrm{eq}},$$

where Δ defines the amplitude of oscillations in the boundary layer of the intersection of discontinuity surfaces during sliding mode and, in its turn, depends of the imperfections of switching devices. As the result, dual variables may be directly measured.

5 Design of Piecewise Smooth Penalty Function

If the conditions of theorems from Sect. 4 are satisfied, the procedure (15.4) provides convergence to X^* which solves the non-linear programming program under consideration. However, it is difficult to verify these conditions because point coordinates x^* are not known in advance. Therefore, as it is usually done in similar situations we shall try to determine conditions that may be stronger than in the theorem, but do not require for verification the knowledge of the coordinates of a point of X^*.

Let us demonstrate as a preliminary that if the permissible domain of x is not empty, vector $G^{\mathrm{T}}u$ is always non-zero in the forbidden domain. Notably, the components u_i of vector u in $G^{\mathrm{T}}u$ may assume either of the two possible values if the distance to the corresponding discontinuity surface differs from zero, or be equal to $u_{i\mathrm{eq}}$ if sliding mode has occurred on the corresponding surface:

$$G^{\mathrm{T}}u = (G')^{\mathrm{T}} u'_{\mathrm{eq}} + (G'')^{\mathrm{T}} u''.$$

If one assumes that this vector is equal to zero in some point of the forbidden domain, from the reasoning used for the proof of Property 3 in Sect. 3 it follows that the convex function $H(x)$ reaches the minimal value in this point. However, this contradicts the definition of penalty function which is non-zero outside the permissible domain and zero on its boundaries.

The norm of vector $G^{\mathrm{T}}u$ depends on the gradients $g^i(x)$ of functions $h_i(x)$ and the penalty function coefficients λ_i. Assume that $\lambda_0 = \min_i \lambda_i$ and that the following estimate holds

$$\| G^{\mathrm{T}}u \| \geqq \lambda_0 g_0, \tag{15.19}$$

where g_0 is a positive number that will be regarded as known. Moreover, assume that the norm of vector grad f is upper-bounded by a number f_0 that is known as well. Let us prove that (15.14) in this case makes possible solution of the original problem of convex programming if the coefficients of penalty function are as follows:

$$\lambda_0 = \min_i \lambda_i > \frac{f_0}{g_0}. \tag{15.20}$$

Demonstrate at first that if (15.20) is satisfied, the state vector always goes out of the forbidden domain. To this end, determine the time derivative of $H(x)$ defined by (15.4):

$$\frac{dH}{dt} = u^{\mathrm{T}}G(-\operatorname{grad} f - G^{\mathrm{T}}u) + \sum_{i=1}^{m+l} h_i(x)\frac{du_i}{dt}. \tag{15.21}$$

If during this motion some functions $h_i(x)$ differ from zero, $u_i = \text{const}$ and $du_i/dt = 0$ according to (15.3); at the occurrence of sliding mode at the intersection of discontinuity surfaces, values du_i/dt that are equal to du_{eq}/dt are, generally, different from zero, but their corresponding functions $h_i(x)$ are equal to zero. This means that all the products $h_i(x)du_i/dt$ are equal to zero, i.e. the last term in (15.21) may be discarded. Function dH/dt may be then estimated by means of conditions (15.19), (15.20):

$$\frac{dH}{dt} \leqq \| G^{\mathrm{T}}u \| (f_0 - \| G^{\mathrm{T}}u \|) \leqq -\lambda_0 g_0 f_0 \left(\frac{g_0}{f_0}\lambda_0 - 1 \right) < 0.$$

Since $H(x) \geqq 0$, it follows from the estimate that this function will be zero within a finite time, i.e. that further motion will take place in the permissible domain where $F(x) = f(x)$ and (15.4) enables determination of the minimum of $F(x)$. Consequently, this procedure at the same time solves the original problem.

6 Linearly Independent Constraints

In this section we also discuss the sufficient conditions which must be satisfied by the penalty function coefficients if the gradients $g^i(x)$ of functions $h_i(x)$ are linearly independent. In order to determine these conditions, we shall take advantage of the methods of scalar control hierarchy and diagonalization described in Sect. 4.5.

Control Hierarchy Method. According to the idea underlying this method, the first component u_1 of vector u is chosen so as to make the state vector leave the domain forbidden by the first constraint independently of the values assumed by the rest of the components. Further motion will proceed either on surface $h_1(x)=0$ in sliding mode, or if the first constraint is inequality, in domain $h_i(x)\leqq 0$. The second component is chosen so as to provide motion towards the domain permitted by the second constraint independently of the values of components $u_3,\dots,u_{m+l}$, etc. As a result, in a control hierarchy organized in this manner the state vector will get into the permissible domain where the minimum of $f(x)$ is located. Let us present relations that will be necessary for implementation of this procedure.

Let Γ^{k-1} be a subspace orthogonal to the subspace spanned on vectors $g^1,\dots,g^{k-1}(2\leqq k\leqq m+l)$. Vectors $g^1,\dots,g^{m+l}$ are linearly independent, i.e. $m+l\leqq n$, therefore, the dimensionality of this subspace is one at least. Denoted by r^k a vector equal to $P_{k-1}g^k$ if $P_{k-1}=I_n-G_{k-1}^{\mathrm{T}}(G_{k-1}G_{k-1}^{\mathrm{T}})^{-1}G_{k-1}$ (it follows from the linear independence of vectors $g^1,\dots,g^{m+l}$ that matrix $(G_{k-1}G_{k-1}^{\mathrm{T}})^{-1}$ exists because $G_{k-1}^{\mathrm{T}}=(g^1,\dots,g^{k-1})$). Vector r^k is the projection of g^k on subspace Γ^{k-1} because $g^k=r^k+G_{k-1}^{\mathrm{T}}(G_{k-1}G_{k-1}^{-1})^{\mathrm{T}}G_{k-1}g^k$ and $G_{k-1}r^k=0$, i.e. $r^k\in\Gamma^{k-1}$ and vector $G_{k-1}^{\mathrm{T}}(G_{k-1}G_{k-1}^{\mathrm{T}})^{-1}G_{k-1}g_k$ belongs to the subspace spanned on vectors $g^1,\dots,g^{k-1}$. It follows from the linear independence of vectors $g^1,\dots,g^k$ that $\|r^k\|>0$.

Note also that if $\bar{r}^k$ is the projection of vector g^k on the subspace orthogonal to the subspace spanned not on all the vectors $g^1,\dots,g^{k-1}$, but on a part of them, then $\|\bar{r}^k\|>\|r^k\|$ or

$$\|\bar{P}_{k-1}g^k\|>\|P_{k-1}g^k\|, \tag{15.22}$$

where $\bar{P}_{k-1}=I_n-\bar{G}_{k-1}^{T}(\bar{G}_{k-1}\bar{G}_{k-1}^{\mathrm{T}})^{-1}\bar{G}_{k-1}$, and matrix $\bar{G}_{k-1}^{\mathrm{T}}$ consists of a part of the columns of G_{k-1}^{T}. Finally, the length of projection of any vector a on subspace Γ^{k-1} does not exceed the length of the vector itself, i.e.

$$\|a\|\geqq\|P_{k-1}a\|. \tag{15.23}$$

Thus, in the case where vectors $g^1,\dots,g^{m+l}$ are linearly independent, the norm of any vector r^k differs from zero. Assume that the lower bound of these values is known:

$$\|r^k\|\geqq r_0,\quad r_0=\text{const},\quad k=1,\dots,m+l \tag{15.24}$$

and, moreover, the upper bounds of vectors g^k and $\operatorname{grad} f$ are known:

$$\begin{aligned}&\|g^k\|\leqq R_0,\quad k=1,\dots,m+l,\\&\|\operatorname{grad} f\|\leqq f_0,\quad R_0, f_0-\text{const}.\end{aligned} \tag{15.25}$$

After these preliminary remarks and assumptions, proceed now directly to the penalty function design. Let state vector be in the forbidden domain. Define then the minimal number k for which the constraint is not satisfied, i.e. control u_k is either λ_k or $-\lambda_k$, and u_i for $i<k$ are zero or $u_{i\mathrm{eq}}$ if sliding mode occurs. Motion equation in this case can be obtained from (15.4) by means of the

equivalent control method

$$\dot{x} = \bar{P}_{k-1}\left(-g^k u_k - \operatorname{grad} f - \sum_{i=k+1}^{m+l} g^i u_i\right). \tag{15.26}$$

In this equation $P_0 = I_n$ (if $k = 1$), the gradients g^i of discontinuity surfaces with numbers $i < k$ along which sliding mode motion takes place are the columns of matrix $\bar{G}_{k-1}^{\mathrm{T}}$ from which $\bar{P}_{k-1}$ is derived, and, finally, u_i for $i \geqq k+1$ are equal to either of the two possible values if the corresponding $h_i(x)$ is not equal to zero or to $u_{i\mathrm{eq}}$ at the occurrence of sliding mode on surfaces $h_i(x) = 0$.

Determine the time derivative of $h_k(x)$ by means of (15.26)

$$\dot{h}_k = -(g^k)^{\mathrm{T}} \bar{P}_{k-1} g^k u_k - (g^k)^{\mathrm{T}} \bar{P}_{k-1} \operatorname{grad} f - \sum_{i=k+1}^{m+l} (g^k)^{\mathrm{T}} \bar{P}_{k-1} g^i u_i.$$

It follows from (15.22) through (15.25) and also from $\bar{P}_{k-1} = \bar{P}_{k-1}^{\mathrm{T}} \bar{P}_{k-1}$ and $|u_{i\mathrm{eq}}| \leqq \lambda_i$ that

$$\dot{h}_k = -\|\bar{r}_k\| u_k + z_k,$$

where

$$|z_k| \leqq R_0 f_0 + R_0^2 \sum_{i=k+1}^{m+l} \lambda_i.$$

Since at the zero time the k-th constraint is violated, u_k is equal to $\pm\lambda_i$ if the constraint is equality, or to λ_k if it is inequality. According to (15.22), (15.23), (15.24), if condition

$$\lambda_k > \frac{R_0}{r_0^2}\left(f_0 + R_0 \sum_{i=k+1}^{m+l} \lambda_i\right) \tag{15.27}$$

is satisfied, h_k and $\dot{h}_k$ have in both cases different signs, the rate of $\dot{h}_k$ differing from zero by a finite value. Consequently, if the constraints with numbers from 1 to $k-1$ are not violated, function $h_k(x)$ will vanish within a finite time, and the state vector will enter the domain permitted by k-th constraint and not leave it later on. If constraint (15.27) is satisfied for all k from 1 to $m+l$, all the constraints will be successively satisfied whereupon the motion will be in the permissible domain where $F(x) = f(x)$. Since procedure (15.4) finds the minimum of $F(x)$, the original problem may be solved if the coefficients λ_k of penalty function (15.3) are chosen according to (15.27).

Importantly, if estimates (15.24) and (15.25) are known, design resting upon the control hierarchy method should be started from its "termination": first, λ_{m+l} is taken greater than $R_0 f_0/\tau_0^2$, next λ_{m+l-1} depending on the found λ_{m+l}, etc. up to λ_1. A remark would be in order that the penalty function generated in this fashion enables also non-convex functions f and h_i to get into the permissible domain, i.e. the penalty function may have local extreme only if the minimized function has them in the permissible domain.

Diagonalization Method. Consideration is given to a special case of the programming problem where the constraints are only equalitities $h_i(x) = 0$ ($i = 1, \ldots, m$). The method consists in selecting such a procedure of varying the vector x that vector u in the equation for vector h would be multiplied by a diagonal (e.g. identity) matrix. In the resulting system, the state vector may be brought to the permissible domain by satisfying the constraints simultaneously (rather than successively as in the control hierarchy method) because each control component affects the rate of distance variation only up to the surface where it undergoes discontinuity.

This approach may be realized if, instead of (15.4), the following algorithm is used:

$$\dot{x} = -\operatorname{grad} f - G^{\mathrm{T}}\bar{u}, \quad \bar{u} = (GG^{\mathrm{T}})^{-1}u, \tag{15.28}$$

where G^{T} is matrix with columns $g^1, \ldots, g^m$; existence of matrix $(GG^{\mathrm{T}})^{-1}$ follows from the linear independence of these columns; u is m-dimensional vector with components $u_1, \ldots, u_m$ undergoing discontinuities on surfaces $h_i(x) = 0$ according to (15.3). By determining the derivative of vector h for algorithm (15.28), obtain $\dot{h} = -G \operatorname{grad} f - u$. Assume that the norms of vectors g^i and $\operatorname{grad} f$ are limited by numbers g_i and f_0, respectively, that will be regarded as known. Then, if the coefficients λ_i are chosen according to inequalities $\lambda_i > g_i f_0$, $i = 1, \ldots, m$, then h_i and their rates $\dot{h}_i$ will simultaneously have different signs and functions $h_i(x)$ will become zero within a finite time.

On each of surfaces $h_i(x) = 0$ sliding mode conditions are satisfied, hence, a stable "in the small" sliding mode will be induced on manifold $h(x) = 0$ which will result in the fulfilment of constraints. By means of the equivalent control method one can determine that in the sliding mode $\dot{x} = -P \operatorname{grad} f$, $P = I_n - G^{\mathrm{T}}(GG^{\mathrm{T}})^{-1}G$, $\dot{f} = -\|P \operatorname{grad} f\|^2$, i.e. that $f(x)$ is decreasing everywhere but in the stationary points where $\operatorname{grad} f$ is a linear combination of vectors $g^1, \ldots, g^m$. According to the Kuhn–Tucker theorem, the stationary point is solution of the convex programming problem with linearly independent constraints [47]. Convergence of (15.28) to the solution of the original problem may be proved by following the reasoning of Sect. 4.

Interestingly, this method does not require a penalty function. We have found a procedure that allows the vector to get into the permissible domain and then steers it to the desired solution. Such an approach was used for solution of programming problems in [81, 82]. In the method of diagonalization as well as in that of hierarchy of control, there is no need that all the functions be convex in order to get into the permissible domain.

Remark. It is appropriate to call attention to the fact that the optimization methods described in this chapter may be used to solve of the system of algebraic equations $h(x) = 0$ where $h^{\mathrm{T}} = (h, \ldots, h_n)$, $x^{\mathrm{T}} = (x_1, \ldots, x_n)$. By considering the positive definite function $H(x)$ (15.3) at $m = n$, $l = 0$ as the function to be minimized, one becomes convinced that a gradient procedure $\dot{x} = -G^{\mathrm{T}}u$ similar to (15.3), (15.4) leads to local decrease of $H(x)$. If the gradients of $h_i(x)$ are linearly independent, the above procedures enable determination of the minimum of $H(x)$. Since in the point of minimum $H(x) = 0$ and $h(x) = 0$, the desired solution of the system of algebraic equations is determined as the result.

Part III. Applications

This part discusses the applied aspects of the design of control algorithms with sliding modes. The classes of plants for which the algorithms might prove effecient are brought out, and some examples of using the algorithms for automatic control in various areas (machines, transport, processes, etc.) are presented.

The experience that has been gained in the applications of control systems relying on sliding modes is a convincing evidence in favour of their effectiveness and universality. This point is also supported by the diversity of controlled plants in terms of their physical nature, types of operators, functional tasks of the control systems. Implementation of the sliding mode algorithms by means of the most common electrical and pneumatic components and actuators has turned out to be simple enough. Notably, for the currently popular electrical actuators with controlled thyristor or transistor power converters, sliding modes are the only possible mode of operation.

It should be stressed that since the time of the relay controllers an opinion has been in existence that it is difficult to realize sliding modes in the discontinuous control systems and that the motion in self-oscillatory mode with finite switching frequency, on the one hand, does not enable high accuracy and the desired dynamic performance and, on the other hand, leads to higher wear of the mechanical parts and greater heat losses in the electrical circuits.

Let us dwell upon the basic reasons that gave rise to this viewpoint. First of all they were due to the inadequacy of switching devices whose delays, lags and hysteresis prevented the attainment of a sufficiently high switching frequency in self-oscillatory modes. Another reason is that the range of structures inducing sliding modes is rather limited even with the ideal switching devices. For instance, at the first stage of the variable structure system theory (where control was formed through the error coordinate and its time derivatives of various orders) [130], the attempts to use sliding modes in practice usually failed because the ideal differentiation is physically non-realizable, and the physical differentiating filters always have polynomials in the denominator of the transfer function. Sliding modes were destroyed namely by these minor lags of filters disregarded in the model. The same effects were caused by neglecting the small time constants of actuators and sensors.

In spite of their seeming convincingness, the doubts as to the realizability of sliding modes has been completely disproved by the technological advances in electronics technology and the advent of new design methods for discontinuous control systems.

The commercially available electronic components enable one to gate powers of several tens kilowatts with frequencies of the order of 10 Khz. It should be readily apparent that these elements enable realization of practically ideal devices forming discontinuous control actions for the majority of controlled plants. It is appropriate to note here that the electronic power converters are inertialess and find increasing use in fast control systems. When using the converters of this type, it seems reasonable to turn to the algorithms with discontinuous control actions, since only the key-mode operation is admissible for them.

As for the doubts about realizability of sliding modes in the presence of discrepancies between the plant operator and the model, they are completely eliminated by the systems with asymptotic observers (Sect. 14.3). While the traditional methods providing the desired dynamic performance by means of dynamic correction filters (including differentiators) lead to self-oscillatory modes, the systems with observers preseve sliding modes. As the analysis carried out in Sect. 14.3 for systems with small dynamic discrepancies between the model and physical system has revealed, in this case only sliding equations are subject to minor variations.

Therefore, from the engineering standpoint implementation of sliding modes is not a problem which is confirmed by their successful practicalization in various applications [150].

Chapter 16

Manipulator Control System

The dynamical properties and control of manipulator or robot arm have been extensively studied [11, 101, 112, 125]. The dynamics of a six-degrees-of-freedom (six joints) manipulator is described by six coupled second-order non-linear differential equations. Physically, the coupling terms represent gravitational torques that depend on link positions, reaction torques due to accelerations, Coriolis and centrifugal torques. The degree of this interaction depends on manipulator's physical parameters such as weight and size of the link, and the weight it carries.

The existing control algorithms are based on non-linear compensations of interactions [101, 112, 125]. Such approaches suffer from the requirement of a detailed manipulator model and load forecasting. Generally, these non-linear compensators are complex and costly in implementation. The control algorithms described below avoid such difficulties by applying discontinuous control algorithms. It is suggested to design manipulator control system in accordance with our basic principle – the use of sliding motions that may occur on discontinuity surfaces in the system state space. The design will be carried out in the space of canonical variables because sliding mode motion in this space is insensitive to parameter variations and disturbances. It is this insensitivity that enables elimination of interactions between various manipulator joints.

The design of discontinuous control system which induces sliding mode can be done without an accurate model. It is sufficient to know only the bounds of plant parameter variations. It is this feature of discontinuous control system that makes them attractive for manipulator control.

1 Model of Robot Arm

The dynamics of a six-joint manipulator is described by

$$D(\theta)\ddot{\theta} = Q(\theta, \dot{\theta}) + G(\theta)g + u, \tag{16.1}$$

where $D(\theta)$ is 6×6 symmetrical matrix; $G(\theta), Q(\theta, \dot{\theta}), u, \theta, \dot{\theta}$ and $\ddot{\theta}$ are column vectors,

$$(G(\theta))^{\mathrm{T}} = (G_1(\theta), \ldots, G_6(\theta)), u^{\mathrm{T}} = (u_1, \ldots, u_m),$$
$$(Q(\theta, \dot{\theta}))^{\mathrm{T}} = (Q_1(\theta, \dot{\theta}), \ldots, Q_6(\theta, \dot{\theta})), \theta^{\mathrm{T}} = (\theta_1, \ldots, \theta_6),$$

g is the gravitational constant; $\theta_i, \dot{\theta}_i, \ddot{\theta}_i$ are, respectively, angular position, velocity and acceleration of joint i. The off-diagonal terms of matrix D in (16.1) represent interactions due to accelerations of other joints, matrix Q represents Coriolis and centrifugal torques, and the components of vector u are input torques applied at various joints.

Assuming that $D^{-1}(\theta)$ exists and defining vector x as

$$x^{\mathrm{T}} = (\theta_1, \ldots, \theta_6, \dot{\theta}_1, \ldots, \dot{\theta}_6) = (p_1, \ldots, p_6, v_1, \ldots, v_6)$$

the following equations can be obtained for the coordinates characterizing manipulator state:

$$\dot{p}_i = v_i, \quad \dot{v}_i = f_i(p, v) + b_i(p)u, \quad i = 1, \ldots, 6, \tag{16.2}$$

where f_i is the i-th component of $f(p, v) = D^{-1}(p)(Q(p, v) + G(p)g), b_i(p)$ is the i-th row of the matrix $B(p) = D^{-1}(p)$.

2 Problem Statements

Let us consider the following problem of control. For the given initial state $p(t_0), v(t_0)$ and desired position p_α at velocity $v_\alpha = 0$ find a control stabilizing manipulator joints in the final position, i.e. $\lim_{t \to \infty} p(t) = p_\alpha$ and $\lim_{t \to \infty} v(t) = 0$. Let position error be $e(t) = p(t) - p_\alpha$. Accordingly, we introduce a new state vector $x_e^{\mathrm{T}} = (e^{\mathrm{T}}, v^{\mathrm{T}})$ with the following equations

$$\dot{e}_i = v_i, \quad \dot{v}_i = f_i(e + p_\alpha, v) + b_i(e + p_\alpha)u, \quad i = 1, \ldots, 6. \tag{16.3}$$

The aim of control is to nullify error $e(t)$.

Another interesting problem in manipulator control is tracking when the manipulator has to follow a desired position path $g(t)$. With such a problem formulation where $x_g^T = (e^T, w^T)$ is state vector, $e(t) = p(t) - g(t)$, $w(t) = v(t) - \dot{g}(t)$ and motion equations are

$$\dot{e}_i = w_i, \quad \dot{w}_i = f_i(e + g, w + \dot{g}) + b_i(e + g) - \ddot{g} \tag{16.4}$$

The goal of control is then to make the vector $(e(t), w(t))$ to tend to zero.

Our primary concern in this chapter is to design a stabilizing discontinuous control. We also shall briefly discuss how the same approach can be applied to tracking.

3 Design of Control

To provide stabilization, control $u(p, v)$ is taken in the form of

$$u_i(p, v) = \begin{cases} u_i^+(p, v) & \text{if} \quad s_i(e_i, v_i) > 0, \\ u_i^-(p, v) & \text{if} \quad s_i(e_i, v_i) < 0, \end{cases} \quad i = 1, \ldots, 6, \tag{16.5}$$

where e_i and v_i are the components of state vector (16.3); the set of equations

$$s_i(e_i, v_i) = c_i e_i + v_i, \quad c_i > 0, \quad i = 1, \ldots, 6 \tag{16.6}$$

defines the set of discontinuity surfaces. Design consists in choosing control functions $u_i^+(p, v)$ and $u_i^-(p, v)$ such that sliding mode occurs on the intersection of planes (16.6).

We first examine the properties of (16.3) in sliding mode. Solving the algebraic equation system $\dot{s}_i = 0, i = 1, \ldots, 6$ with respect to u, we see that the equivalent control exists and is unique:

$$u_{eq} = -D(p)(f(p, v) + Cv), \tag{16.7}$$

where $C = \text{diag}(c_1, \ldots, c_6)$. Following the equivalent control method, the motion differential equations are obtained that represent sliding mode:

$$\dot{e}_i = -c_i e_i, \quad i = 1, \ldots, 6. \tag{16.8}$$

The Eq. (16.8) involves six uncoupled first-order linear differential equations representing each the dynamics of one degree of freedom when system (16.3) is in sliding mode. Obviously, non-linear interactions of manipulator subsystems are eliminated completely. Furthermore, the sliding-mode dynamics depends only on parameters c_i which can be chosen arbitrarily. In sliding mode, therefore, the system is insensitive to the interactions of joints and to load variations. Recognizing that $-c_i$ $(i = 1, \ldots, 6)$ are the eigenvalues of (16.8), the choice of positive coefficients in (16.6) guarantees asymptotic stability of (16.3) in sliding mode. The sliding mode motion may be effectively speed up by choosing large c_i.

In order to obtain in the multi-variable discontinuous control system conditions imposed on u_i^+ and u_i^- in (16.5) which ensure the existence of sliding mode on the interaction of discontinuity surfaces (16.6) we use the design method with scalar control hierarchy (Sect. 4.5). This method is chosen because of its robustness – it boils down to the successive design of scalar control systems whose conditions for sliding mode to exist have the form of inequalities.

Recall briefly the procedure of the hierarchy of control method. We introduce the notion of "sliding mode on $s_i = 0$" which means that $e_i(t)$ satisfies equations

$$\dot{e}_i = -c_i e_i. \tag{16.9}$$

By a hierarchy of switching surfaces, we mean that sliding mode occurs earlier on those discontinuity surfaces which are higher in the hierarchy. The hierarchy of control methods is then described by the following procedure:

Step 1. Suppose the hierarchy $s_1 \to s_2 \to s_3 \to s_4 \to s_5 \to s_6$ is assumed. The arrows indicate hierarchy direction.

Step 2. Suppose that sliding mode occurs on the surfaces $s_j = 0, j = 1, \ldots, i-1$, $1 \leqq i \leqq 6$.

Step 3. For the surface $s_i = 0$, find u_i^+ and u_i^- such that

$$s_i \dot{s}_i < 0 \tag{16.10}$$

outside the discontinuity surfaces.

This inequality guarantees that the state space trajectories of (16.3) which lie on the intersection of surfaces $s_j = 0, j = 1, \ldots, i-1$, move towards the surface $s_i = 0$. After getting on $s_i = 0$, the system state vector varies along the intersection of $s_1 = 0, \ldots, s_i = 0$. Condition (16.10) implies that u_i^+ and u_i^- satisfy

$$b_i^i(p) u_i^+ < -\min_r \left(c_i v_i + f_i(p, v) + (d_i(p))^{\mathrm{T}} r_{\mathrm{eq}}^{i-1} + (h_i(p))^{\mathrm{T}} r^{i+1}\right), \tag{16.11}$$

$$b_i^i(p) u_i^- > -\max_r \left(c_i v_i + f_i(p, v) + (d_i(p))^{\mathrm{R}} r_{\mathrm{eq}}^{i-1} + (h_i(p))^{\mathrm{T}} r^{i+1}\right), \tag{16.12}$$

where

$$(d_i(p))^{\mathrm{T}} = (b_i^1, b_i^2, \ldots, b_i^{i-1}), (h_i(p))^{\mathrm{T}} = (b_i^{i+1}, \ldots, b_i^6),$$
$$(b_i(p))^{\mathrm{T}} = ((d_i(p))^{\mathrm{T}}, b_i^i(p), (h_i(p))^{\mathrm{T}}),$$
$$(r_{\mathrm{eq}}^{i-1})^{\mathrm{T}} = (u_{\mathrm{eq}}^1, \ldots, u_{\mathrm{eq}}^{i-1}), (r^{i+1})^{\mathrm{T}} = (u_{i+1}, \ldots, u_6),$$

r_{eq}^{i-1} is equivalent control which is function of $u_k, k = i, \ldots, 6$. It is obtained by solving the algebraic system $\dot{s}_j = 0, j = 1, \ldots, i-1$, for $u_1, \ldots, u_{i-1}$.

Step 4. Replace i by $i-1$. If $i > 0$, go to Step 2. The procedure begins with $i = 6$ and ends with $i = 1$.

As it follows from this procedure, sliding mode initially occurs on the plane $s_1 = 0$, then on the intersection of planes $s_1 = 0$ and $s_2 = 0$ and so on until it occurs on the intersection of all the discontinuity surfaces. Then sliding mode is said to take place on the manifold $s = 0, s^T = (s_1, \ldots, s_6)$.

We see that the sliding mode existence conditions (16.10) have the form of (1.9) like the existence conditions in single-input systems (Sect. 1.1). This fact reveals the basic idea behind the scalar hierarchy of control method which is to replace the design of multi-variable system by a sequence of single-input problems.

It should be stressed that the sliding mode existence conditions have the form of (16.11), (16.12), and it is, therefore, sufficient to know only the bounds of the uncertainties in the manipulator physical parameters in order to determine u_i^+ and u_i^-. This is the robustness as mentioned in the beginning of this chapter. We also see that each control component u_i is determined independently from (16.11) and (16.12). Moreover, the motion speed before reaching switching plane $s_i = 0$ is defined by the absolute values of u_i^+ and u_i^-. Stated differently, $|\dot{s}_i|$ is proportional to the difference between the right and left sides of (16.11) for $s_i > 0$ (or (16.12) for $s_i < 0$).

4 Design of Control System for a Two-Joint Manipulator

We shall follow the design procedure outlined in the last section for a two-degrees-of-freedom manipulator diagrammed in Fig. 22. Under the assumption of lumped equivalent masses and massless links, manipulator dynamics is representable [166] as

$$\underbrace{\begin{bmatrix} \alpha_{11}(\varphi) & \alpha_{12}(\varphi) \\ \alpha_{12}(\varphi) & \alpha_{22}(\varphi) \end{bmatrix}}_{D(\varphi)} \begin{bmatrix} \ddot{\theta} \\ \ddot{\varphi} \end{bmatrix} = \underbrace{\begin{bmatrix} \beta_{12}(\varphi)\dot{\theta}^2 + 2\beta_{12}(\varphi)\dot{\theta}\dot{\varphi} \\ -\beta_{12}(\varphi)\dot{\varphi}^2 \end{bmatrix}}_{Q(\theta, \dot{\theta}, \varphi, \dot{\varphi})} + \underbrace{\begin{bmatrix} \gamma_1(\theta, \varphi) \\ \gamma_2(\theta, \varphi) \end{bmatrix}}_{G(\theta, \varphi)} g + \begin{bmatrix} u_1 \\ u_2 \end{bmatrix}. \tag{16.13}$$

where θ and φ are defined by the manipulator geometry (Fig. 22), g is the gravitational constant, and u_i are the input torques. Let us assume that $D^{-1}(\varphi)$ exists, and let θ_α and φ_α be the desired positions. To solve the stabilization problem, define the state vector $x^T = (e_1, x_2, e_3, x_4) = (\theta - \theta_\alpha, \dot{\theta}, \varphi - \varphi_\alpha, \dot{\varphi})$. Equation (16.13) in state space becomes

$$\dot{e}_1 = x_2, \tag{16.14}$$

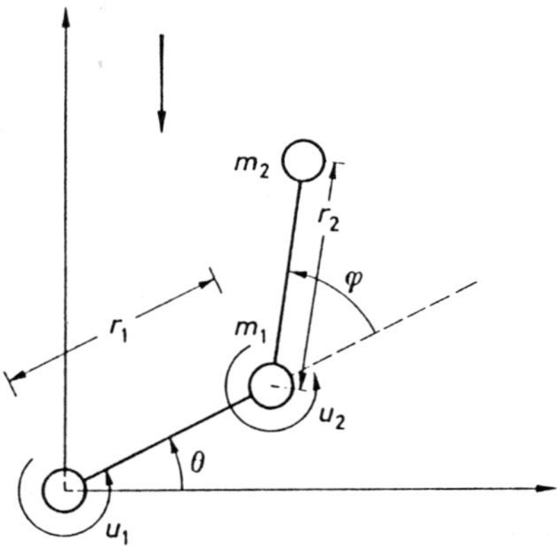

Fig. 22

$$\dot{x}_2 = \frac{\alpha_{22}}{\alpha}(\beta_{12}(x_3)x_2^2 + 2\beta_{12}(x_3)x_2x_4 + \gamma_1(x_1, x_3)g + u_1)$$

$$- \frac{\alpha_{12}(x_3)}{\alpha}(-\beta_{12}(x_3)x_4^2 + \gamma_2(x_1, x_3)g + u_2)$$

$$\equiv f_1(x) + b_1^1(x_3)u_1 + b_1^2(x_3)u_2, \tag{16.15}$$

$$\dot{e}_3 = x_4, \tag{16.16}$$

$$\dot{x}_4 = -\frac{\alpha_{12}(x_3)}{\alpha}(\beta_{12}(x_3)x_2^2 + 2\beta_{12}(x_3)x_2x_4 + \gamma_1(x_1, x_3) + u_1)$$

$$+ \frac{\alpha_{11}(x_3)}{\alpha}(-\beta_{12}(x_3)x_4^2 + \gamma_2(x_1, x_3)g + u_2)$$

$$\equiv f_2(x) + b_2^1(x_3)u_1 + b_2^2(x_3)u_2 \tag{16.17}$$

with

$$x_1 = \theta = e_1 + \theta_\alpha, \quad x_3 = \varphi = e_3 + \varphi_\alpha,$$

$$\alpha = \alpha_{11}(x_3)\alpha_{22} - \alpha_{12}^2(x_3), \tag{16.18}$$

$$\alpha_{11}(x_3) = (m_1 + m_2)r_1^2 + m_2r_2^2 + 2m_2r_1r_2\cos x_3 + I_1, \tag{16.19}$$

$$\alpha_{22} = m_2r_2^2 + I_2, \tag{16.20}$$

$$\alpha_{12}(x_3) = m_2r_2^2 + m_2r_1r_2\cos x_3, \tag{16.21}$$

$$\gamma_1(x_1, x_3) = -(m_1 + m_2)r_1\cos x_3 + m_2r_2\cos(x_1 + x_3), \tag{16.22}$$

$$\gamma_2(x_1, x_3) = -m_2 r_2 \cos(x_1 + x_3), \tag{16.23}$$

$$\beta_{12}(x_3) = m_2 r_1 r_2 \sin x_3, \tag{16.24}$$

where m_i, r_i and $I_i, i = 1, 2$, denote the point mass, link length and additional constant moment of inertia with respect to axes of rotation. It is assumed that only a bounded domain of the state space is of interest for manipulator operations.

In the hierarchy of control method outlined in the last, section u_i^+ and u_i^- are chosen so as to satisfy (16.11), (16.12). Although constant u_i^+ and u_i^- can be always chosen so as to satisfy (16.11) and (16.12), the state feedback control functions $u_i^+(x)$ and $u_i^-(x)$ are preferred. Recognizing that for a bounded domain of x for $\alpha \neq 0$ there always will be an upper estimate for all the functions in (16.14) through (16.24)

$$N|x| + M, \quad N, M - \text{const}, \quad |x| = |e_1| + |x_2| + |e_3| + |x_4|, \tag{16.25}$$

control action is chosen in the form of

$$u_i(x) = -(\alpha_i^1 |e_1| + \alpha_i^2 |x_2| + \alpha_i^3 |e_3| + \alpha_i^4 |x_4| + \delta_i) \operatorname{sign} s_i, \alpha_i^j, \delta_i = \text{const}, \quad i = 1, 2. \tag{16.26}$$

Denote

$$u_i(x) = \begin{cases} u_i^+(x) & \text{at} \quad s_i > 0 \quad (16.27) \\ u_i^-(x) & \text{at} \quad s_i > 0 \quad (16.28) \end{cases}$$

with discontinuity surfaces

$$s_1(e_1, x_2) = c_1 e_1 + x_2 = 0, \quad c_1 > 0, \tag{16.29}$$

$$s_2(e_2, x_4) = c_2 e_2 + x_4 = 0, \quad c_2 > 0. \tag{16.30}$$

Assume that hierarchy of switching surfaces is $s_1 \to s_2$, then (16.11), (16.12) become

$$h_2(x_3) u_2^+ < -\left(c_2 x_4 - \frac{b_2^1(x_3) b_1^2(x_3)}{b_1^1} (c_1 x_1 + f_1) + f_2 \right), \tag{16.31}$$

$$h_2(x_3) \bar{u}_2 > -\left(c_2 x_4 - \frac{b_2^1(x_3) b_1^2(x_3)}{b_1^1} (c_1 x_1 + f_1) + f_2 \right), \tag{16.32}$$

$$b_1^1 u_1^+ < -\min_{u_2} (c_1 x_2 + f_1 + b_1^2(x_3) u_2), \tag{16.33}$$

$$b_1^1 u_1^- > -\max_{u_2} (c_1 x_2 + f_1 + b_1^2(x_3) u_2), \tag{16.34}$$

where

$$h_2(x_3) = b_2^2(x_3) - \frac{b_2^1(x_3) b_1^2(x_3)}{b_1^1}. \tag{16.35}$$

Signs of b_1^1 and $h_2(x_3)$ must be known in order to determine u_i^+ and u_i^-. The sign of $h_2(x_3)$ is independent of x_3 since $b_1^1 > 0$ for all x_3, $h_2(x_3)$ is continuous function and $h_2(x_3) = \det D^{-1}/b_1^1 \neq 0$ [1]. It remains to choose coefficients α_i^j in (16.26) so that (16.31) through (16.34) are satisfied. To this end, it is sufficient to know the lower bounds of b_1^1 and $h_2(x_3)$, the upper bounds of $b_1^1, b_1^2(x_3) = b_2^1(x_3)$, f_1 and f_2. According to our hierarchy, sliding mode occurs first on the discontinuity surface $s_1 = 0$ and then on the intersection of $s_1 = 0$ and $s_2 = 0$.

5 Manipulator Simulation

Hybrid simulation has been carried out for the sliding motion of manipulator of Fig. 22 [166]. The purpose of simulating the system in hybrid environment was twofold: first, to observe the non-ideal sliding motions due to the delays introduced by the hardware realizing the control law and by other imperfections, and second, to demonstrate that this control algorithm is suitable to be implemented on computers.

Under the assumption that $\beta_{12}(x_3)$ is small, i.e. that Coriolis and cetrifugal effects are negligible, the variable α in (16.18) may be regarded as constant. The system (16.14) through (16.17) was simulated on an AD5 analogue computer with the following set of parameters:

$m_1 = 0.5\,\mathrm{kg}$, $m_2 = 6.25\,\mathrm{kg}$, $r_1 = 1.0\,\mathrm{m}$, $r_2 = 0.8\,\mathrm{m}$, $I_1 = 5\,\mathrm{kg\,m^2}$,

$I_2 = 5\,\mathrm{kg\,m^2}$. The excessive ratio between m_1 and m_2 is used to emphasize the load effect; u_1 and u_2 are control components (external torques). For implementation convenience $\cos y$ was approximated by piecewise linear function $\cos y \approx 1 - |y|/(\pi/2), |y| \leqq \pi$. The bounded region of the manipulator state space is defined by $|e_1| \leqq 180°$, $|e_3| \leqq 180°$, $|x_2| \leqq 180°/\mathrm{sec}$, and $|x_4| \leqq 180°/\mathrm{sec}$.

The discontinuous control was implemented on a PDP 11/40 minicomputer interfaced to AD5 and has the following form

$$u_1(x) = -(\alpha_1^2|x_2| + \alpha_1^s|x_s| + \alpha_1^4|x_4| + \delta_1)\,\mathrm{sign}\, s_1, \tag{16.36}$$

$$u_2(x) = -(\alpha_2^2|x_2| + \alpha_2^s|x_s| + \alpha_2^4|x_4| + \delta_2)\,\mathrm{sign}\, s_2, \quad \alpha_1^s, \alpha_2^s = \mathrm{const}, \tag{16.37}$$

$$x_s = x_1 + x_3 (x_1 = e_1, x_3 = e_2) \tag{16.38}$$

with s_1, s_2 and x_s as defined by (16.29), (16.30) and (16.38). Control hierarchy was $s_1 \to s_2$. The magnitudes of the relay components δ_i in (16.36), (16.37), are chosen so as to fulfil the inequalitites (16.11), (16.12). The implementation of (16.36), (16.37) involves determination of the signs of x_i and s_i. Rewriting (16.36)

[1] For manipulators with more degrees of freedom, the signs of $h_i(x)$ are independent of x like in (16.31) and (16.32) owing to the positive definiteness of D in (16.1).

and (16.37) as

$$u_1(x) = u_1^2(x_2) + u_1^s(x_s) + u_1^4(x_4) + u_1^r, \tag{16.39}$$

$$u_2(x) = u_2^2(x_2) + u_2^s(x_s) + u_2^4(x_4) + u_2^r, \tag{16.40}$$

where

$$u_i^j(x_j) = \begin{cases} -\alpha_i^j x_j & \text{at} \quad x_j s_i > 0, \quad j = 2, 4, s, \\ \alpha_i^j x_j & \text{at} \quad x_j s_i < 0, \quad i = 1, 2, \quad \alpha_i^j > 0, \end{cases}$$

$$u_i^r = \begin{cases} -\delta_i & \text{at} \quad s_i > 0, \\ \delta_i & \text{at} \quad s_i < 0, \quad i = 1, 2, \quad \delta_i > 0. \end{cases} \tag{16.41}$$

The control laws (16.39), (16.40) may be, therefore, implemented as a sum of piecewise linear terms and a relay term. The gains and relay signals are switched depending on the signs of $s_i(x)$. Computation of $s_i(x)$ and determination of the signs of s_i and x_i were performed by a FORTRAN program run on the PDP 11 computer. Vector x and control actions are effected through analog-to-digital and digital-to-analog converters, respectively.

We describe only one simulation algorithm which is representative for this experiment. The manipulator is idle in some initial configuration. It is to be moved to a different configuration be applying torques defined by the control laws (16.39), (16.40). Various initial and final configurations were experimented with, and it was observed that the manipulator was always regulated to the desired positions; in particular, from the initial state $(x(t_0))^T = (-159.5°; 0; -69°; 0)$ to the desired configuration with all the coordinates being equal to zero. The final configuration where x is zero deserves attention because the maximal torques are required in order to maintain the manipulator in this configuration under fixed load.

Switching surfaces

$$s_1(x) = 2x_1 + x_2 = 0, \tag{16.42}$$

$$s_2(x) = 2,5x_3 + x_4 = 0, \tag{16.43}$$

with hierarchy $s_1 \to s_2$ were experimented here. The ideal sliding-mode equations were

$$\dot{x}_1 = -2x_1, \tag{16.44}$$

$$\dot{x}_3 = -2,5x_3. \tag{16.45}$$

In order to dramatize the effect of non-ideal switching on sliding mode, a delay of approximately 20 ms was introduced. The effect of torque chattering was observed in the velocity graphs of $x_2(t)$ and $x_4(t)$. The position trajectories $x_1(t)$ and $x_2(t)$ were fairly smooth which may be attributed to the fact that the manipulator acts as a low-pass filter with respect to each discontinuous input control. The beginning of control torque chattering may be interpreted as the beginning of sliding mode. It was observed that shortly after the start of sliding mode on plane $s_1 = 0$, sliding mode was initiated on plane $s_2 = 0$ as intended by the design. The time constants of the non-ideal sliding mode were approximately 0.5 and 0.4 sec for $s_1 = 0$ and $s_2 = 0$, respectively. Compared to (16.44), (16.45), this result confirms that the non-ideal sliding mode is close to the ideal one predicted by the theory, i.e. that despite the chattering phenomenon, the speed of sliding mode is defined by the coefficients of plane equations. For sufficiently small delay t_α in the switching device during sliding mode, the following relations are fulfilled

$$|s_1| \leqq \Delta_1(t_\alpha, x, u_1, u_2),$$

$$|s_2| \leqq \Delta_2(t_\alpha, x, u_1, u_2), \tag{16.46}$$

$$\Delta_i \approx \dot{s}_i t_\alpha. \tag{16.47}$$

The magnitude of Δ_i hence is dependent on the gains α_i^j and relay signal δ_i. Larger α_i^j and δ_i speed up the motion towards the discontinuity surfaces, but also increase Δ_i. At the origin we have

$$\Delta_i(t_\alpha, 0, \delta_1, \delta_2) = (k_i^1 \delta_1 + k_i^2 \delta_2) t_\alpha, \tag{16.48}$$

where k_i^1 and k_i^2 depend on the physical parameters of the manipulator arm and on the final desired position. The accuracy requirements can be met by a selection of δ_1, δ_2 and t_α.

6 Path Control

As it was already noted, path control is another interesting question in manipulator control. By comparing the motion equations for the case where the manipulator has to track a desired trajectory (16.4) with those of stabilization system (16.3), we observe that the only difference is the presence of the additional term $\ddot{g}_i(t)$ in (16.4). The design of discontinuous control system for the trajecgtory problem is basically the same as in the case of stabilization. If we assume the function $\ddot{g}_i(t)$ to be known, the discontinuous control

$$u_i(x) = -(\alpha_i^1|e_1| + \alpha_i^2|w_2| + \alpha_i^3|e_3| + \alpha_i^4|w_4| + \alpha_i^5|\ddot{g}_i| + \delta_i)\operatorname{sign} s_i, \quad i = 1, 2. \tag{16.49}$$

with

$$e_i = \theta_i - g_i, \tag{16.50}$$

$$w_j = \dot{\theta}_{j-1} - \dot{g}_{j-1}, \quad \theta_1 = \theta, \quad \theta_3 = \varphi, \quad i = 1, 3, \quad j = 2, 4, \tag{16.51}$$

solves the tracking problem for the two-joint manipulator. The parameters in (16.49) are chosen as previously so that sliding mode occurs on the intersection of discontinuity surfaces. This system belongs to the class of combined control systems which is commonly adopted for tracking problems.

7 Conclusions

The hybrid simulation has borne out the fact that discontinuous controls are applicable to manipulator control system design. By introducing sliding modes, the discontinuous control algorithms compensate for the non-linear dynamic interaction of manipulator joints. Comparing the proposed algorithms with the existing control laws which emphasize complex non-linear compensation, we see that their implementation is simplier and requires less *a priori* knowledge of the manipulator physical parameters since when designing the control only the inequality-type conditions (16.11), (16.12) must be satisfied. However, it is clear that the type of control discussed here leads to discontinuous control signals with rapid sign changes similar to amplitude/pulse modulated signals.

Understandably, the effects of such control signals on all the manipulator elements should be taken into consideration. Despite the discontinuous nature of control signals, the simulated joint position trajectories were observed to be smooth. This indicates to the fact that the manipulator itself filters out the high-frequency component introduced by the physical imperfections of switching devices. These results were obtained with the "delay" imperfections of switching devices. Other types of non-idealities may result in non-ideal sliding modes that differ from those observed in our experiments but are still close enough to the ideal sliding modes.

Chapter 17

Sliding Modes in Control of Electric Motors

1 Problem Statement

This chapter addresses one of the most challenging applications of sliding modes, the control of electric motors. Its attractiveness is due to the wide use of motors, the advances of electronics in the area of controlled power converters, the insufficience of linear control methodology for essentially non-linear high-order plants such as a.c. motors.

Until recently, in applications where speed control over a wide range was emphasized, d.c. drives were most common because of the relative simplicity of their mathematical model and the possibility to exercise control by small powers in the excitation channel. However, owing to such disadvantages as commutators, great moments of inertia, high production cost and maintenance charges, design of a.c. motor-operated drives has always been topical. The advent of completely controllable thyristors and transistors able to control in the switching mode tens and even hundreds of kilowatts has provided the ground for attacking this problem. But the truth is that the potentialities of power convertors were not used to full advantage owing to the orientation to frequency control, correction of static characteristics, and linear controllers. Within the framework of these approaches, it is impossible to find satisfactory solutions, especially under strict resuirements to the accuracy and range of speed control, because of incomplete process information and uncontrollable disturbances, non-linear cross-couplings between control loops, variations of the local properties of the motor at transition from one mode to another. In this situation, the desire to confine the power convertor functions only to the pulse-width modulated amplification of continuous input signals which are the desired averages of motor's voltages and currents seems to be unjustified. It seems more natural to deisgn new control algorithms so as to make the best use of the convertor's essentially non-linear discontinuous characteristics. The presence of discontinuities is accounted for by the switching mode performance of the power supply, i.e. by the nature of transistor or thyristor convertors.

Thus, discontinuous control actions are "prescribed" to us. Let us make use of them in order to induce in the system state space sliding modes on surfaces whose equations will define deviations from the desired mode. As it was demonstrated in Parts I and II, the order of sliding mode equations is lower as compared with that of the original system, the motion itself may be invariant to parametric and additive disturbances, the existence conditions of sliding mode have usually the form of inequalities and, therefore, design of appropriate algorithms does not require exact information about the plant operator.

In what follows, we shall demonstrate the possibilities of using sliding modes in systems with non-linear control laws for the purpose of overcoming a wide scope of problems occurring in the design of various electric drives.

2 Control of d.c. Motor

For the beginning, let us consider a most simple (in terms of control) d.c. motor with autonomous excitation whose behaviour with respect to the angular shaft speed is described by a linear second-order equation

$$T_m T_e \ddot{n} + T_m \dot{n} + n = k_1 u - k_2 \dot{M}_l - k_0 M_l,$$

where u is the supply voltage of rotor circuit, T_m and T_e are respectively, the electromechanical and electrical time constants, M_l is the load torque, k_1, k_2, k_0 are constant coefficients. Parameters T_m, T_e, k_0, k_1, k_2 depend on the moment of inertia J reduced to the motor shaft, inductance L and resistance r of rotor circuit, and on the proportionality factor k_m between the rotor current i and the torque developed by the motor M:

$$M = k_m i. \tag{17.1}$$

It is assumed that speed is controlled by varying the voltage u, which may assume one of two possible values u_0 or $-u_0$. Such a control is implemented by a d.c. source and controllable convertors.

Let $n_z(t)$ be the desired speed profile. Write motor equation in the coordinates of error $x_1 = n_z - n$ and its derivative:

$$\dot{x}_1 = x_2, \quad \dot{x}_2 = -\frac{1}{T_e} x_2 - \frac{1}{T_m T_e} x_1 - \frac{k_1}{T_e T_m} u + f(t),$$

where

$$f(t)=\ddot{n}_z+\frac{1}{T_e}\dot{n}_z+\frac{1}{T_eT_m}n_z+\frac{k_2}{T_eT_m}\dot{M}_l+\frac{k_0}{T_eT_m}M_l.$$

For such a system represented in the canonical form (Sect. 7.4), sliding mode motion along the straight line

$$s=cx_1+x_2=cx_1+\dot{x}_1=0 \tag{17.2}$$

is invariant to load torques and reference input, and to the variations of motor parameters, of which most representative is the variation of moment of inertia J in the course of operation. After the start of sliding mode, the error coordinate exponentially tends to zero with time constant $1/c$.

To determine the conditions for the sliding modes to exist, let us write the motion equation with respect to coordinate s:

$$\dot{s}=cx_2-\frac{1}{T_e}x_2-\frac{1}{T_eT_m}x_1-\frac{k_1}{T_eT_m}u+f(t)$$

If

$$u=u_0\,\mathrm{sign}\,s \tag{17.3}$$

and inequality

$$\frac{k_1}{T_eT_m}u_0>\left|cx_2-\frac{1}{T_e}x_2-\frac{1}{T_eT_m}x_1+f(t)\right| \tag{17.4}$$

is satisfied, s and $\dot{s}$ have different signs and the state vector will reach the straight line $s=0$ within a finite time. A value of u_0 such that control (17.3) induces sliding mode and, consequently, the error for a given class of inputs $M_l(t)$ and $n_z(t)$ vanishes, may be determined through (17.4). Note that the class of such functions is sufficiently wide: (17.4) may be satisfied if these functions are bounded together with their derivatives (first derivative of $M_l(t)$ and second derivative of $n_z(t)$).

To implement control (17.3), one has to measure the time derivative of motor angular speed since $x_2=\dot{n}_z-\dot{n}$ (it will be assumed below that the rate of reference input variation $\dot{n}_z$ may be determined). The time constants of the existing devices for speed measurement and subsequent differentiation are negligible as compared with those characterizing decay of transients in control processes.[1] The sliding mode motion features fast switchings and its corresponding high-frequency component may be comparable with the rates of proper motions in measuring and differentiation devices. As it was shown in Sect. 5.3, the distorting contribution of small dynamic imperfections to the average motion is insignificant, but the frequency of voltage switching may be significantly reduced. Apart from the reduction in accuracy, this leads to

[1] In the case of sliding mode control, the process rate is determined by substituting the equivalent control for the discontinuous one (Sect. 2.4).

significant heat losses because, owing to the low frequency, the amplitude of current oscillations about the mean value may be rather high.

Let us at first demonstrate qualitatively how by measuring current i in addition to the velocity one can increase the switching frequency and, thus, prevent this undesirable effect and maintain the low sensitivity to external actions. Current measurements present no special engineering problem. Keeping in mind that the motor torque is proportional to rotor current (17.1), we can determine the relation between current and acceleration:

$$= \frac{J}{k_m}\frac{dn}{dt} - \frac{1}{k_m} M_l. \tag{17.5}$$

It is evident that, motor parameters being known, the rotation speed derivative may be computed to an accuracy of load torque through the measurements of the rotor current. Represent it as a sum of average and high-frequency component: $i = i_{\text{av}} + i_f$. The possible frequency range of load torque is much lower than the sliding mode frequency (which may be chosen for electric motors of the order of several kHz). This means that a time constant τ of the inertial element may be found

$$\tau \dot{z} + z = i$$

such that its output z will only slightly differ from i_{av} or from

$$\frac{J}{k_m}\frac{dn_{\text{av}}}{dt} - \frac{1}{k_m} M_l,$$

where n_{av} is the average speed component (i.e. speed without the high-frequency component). The value of dn_{av}/dt may be estimated without essential distortions by mean of an implementable differentiating element with transfer function $W_{\text{av}}(p) = p/(\tau p + 1)$ if the directly measured velocity n is fed into its input. The high-frequency component of acceleration dn_f/dt may be determined by subtracting z from the rotor current

$$\frac{J}{k_m}\frac{dn_f}{dt} = i - z.$$

Stated differently, dn_f/dt may be estimated by means of an element with transfer function

$$W_f(p) = \frac{k_m}{J}\frac{\tau p}{\tau p + 1}$$

with input i.

The above reasoning shows how function s must be formed if coordinate x_2 is estimated by means of inertial measuring and differentiating devices which introduce essential distortions into the high-frequency component:

$$s = c(n_z - n) + \dot{n}_z - \frac{dn_{\text{av}}}{dt} - \varepsilon(i - z),$$

where n *and* dn_{av}/dt are the measured and estimated speed and acceleration, ε is a positive coefficient. It should be noted that for sliding mode to commence, $\dot{s}$ should have discontinuities at the times of sign alteration of s. This condition is satisfied because di/dt depends linearly on d^2n/dt^2 (17.5) which, in its turn, depends on the discontinuous control. The average value of function $i-z$ is close to zero; therefore, at the start of sliding mode, i.e. at $s=0$, motion is still defined by coefficient c independently of the load.

Now we demonstrate that the observer of Sect. 14.1 leads to the just discussed dynamic system used for estimation of acceleration dn/dt. Assume as before that the load torque M_l is unknown, varies slowly and may be regarded as constant, and obtain a second order system with state vector (M_l, n) and current i as an input:

$$\frac{dM_l}{dt}=0, \quad \frac{dn}{dt}=-\frac{1}{J}M_l+\frac{k_m}{J}i.$$

If we succeed in estimating M_l, acceleration will be determined from the second equation. Bearing in mind that n may be measured, let us design an asymptotic first-order Luenberger observer. Introduce a new variable M_l' similar to x' in (14.15): $M_l'=M_l+L_1 n$, $L_1=\text{const}$, and write the motion equations in terms of M_l' and n:

$$\frac{dM_l'}{dt}=-\frac{L_1}{J}M_l'+\frac{L_1^2}{J}n+\frac{L_1 k_m}{J}i,$$

$$\frac{dn}{dt}=-\frac{1}{J}M_l'+\frac{L_1}{J}n+\frac{k_m}{J}i.$$

Observer's equation, correspondingly, is

$$\frac{d\bar{M}_l'}{dt}=-\frac{L_1}{J}\bar{M}_l'+\frac{L_1^2}{J}n+\frac{L_1 k_m}{J}i,$$

and dn/dt is estimated as follows

$$\frac{d\bar{n}}{dt}=-\frac{1}{J}\bar{M}_l'+\frac{L_1}{J}n+\frac{k_m}{J}i.$$

Mismatch $\hat{M}_l'=M_l'-\bar{M}_l$, obviously, tends asymptotically to zero because it is the solution of

$$\frac{d\hat{M}_l'}{dt}=-\frac{L_1}{J}\bar{M}_l',$$

if $L_1>0$, the value of L_1 defining the rate of decay of $\bar{M}_l'$. Thus, the estimate $\bar{M}'$ tends to M_l' and, consequently, we are able to determine both the load torque and acceleration dn/dt which is required for generation of switching function s.

It follows directly from the observer equation that it may be regarded as a first-order dynamic system with scalar output $d\bar{n}/dt$ and two inputs n and i. The transfer functions of each input are, respectively,

$$W_n(p) = \frac{p}{\tau p + 1}, \quad W_i(p) = \frac{k_m}{J} \frac{\tau p}{\tau p + 1}, \quad \tau = \frac{J}{L_1}.$$

As one may see, they coincide with the transfer functions $W_{\text{av}}(p)$ and $W_f(p)$ that were used for determination of the average and high-frequency components of acceleration dn/dt.

It should be said in conclusion how L_1 in the observer is to be chosen. The increase of L_1 speeds up the decay of error between the true acceleration and its estimate, but at the same time this may deteriorate system's noise immunity since the element with transfer function $W_n(p)$ (or $W_{\text{av}}(p)$) becomes an ideal differentiator at $L \rightarrow \infty$. Therefore, the choice of parameter L_1 should represent a trade-off between these two contradictory requirements. The reader may refer, for example, to monographs [6, 92] where the methods used in the design of r.m.s.-optimal observers or Kalman filters are presented in full detail.

3 Control of Induction Motor

The most simple, reliable and economic of all motors, the induction motor is commonly used in electric drives where speed need not be controlled within a wide range. However, in the controlled drives the induction motor proved to be inconvenient because it seems to be the most complicated motor in terms of controllability. The point is that it is described by an essentially non-linear high-order system of differential equations, and to support its normal operation through using only the power supply channel one has to simultaneously control several interrelated variables such as, for example, motor shaft rotation speed and magnetic flux (or angular position, acceleration, torque, power, etc.).

Let us consider the control of rotation speed of three-phase induction motor with squirrel-cage rotor. Non-saturated symmetrical induction motor is described in the fixed coordinate system (α, β) by a system of fifth-order non-linear differential equations [31] with respect to rotor velocity n, the components of rotor magnetic flux ψ_α, ψ_β and of stator current i_α, i_β:

$$\frac{dn}{dt} = \frac{1}{J}(M - M_l), \quad M = \frac{x_H}{x_R}(i_\alpha \psi_\beta - i_\beta \psi_\alpha),$$

$$\frac{d\psi_\alpha}{dt} = -\frac{r_R}{x_R}\psi_\alpha - n\psi_\beta + \frac{r_R x_H}{x_R} i_\alpha,$$

$$\frac{d\psi_\beta}{dt} = -\frac{r_R}{x_R}\psi_\beta + n\psi_\beta + \frac{r_R x_H}{x_R} i_\beta, \tag{17.6}$$

$$\frac{di_\alpha}{dt} = \frac{x_R}{x_S x_R - x_H^2}\left(-\frac{x_H}{x_R}\frac{d\psi_\alpha}{dt} - r_S i_\alpha + u_\alpha\right),$$

$$\frac{di_\beta}{dt} = \frac{x_R}{x_S x_R - x_H^2}\left(-\frac{x_H}{x_R}\frac{d\psi_\beta}{dt} - r_S i_\beta + u_\beta\right),$$

where M and M_l are, respectively, the torque developed by the motor and load torque; r_S and r_R are the resistances of, respectively, stator and rotor windings; x_R and x_S are the inductances of rotor and stator, x_H is the mutual inductance of rotor and stator, J is the moment of inertia as reduced to the motor shaft. The Eqs. (17.6) may be represented as the Cauchy form if $d\psi_\alpha/dt$ and $d\psi_\beta/dt$ in the last two equations are replaced by the right-hand sides of the second and third equations, respectively.

Let the induction motor be supplied from a source of d.c. voltage $\pm u_0$ through a thyristor voltage inverter with forced switching or through a transistor convertor operating in switching mode. Each of the inverter's output voltages u_R, u_S, u_T of phases R, S, T connected to the respective phases of the motor may be at each instant made equal either to u_0 or $-u_0$, the stator voltage components are related to the convertor phase voltages by a linear expression

$$\begin{bmatrix} u_\alpha \\ u_\beta \end{bmatrix} = \frac{2}{3}\begin{bmatrix} e_{R\alpha} & e_{S\alpha} & e_{T\alpha} \\ e_{R\beta} & e_{S\beta} & e_{T\beta} \end{bmatrix}\begin{bmatrix} u_R \\ u_S \\ u_T \end{bmatrix}, \tag{17.7}$$

where $(e_{R\alpha}, e_{R\beta})$, $(e_{S\alpha}, e_{S\beta})$, $(e_{T\alpha}, e_{T\beta})$ are the components of unit vectors of phases R, S, T, respectively. In particular, if the axis of phase R is oriented along the axis α,

$$e_R^T = (1, 0), \quad e_S^T = \left(-\frac{1}{2}, \frac{\sqrt{3}}{2}\right), \quad e_T^T = \left(-\frac{1}{2}, -\frac{\sqrt{3}}{2}\right).$$

(17.6) and (17.7) describe operation of the motor and power convertor to a precision of the zero component of stator current. If this component occurs, the requirement is imposed that the sum of mean convertor voltages u_R, u_S, u_T be equal to zero, i.e. that they form a three-phase group [31]. Another requirement often imposed on the stationary mode of induction motor is that the rotor flux $\|\psi\| = (\psi_\alpha^2 + \psi_\beta^2)^{1/2}$ be maintained at a given level $\psi_z(t)$ which is close, for instance, to the saturation of magnetic circuit [106]. Thus, the task of control is to generate switching control actions u_R, u_S, u_T so that rotor speed is equal to the reference input $n_z(t)$, the above requirements being satisfied.

Let us form the following functions that characterize the deviations from the desired mode:

$$s_1 = c_1(n - n_z) + \frac{d}{dt}(n - n_z), \quad c_1 = \text{const},$$

$$s_2 = c_2(\|\psi\| - \psi_z) + \frac{d}{dt}(\|\psi\| - \psi_z), \quad c_2 = \text{const}, \tag{17.8}$$

$$s_3 = \int_0^t (u_R + u_S + u_T)d\lambda.$$

In order to solve the problem of control, it suffices to nullify functions (17.8). Indeed, $s_1 = 0$ may be regarded as a differential equation for the mismatch between the actual and desired rotor speeds which exponentially tends to zero at $c_1 > 0$ with time constant $1/c_1$. At $c_2 > 0$ equation $s_2 =$ describes an aperiodic decaying variation of the mismatch between the actual and desired rotor fluxes. Finally, if s_3 is zero, the symmetry condition for time-averaged convertor voltages u_R, u_S, u_T is satisfied.

Let us seek after simultaneous equality to zero of functions (17.8) by inducing sliding mode at the intersection of surfaces $s_1 = 0$, $s_2 = 0$, $s_3 = 0$ by means of three-dimensional control $u^{\mathrm{T}} = (u_R, u_S, u_T)$ whose components, phase voltages, are discontinuous. This problem is equivalent to that of stability of motion projection on a subspace $s(s^{\mathrm{T}} = (s_1, s_2, s_3))$ described by differential equation

$$\frac{ds}{dt} = F + Du, \tag{17.9}$$

where $F^{\mathrm{T}} = (f_1, f_2, 0)$ and matrix D are independent of control vector and in virtue of (17.6) may be found through differentiation of s_1, s_2, s_3 (17.8). The components f_1 and f_2 of vector F are continuous functions of the motor parameters, angular speed, acceleration, magnitude of rotor flux and its time derivative, matrix D is defined by

$$D = \begin{bmatrix} D_1 \\ d \end{bmatrix},$$

$$D_1 = \frac{kx_H}{x_S x_R - x_H^2} \begin{bmatrix} \frac{1}{J} & 0 \\ 0 & \frac{r_R}{\|\psi\|} \end{bmatrix} \begin{bmatrix} \psi_\beta & -\psi_\alpha \\ \psi_\alpha & \psi_\beta \end{bmatrix} \begin{bmatrix} e_{R\alpha} & e_{S\alpha} & e_{T\alpha} \\ e_{R\beta} & e_{S\beta} & e_{T\beta} \end{bmatrix},$$

$$d = (1, 1, 1), \quad k = \text{const}.$$

The convergence of the non-linear system (17.9) to the origin may be determined only for particular matrices D (see Sect. 4.3–5). In order to make solvable the problem of motion stability in subspace s, let us use one of the invariant transformations of equations of discontinuity surfaces (see Sect. 6.2). Take matrix Ω in transformation

$$s^* = \Omega s \tag{17.10}$$

as follows

$$\Omega = D^{-1} = \frac{1}{3}\begin{bmatrix} \nu(e_{R\alpha}\psi_\beta - e_{R\beta}\psi_\alpha) & \mu(e_{R\alpha}\psi_\alpha + e_{R\beta}\psi_\beta) & 1 \\ \nu(e_{S\alpha}\psi_\beta - e_{S\beta}\psi_\alpha) & \mu(e_{S\alpha}\psi_\alpha + e_{S\beta}\psi_\beta) & 1 \\ \nu(e_{T\alpha}\psi_\beta - e_{T\beta}\psi_\alpha) & \mu(e_{T\alpha}\psi_\alpha + e_{T\beta}\psi_\beta) & 1 \end{bmatrix}, \tag{17.11}$$

with

$$\nu = \frac{3J(x_S x_R - x_H^2)}{x_H \| \psi \|^2}, \quad \mu = \frac{3}{2}\frac{x_S x_R - x_H^2}{x_H x_R \| \psi \|^2}.$$

Matrix Ω exists because everywhere $\det D \neq 0$ with the exception of points $\| \psi \| = 0$. Then the equation of motion projection on subspace s^* is

$$\dot{s}^* = \Omega F + \frac{d\Omega}{dt}\Omega^{-1}s^* + u. \tag{17.12}$$

Let the components of control vector u have discontinuities on surfaces s_1^*, s_2^*, s_3^*:

$$u = -u_0 \operatorname{sign} s^*. \tag{17.13}$$

At sufficiently great u_0, there always will be in subspace a domain including the origin such that for each component of vector

$$F_0 = \Omega F + \frac{d\Omega}{dt}\Omega^{-1}s^* \tag{17.14}$$

condition

$$u_0 > |F_{0i}|, \quad i = 1, 2, 3 \tag{17.15}$$

is satisfied.

Relations (17.12) through (17.15) mean that the signs of functions of s_i^* and their rates will be opposite and sliding mode will occur each surface $s_i^* = 0$. Thus, equality of s^* or, in virtue of non-singularity of Ω, of s to zero is provided, which solves the problem of control.

The resulting control algorithm of system (17.8), (17.10), (17.11) for the class of reference inputs and disturbances (17.14), (17.15) ensures reproducibility of reference inputs without dynamic tracking error.

Observe that by means of (17.10) the matrix preceding control in (17.12) may be made not only diagonal, but also positively definite, symmetrical or Hadamard matrix. For equations with matrices of this type, Chap. 4 describes methods for determination of sliding mode stability conditions through the piecewise smooth and quadratic Lyapunov functions. Moreover, sliding modes may be induced also by means of the hierarchy of control method.

Implementation of this algorithm relies on information about the components ψ_α and ψ_β of rotor flux as well as about motor shaft angular speed and acceleration. In special induction motors with built-in sensors ψ_α and ψ_β are directly measured. Magnetic flux may be estimated through the stator phase currents and angular speed that usually are measurable. It follows from the consideration of the equations with respect to ψ_α and ψ_β in (17.6) that the rotor circuit model or observer of rotor flux components may be designed in the form of a system consisting of two integrators with non-linear cross-coupling, three input variables n, i_α and i_β (the latter two being related to phase currents i_R, i_S, i_T by relations similar to (17.7)) and rotor flux components ψ_α and ψ_β as outputs:

$$\frac{d\bar{\psi}_\alpha}{dt} = -\frac{r_R}{x_R}\bar{\psi}_\alpha - n\bar{\psi}_\beta + \frac{r_R x_H}{x_R} i_\alpha,$$

$$\frac{d\bar{\psi}_\beta}{dt} = -\frac{r_R}{x_R}\bar{\psi}_\beta + n\bar{\psi}_\alpha + \frac{r_R x_H}{x_R} i_\beta,$$

where $\bar{\psi}_\alpha$ and $\bar{\psi}_\beta$ are estimates of ψ_α and ψ_β, respectively. Equations with respect to mismatches $\hat{\psi}_\alpha = \psi_\alpha - \bar{\psi}_\alpha$ and $\hat{\psi}_\beta = \psi_\beta - \bar{\psi}_\beta$ are

$$\frac{d\hat{\psi}_\alpha}{dt} = -\frac{r_R}{x_R}\hat{\psi}_\alpha - n\hat{\psi}_\beta, \quad \frac{d\hat{\psi}_\beta}{dt} = -\frac{r_R}{x_R}\hat{\psi}_\beta + n\hat{\psi}_\alpha.$$

Let us check by means of the Lyapunov function $v = \frac{1}{2}(\hat{\psi}_\alpha^2 + \hat{\psi}_\beta^2)$ that the equilibrium state in space $\hat{\psi}_\alpha, \hat{\psi}_\beta$ is asymptotically stable in the large. The time derivative of v on the trajectories of non-linear system of equations with respect to $\hat{\psi}_\alpha$ and $\hat{\psi}_\beta$

$$\dot{v} = -\frac{r_R}{x_R}(\hat{\psi}_\alpha^2 + \hat{\psi}_\beta^2) = -\frac{2r_R}{x_R} v$$

is a negative definite function. The mismatches $\hat{\psi}_\alpha$ and $\hat{\psi}_\beta$, therefore, tend to zero and, correspondingly, estimates $\bar{\psi}_\alpha$ and $\bar{\psi}_\beta$ asymptotically approach the true values of rotor flux components. (Convergence rate is defined by r_R/x_R). The time derivatives of ψ_α and ψ_β that are also required for implementation of s_2 in (17.8) may be measured at the input of model integrating elements.

In the environment of measuring and differentiating devices with dynamic imperfections, s_1 in (17.8) is generated, like in the case of d.c. motor, with the aid of first-order asymptotic observer. A first-order observer with input $i_\alpha\psi_\beta - i_\beta\psi_\alpha$ (recall that ψ_α and ψ_β are already known and the current components i_α and i_β may be found through the measured phase currents), that enables determination of load torque M_l, may be designed by means of an equation similar to (17.5) and obtained through (17.6):

$$i_\alpha\psi_\beta - i_\beta\psi_\alpha = \frac{x_R}{x_H} J \frac{dn}{dt} - \frac{x_R}{x_H} M_l$$

From this equation acceleration dn/dt is determined, of course to an accuracy of the decaying transition component. By generating s_1 through the output of this auxiliary dynamic system one may induce sliding mode motion which is defined basically by c_1 and is independent of the load torque. The structure of this observer may be found in Sect. 2.

Obviously, this approach me be used for controlling not only the rotor speed, but also other induction motor coordinates such as rotor position and angular acceleration, electric torque developed by the motor, and stator current. The only distinction is that the deviations from the desired mode should be characterized by the values of switching functions depending on particular control objectives.

4 Control of Synchronous Motor

The current trend to a.c. motors that in contrast to d.c. ones do not need collectors has made topical the use of synchronous motors of medium and high powers in controlled drives [31]. Both induction and synchronous motors offer a challenge to control because of the high order and non-linearity of their differential equations. Induction motors seem superior as regards simplicity of design and cost. But the synchronous motor-based drive features essentially wider scope of functional abilities because, in addition to the power supply channel, the rotor circuit is also controllable. This fact seems to account for the significant research efforts in the area of synchronous motor applications.

This section considers the possibility of designing non-linear control algorithms for synchronous motors by using sliding modes.

Using an orthogonal system (d, q) rotating together with rotor, let us consider the differential equations of synchronous motor written in relative variables [31]:

$$e_d = ri_d + x_d \frac{di_d}{d\tau} + x_{\mathrm{ad}} \frac{di_f}{d\tau} - x_q i_q \frac{d\gamma}{d\tau},$$

$$e_q = ri_q + x_q \frac{di_q}{d\tau} + (x_d i_d + x_{\mathrm{ad}} i_f) \frac{d\gamma}{d\tau},$$

$$e_f = r_f i_f + x_{\mathrm{ad}} \frac{di_d}{d\tau} + x_f \frac{di_f}{d\tau},$$

$$M = ((x_d - x_q) i_d + x_{\mathrm{ad}} i_f) i_q,$$

$$J \frac{dn}{dt} = M - M_l,$$

$$n = \frac{d\gamma}{d\tau}, \tag{17.16}$$

where i_d, i_q are the components of stator winding current; i_f is the excitation winding current; e_d, e_q are the components of the stator winding voltage; e_f is the winding excitation voltage; r, x_q, x_d, are resistance and inductances of stator winding; r_f and x_f are the resistance and inductance of excitation winding; x_{ad} is the mutual inductance of stator and excitation winding; n is the angular speed; M is the electric torque; M_l is the load torque; γ is the angle between rotor axis and that of stator phase R; J is the moment of inertia. Variables e_0, e_d, e_q, i_0, i_d, i_q are related to phase variables e_R, e_S, e_T, i_R, i_S, i_T as follows

$$E = CE', \quad I = CI',$$

$$E' = \begin{bmatrix} e_R \\ e_S \\ e_T \end{bmatrix}, \quad I' = \begin{bmatrix} i_R \\ i_S \\ i_T \end{bmatrix}, \quad E = \begin{bmatrix} e_0 \\ e_d \\ e_q \end{bmatrix}, \quad I = \begin{bmatrix} i_0 \\ i_d \\ i_q \end{bmatrix}. \tag{17.17}$$

where

$$C = \begin{bmatrix} \frac{1}{3} & \frac{1}{3} & \frac{1}{3} \\ \frac{2}{3}\cos\gamma & \frac{2}{3}\cos\left(\gamma - \frac{2\pi}{3}\right) & \frac{2}{3}\cos\left(\gamma + \frac{2\pi}{3}\right) \\ -\frac{2}{3}\sin\gamma & -\frac{2}{3}\sin\left(\gamma - \frac{2\pi}{3}\right) & -\frac{2}{3}\sin\left(\gamma + \frac{2\pi}{3}\right) \end{bmatrix}.$$

This mathematical model of synchronous motor was formed under the usual assumptions such as the symmetry of phase windings and freedom from voltage e_0 and current i_0 in the zero phase-sequence (symmetry condition), no-saturation condition [31]. In the case under consideration, power is supplied from a thyristor voltage convertor with forced switching. Each of the phase voltages e_R, e_S, e_T may assume at any time instant value u_0 or $-u_0$. Voltage e_f on the excitation winding may be e_{f_0} or $-e_{f_0}$.

Discontinuous functions e_R, e_S, e_T and e_f will be regarded as the components of the four-dimensional control vector u designed in the form of

$$u = V \operatorname{sign} s^*, \tag{17.18}$$

where $u^{\mathrm{T}} = (e_R, e_S, e_T, e_f)$, $V = \operatorname{diag}(u_0, u_0, u_0, e_{f_0})$, $(\operatorname{sign} s^*)^{\mathrm{T}} = (\operatorname{sign} s_R, \operatorname{sign} s_S, \operatorname{sign} s_T, \operatorname{sign} s_f)$ functions s_R, s_S, s_T and s_f are to be chosen depending on the control objectives.

Usually, the major controlled variable of the drive is rotor speed that should be equal to the reference input $n_z(t)$. As in the case of induction motor, the symmetry requirement is common which implies that

$$\int_0^{\tau} (e_R + e_S + e_T)\, d\lambda = 0.$$

Since there are four control actions in synchronous motor e_R, e_S, e_T, e_f, two more state variables may be controlled in addition to maintaining symmetry and given speed. For example, one may support the maximal developed torque under the fixed magnitude of stator current, or desired $\cos\varphi$, (φ is the angle between the current and voltage vectors) etc. Assume that these particular requirements may be recalculated into reference input i_{dz} and i_{fz} for the component of stator and excitation winding currents i_d and i_f, respectively. It will be illustrated below by means of an example how these reference inputs may be formed).

The task of control, thus, is to form such a low of switching control actions e_R, e_S, e_T, e_f that supports rotor velocity n equal to the reference input $n_z(\tau)$ provided that requirements $i_d = i_{dz}$, $i_f = i_{fz}$ are followed and the symmetry condition is satisfied.

Let us form functions characterizing the deviation from the desired mode:

$$s_0 = \int_0^\tau (e_R + e_S + e_T)d\lambda,$$

$$s_1 = c_1(n_z - n) + \frac{d}{d\tau}(n_z - n), \quad c_1 = \text{const}, \tag{17.9}$$

$$s_2 = i_{dz} - i_d, \quad s_3 = i_{fz} - i_f.$$

For the stated problem to be solvable, it suffices that the functions (17.19) be equal to zero. If $s_0 = 0$, this implies that the symmetry condition is obeyed. Condition $s_1 = 0$ may be regarded as a differential equation of the mismatch between the reference and actual rotor speeds whose solution will exponentially tend to zero with time constant $1/c_1$ at $c_1 > 0$. Conditions $s_2 = 0$ and $s_3 = 0$ imply that, respectively, i_d and i_f are equal to their reference inputs i_{dz} and i_{fz}. Simultaneous equality to zero of the functions (17.19) may be provided by inducing sliding mode in the system state space on the intersection of surfaces $s_0 = 0$, $s_1 = 0$, $s_2 = 0$, $s_3 = 0$.

Control vector u is defined, as before, from the equations of system motion projection on subspaces ($s^{\mathrm{T}} = (s_0, s_1, s_2, s_3)$):

$$\frac{ds}{d\tau} = F + Du,$$

where vector F and matrix D may be found through differentiation of s_0, s_1, s_2, s_3 (17.19) in virtue of (17.16). Components of F and D are continuous functions of motor parameters and coordinates; matrix D, like in the case of induction motor, is non-singular everywhere with the exception of the points where the motor flux that defines its torque is equal to zero.

The further design procedure is similar to that for induction motor as described in Sect. 3. Assuming that $\det D \neq 0$, we take an invariant transformation of the discontinuity surface equations

$$s^* = \Omega s, \quad \Omega = D^{-1}, \tag{17.20}$$

$$\frac{ds^*}{d\tau} = \Omega F + \frac{d\Omega}{dt}\Omega^{-1}s^* + u. \tag{17.21}$$

As follows from (17.18) and (17.21), all the components of s^* and their rates will have opposite signs if

$$u_0 > |F_{0i}|, \quad i = 1,2,3, \quad e_{f_0} > |F_{04}|, \tag{17.22}$$

where F_{01}, F_{02}, F_{03}, F_{04} are the components of vector

$$F_0 = \Omega F + \frac{d\Omega}{d\tau}\Omega^{-1}s^*.$$

Hence, for the class of external actions (17.22) reference inputs are reproduced by the control algorithm (17.18), (17.19), (17.20) with zero dynamic error.

Let us dwell now upon issues concerning optimization of the operation modes of synchronous motors. In addition to the requirements of control accuracy and dynamic quality, a number of requirements due to static operating conditions are imposed on the system such as maintaining the maximal eletric torque under fixed magnitude of stator current, or desired $\cos\varphi$, etc. In the above algorithms, the desired static modes are defined by reference inputs n_z, i_{dz} and i_{fz}. Therefore, optimization problems must be reformulated in terms of these variables.

Let us consider several of the most important problems under an assumption that excitation flux is constant ($\psi_{\mathrm{ad}} = \mathrm{const}$). This class is represented by synchronous motors with magneto-electric excitation and with fixed excitation flux for which $\psi_{\mathrm{ad}} = i_f x_{\mathrm{ad}}$. In this case, if one still has to maintain speed n_z, only one degree of freedom is left for optimization, that of reference input for the stator current component i_d.

Consider first maintaining maximal electric torque under fixed magnitude of stator current. Electric torque is representable as

$$M = \psi_{\mathrm{ad}} i_{\mathrm{st}} \sin\alpha + (x_d - x_q) i_{\mathrm{st}}^2 \tfrac{1}{2} \sin 2\alpha, \tag{17.23}$$

where i_{st} is the stator current magnitude equal to $\sqrt{i_d^2 + i_q^2}$, α is the angle between the direction of stator current and axis d. Equation

$$\frac{dM}{d\alpha} = \psi_{\mathrm{ad}} i_{\mathrm{st}} \cos\alpha + (x_d - x_q) i_{\mathrm{st}}^2 \cos 2\alpha = 0,$$

$$\cos\alpha = -\frac{\psi_{\mathrm{ad}}}{4(x_d - x_q) i_{\mathrm{st}}} + \sqrt{\left(\frac{\psi_{\mathrm{ad}}}{4(x_d - x_q)^2}\right)^2 + \tfrac{1}{2}}.$$

defines the condition of electric torque maximum.

It is often the case in practice that condition $\psi_{\mathrm{ad}} i_{\mathrm{st}} \gg \frac{1}{2}(x_d - x_q)^2 i_{\mathrm{st}}^2$ is satisfied, and one may assume with reasonable accuracy that for $\psi_{\mathrm{ad}} \neq 0$ and $i_{\mathrm{st}} \neq 0$ the electric torque reaches its maximum at $\alpha = \pi/2$, i.e. the stator current vector goes along axis q or

$$i_{dz} = 0.$$

Among other requirements that may be satisfied through the appropriate choice of reference inputs are maximization of efficiency, minimization of stator current magnitude under fixed shaft torque, provision of the desired $\cos\varphi$, constraints on currents, etc.

Presented in this chapter concept of control resting upon the deliberate introduction of sliding modes is applicable also to other kinds of electromechanical transformers of power such as electric generators, a.c. electronic motors, doubly fed motors.

The general design methodology for the class of control systems under consideration consists, first, in choosing the desired nature of control processes and functions characterizing the deviation from the desired operational conditions, and, second, in designing sliding motions so that the deviations be equal to zero. The number of independent control actions (dimensionality of control vector) and, consequently, the number of sliding surfaces characterizing the deviations from the desired mode may differ depending on the particular type of motor and on the nature applications.

Chapter 18
Examples

1 Electric Drives for Metal-Cutting Machine Tools

The increased requirements to the static and dynamic accuracy of the machine tools having electric drives with transistor or thyrisotor converters cannot be met within the framework of the classical methods of linear theory, and in order to design control systems for the installations of this sort it is necessary to employ methods taking into account the discontinuous nature of control actions and non-linearity of the equations of electric motors. Below, a number of electric drives is described having sliding mode control and intended for machine tools.

Induction Motor-Based Feed Drive for Numerically Controlled Machines

The drive uses a commercial squirrel-cage induction motor with transistor voltage inverter powered from the industrial network through a transformer and uncontrollable rectifier. The required data are sensed by a transducer of phase currents and photo-pulse transducer of shaft motion. The reference input signal of the drive is the unitary code of displacement generated by the numerical controller. Since in the drive in question, controlled variables are the angular position of the motor shaft and magnetic flux, therefore

$$s_1 = c_{11}(\theta - \theta_z) + c_{12}\frac{d}{dt}(\theta - \theta_z) + \frac{d^2}{dt2}(\theta - \theta_z) \tag{18.1}$$

is used instead of s_1 in (17.8) where θ and θ_z are the actual and given angular positions of the shaft, respectively; c_{11} and c_{12} are constant parameters that together with c_2 in (17.8) define system motion in sliding mode. In order to provide sliding motion, vector control is used with transformation of the motor current vector (similar to (17.10) from the fixed coordinate system into the system oriented along the rotor flux vector [16, 65, 124]. Although the sixth order motion equation system of the induction motor is essentially non-linear, the mismatches of position and flux decay in sliding mode according to the independent homogeneous second- and first-order linear Eqs. (18.1) and (17.8).

According to the methods of Chap. 17, the motor magnetic flux and switching function s_2 in (17.8) are computed through a dynamic model of the electromagnetic rotor circuit, and the instantaneous values of the switching function s_1 (18.1) are restored through the measured angular error and stator current components by the linear Luenberger state observer designed under the assumption that load torque is piecewise-constant time function. The prototypes of tracking drives [152] designed and tested jointly by the Institute of Control Sciences, Research and Production Amalgamation ENIMS and Production Amalgamation LEMZ have nomial torque 21 Nm and speed range of $\pm 10^3$ rpm and tolerate a double short-time overload. The accuracy of positioning and tracking is ± 1 pulse of the displacement transducer (with 10^4 pulses per revolution), the reversal time is 10^{-2} s, for the frequency ± 60 rpm, and the bandwidth is 80 Hz.

Controlled Induction-Motor for Spindle Drive

The development of a controlled drive with no mechanical motion transducer mounted on the motor shaft is an urgent problem from the standpoint of improving drive performance and reliability. The drive is built around a special-purpose three-phase squirrel cage induction motor of the motor-spindle type. The range of controlled speed is 40 through 8,000 rpm. Up to about 10^3 rpm the motor operatres with constant rotor flux and, therefore, with constant maximal torque (50 Nm), at higher speeds the flux is reduced in order to maintain a constant power (up to 5.5 KW).

Information is captured through the sensors of instantaneous currents and voltages of the stator windings. The rotor speed and the components of the vector of rotor flux are computed by the model of motor electromagnetic circuits – non-linear state observer belonging to the class of discontinuous parameter systems [63].

The observer simulates the differential equations that describe the electromagnetic processes in the stator and rotor windings of the motor (17.6). The measured instantaneous voltage u and modelled stator current i^* are model's input and output, respectively. The model is controlled by the components of difference $\Delta i = i^* - i$ of modelled i^* and measured i stator currents. To this end, two discontinuous parameters are introduced into the model: n^* representing the unknown (on the average) rotor angular speed and μ that is an auxiliary parameter for stabilization of computations. The observer's equations with respect to the modelled stator current $i^* = i^*_\alpha + ji^*_\beta$ and rotor flux $\psi^* = \psi^*_\alpha + j\psi^*_\beta$ represented in the complex form are as follows:

$$\frac{di^*}{dt} = \frac{x_R}{x_S x_R - x_H^2}\left(\frac{x_H}{x_R}\left(\frac{r_R}{x_R} - \mu - jn^*\right)\psi^* - \left(\frac{r_R x_H^2}{x_R^2} + r_S\right)i^* + u\right),$$

$$\frac{d\psi^*}{dt} = -\left(\frac{r_R}{x_R} - \mu\left(1 + \frac{k}{\dfrac{r_R}{x_R} - jn^*_{eq}}\right) - jn^*\right)\psi^* + \frac{r_R}{x_R}i^*,$$

$$\mu = \begin{cases} \mu_0 & \text{with } (\Delta i_\alpha \psi^*_\alpha + \Delta i_\beta \psi^*_\beta) > 0 \\ -\mu_0 & \text{with } (\Delta i_\alpha \psi^*_\alpha + \Delta i_\beta \psi^*_\beta) < 0, \end{cases}$$

$$\mu_0,\ n - \text{const},$$

$$n^* = \begin{cases} n_0 & \text{with } (\Delta i_\beta \psi^*_\alpha + \Delta i_\alpha \psi^*_\beta) > 0 \\ -n_0 & \text{with } (\Delta i_\beta \psi^*_\alpha + \Delta i_\alpha \psi^*_\beta) < 0, \end{cases}$$

(the equations are similar to the corresponding motor Eqs. (17.6) with respect to $i = i_\alpha + ji_\beta$ and $\psi = \psi_\alpha + j\psi_\beta$), where n^*_{eq} is the equivalent value of discontinuous parameter n^* (derived, for example, by a low-pass filter), k – constant, $k > 0$. In this system, sliding mode occurs on the surfaces defined by the zero mismatch between the modelled and measured currents $\Delta i_\alpha = 0$ $\Delta i_\beta = 0$. In the physical sliding mode, the parameters μ and n^* vary with high frequency so that their equivalent values tend, respectively, to zero and n, and ψ^* tends to the actual rotor flux. The convergence rates of the modelled rotation speed and rotor flux are defined by k (notably, reconstruction of the motor speed and flux through the observed currents and voltages is possible only if the motor field is varying, $\psi(t) \neq \text{const}$ [63]).

The off-line trials of the device computing the angular velocity and flux of induction motor (designed by the Institute of Control Sciences in collaboration with the Ivanovo Power Institute) and the breadboard of controlled drive (designed by the Institute of Control Sciences and the Research and Production Amalgamation ENIMS) have confirmed its operability in compliance with the specifications. The motor-spindle drive is being tested now in the field environment.

Controlled Transistor–Thyristor Drive with Synchronous Motor

This drive is intended for the tracking system controlling the feed of working tools in numerically controlled precision machine tools. It is built around a three-phase synchronous motor with permanent-magnet excitation and thyristor commutator. The synchronous motor, thus, operates as the electronic d.c. motor. The d.c. circuit has a transistor convertor for control of current (or motor torque) by means of high-frequency modulation of the output voltage. The drive has a rotor position transducer used for generation of commands to switch the thyristors and a tacho generator as transducer of angular velocity.

For normal operation of the modulator, its input must be bounded not only by level, but by rate of variation as well. This requirement may be met by serial connection of a relay element with output u equal to $\pm u_0$ and integrating one that generate the modulator input signal. The introduction of integrating element increases the order of motor equations by one, and the second derivative of error with respect to velocity must be used in the switching function of (17.19) in order to induce sliding mode:

$$s_1 = c_1(n_z - n) + c_2 \frac{d}{dt}(n_z - n) + \frac{d^2}{dt^2}(n_z - n). \tag{18.7}$$

An asymptotic state observer is used for restoration of the instantaneous values of the relay element switching function. The modulator input and signal of rotation speed error are the observer's inputs.

The bench tests of the drive designed by the Research and Production Amalgamation ENIMS have shown up its dynamic performance even in the heaviest modes of slow speeds and pulse load. The quality of factors of the transients proved to be close to the factors of the "ideal" drive with PID control law that cannot be implemented because of the mismatch between the model and real process on high frequencies and insufficient information about process operator and state. The designed drive has identical characteristics over all the control range, low sensitivity to load disturbances, and reversal time under the nominal load 3 to 5 times smaller than that of standard drives. In addition, it has low sensitivity to motor parameter variations.

Conclusion

In spite of the existing trend towards a.c. drives, d.c. motors are still in general use in machine tools. Since they are usually controlled by transistor or thyristor converters operating in key mode, sliding mode algorithms can prove effective also for d.c. drives.

Such a drive may be exemplified by a modification of the commercial drive ETZI that has the same size, power unit (controlled rectifier and sampled-data phase control) and transducer (tacho generator), i.e. only the controller board has been changed which facilitates its introduction into industry. The d.c. drive controller [64] is smaller to that of synchronous drive: continuous reference input of sampled-data phase control system is generated by relay and integrating elements, sliding mode is induced, and an observer is employed. The effectiveness of this drive has been confirmed both by off-line tests and field operation of the breadboard. At low rotation frequencies, the drive has reverse time from 0.2 to 0.4 s which is by a factor of 5 through 7 smaller than with the existing ETZI drive with PI controller. Its high dynamic performance is comparable with that of d.c. drives using pulse-width modulation convertors. Therefore, this modification is fit for machines of higher precision class.

Sliding modes are used also in the current loops of cascaded control systems of d.c. [70] and a.c. [121] drives. The relay current control system for induction and synchronous motors [121] and the control of d.c. motor rotational speed [38, 59] are based on similar design principles. The results obtained in [38, 59] were used in the microprocessor manipulator control system [37, 59].

2 Vehicle Control

The designs of control systems for transportation facilities were oriented to electrically driven vehicles with autonomous and external power supplies. In the last chapter the sliding mode algorithms in the control of electrical drives were shown to be prospective. It must be noted here only that, in contrast to the machine tool drives, in transport applications (especially in autonomous vehicles) power consumption criteria prevail over those of accuracy. In particular, energy regeneration into the power supply should be usually envisioned in them.

Battery-Driven Vehicle

The control system for one-motor d.c. vehicle drive has been designed by the VAZ motor-factory (Togliatti) in collaboration with Institute of Control Sciences (Moscow). In order to describe the control algorithm, let us write the d.c. motor equations

$$J\frac{dn}{dt} = k_1 i_1 i_2 - M_l,$$

$$L_1\frac{di_1}{dt} = -R_1 i_1 - k_2 i_2 n + u_1$$

$$L_2\frac{di_2}{dt} = -R_2 i_2 + u_2, \qquad (18.8)$$

where n is the shaft rotational velocity; i_1, u_1, i_2, u_2 are the currents and voltages of rotor and excitation windings, respectively; M_l is the load torque; J is the moment of inertia; k_1, k_2, L_1, L_2, R_1 and R_2 are motor parameters.

The discontinuity voltages u_1 and u_2 are chosen so as to make the rotor and excitation currents to track in sliding mode the reference inputs i_{1z} and i_{2z}:

$$u_1 = u_{10}\,\mathrm{sign}(i_{1z} - i_1),$$

$$u_2 = u_{20}\,\mathrm{sign}(i_{2z} - i_2), \quad u_{10}, u_{20} - \text{const.} \tag{18.9}$$

The reference input i_{1z} is defined by the driver's command (with allowance for the natural constraints on the rotor current) and then recalculated into the excitation current reference input:

$$i_{2z} = k i_{1z}, \tag{18.10}$$

$$\dot{k} = \lambda_0(\Delta - (i_{1z} - i_1)) - \lambda_1 \mathrm{Sg}(k - k_{\max}) + \lambda_2 \mathrm{Sg}(k_{\min} - k), \tag{18.11}$$

where λ_0, λ_1, λ_2 and Δ are positive parameters, $\lambda_1 > \lambda_0\Delta$, $\lambda_2 > \lambda_0\Delta$; $\mathrm{Sg}\,a = 1$ for $a > 0$, $\mathrm{Sg}\,a = 0$ for $a < 0$; $k_{\max}$ is the proportionality factor, providing minimal heat loses, $k_{\min}$ is the proportionality factor providing the maximal torque under the voltage constraint on u_1.

At the initial stage of low-speed motion when $k_2 i_2 n$ of the equation in (18.8) is small, the voltage u_{10} in the control (18.9) suffices for maintaing the equality of reference and true values of rotor current. As the result, $i_{1z} = i_1$, $\dot{k} = \lambda_0\Delta > 0$, the parameter k becomes $k_{\max}$ and is maintained at this level in sliding mode, and the vehicle motor develops torque as defined by the driver (and equal to $k_{\max} i_{1z} i_{2z}$) at minimal losses. As the vehicle accelerates, the relation $i_{1z} = i_1$ cannot be fulfilled, $i_{1z} - i_1$ in (18.11) exceeds Δ which results in the decrease of the parameter k in (18.10) or excitation current[1] (and flux), and in the steady state the system will maintain the given torque at a new value of k ($k_{\min} < k < k_{\max}$). If k drops down to $k_{\min}$, the motor will be unable to maintain the desired torque because of physical limitations, and the motion will have maximal acceleration until the desired speed is attained following which the torque setting will be reduced.

The drive realizing the above control algorithm has already been tested on a vehicle. The first 10 units of the unified electric drive have been manufactured for various types of battery-driven vehicles that will be commercially produced in the USSR.

Apart from the design of traditional one-drive vehicles oriented to quantity production it is noteworthy to mention that the VAZ motor-factory in collaboration with the Institute of Control Sciences has designed and tested a vehicle with two wheels driven by induction motors. In this design each of the two wheels is directly attached to an individual special-purpose induction motor.

This basically new technological approach to the design of cars enables one to improve significantly their controllability, safety and cross-country

[1] The reference and true currents in the excitation winding are assumed to be always equal.

capability because the automonous control of each motor-driven wheel, eliminates wheel slippage and, thus, introduces into the control system qualitatively new properties that cannot be obtained in vehicles with a mechanical differential.

The tests have confirmed that the sliding mode control system for motor-driven wheels is the most promising from the viewpoint of controllability, cost, power consumption, simplicity of system adjustment and maintenance.

Tram Car with Induction Motors

The tram car with commercial controllable squirrel-case induction motors seems to have the edge over the d.c. motor-driven tram. The induction motor is smaller and lighter than the corresponding d.c. motor, and the absence of commutator and brushes offers a substantial advantage because all the problems of sliding contact maintenance are eliminated. By choosing an adequate control algorithm one can overcome the difficulties involved into induction motor control that are due to its multi-variable and non-linear nature. Discontinuous controls and sliding modes seem to be natural for the motor fed by a converter with key-mode operation.

The results obtained through a joint Soviet-Yugoslav venture were applied to the design of a prototype of induction motor-driven tram at "Energoinvest" (Sarajevo, Yugoslavia). The system relies upon a modification of the control algorithm described in [65, 124].

Comparison of various induction motors has revealed that the four-pole one has the least mass [116]. The designed system fulfills the requirements usually presented to the traction motor: maximal torque for the speed range $0 \leqq v \leqq v_{max}/2$ and constant power for $v_{max}/2 \leqq v \leqq v_{max}$ (for the tram $v_{max} = 60$ kph).

Four commercial 37 KW induction motors were powered by a modified McMurray inverter that features low weight and cost [106].

The trials carried out on Saraevo municipal lines have demonstrated that the induction motor-based drive features the same performance as that with d.c. motors and can be used to advantage instead of the existing d.c. drives.

3 Process Control

Chemical Fiber Drafting

Production of the majority of chemical fibers rests upon the drafting of melted or softened material followed by winding on the taker-in. In this process, control is applied to so called "onion" from which the fiber is formed, the fiber diameter is the controlled parameter, and the drafting speed v is the control action that can be varied by a d.c. motor. The "onion" is subject to a number of controllable ("onion" temperature, speed of workpiece feed or of material loading) and uncontrollable (variations in the chemical composition and form of the workpiece; fluctuations of air flows at the spot of fiber formation) disturbances affecting the variation in thickness which is the basic product parameter.

The need to remove the fiber diameter sensor from the high-temperature zone creates a significant difficulty for the design of precise control system because this gives rise to transport delay in the measuring circuit and essentially complicates the design of control.

Control systems for processes of this sort usually are cascaded (Fig. 23): using the mismatch between the desired diameter d_0 and the true value as measured by the sensor D_d with delay τ, the controller C_d generates the setting of angular speed η_z for the internal loop comprising the controller C_n which makes the motor speed to track the setting. The standard linear methods of control cannot provide the desired dynamic performance and accuracy of the system and eliminate the mutual dynamic influence between the loops which makes controller adjustment unacceptably complicated.

The internal-loop tracking system relies upon the sliding mode algorithm of Sect. 17.2 and employs rotor current transducer D_I in addition to the speed transducer D_n. It performs the ideal tracking of the reference input n_z by motor speed. As the result, the drive dynamics is practically eliminated and the problem boils down to the control of a lower-order plant with transfer function of the "inertial element + delay" type and linear fiber drafting speed as a control. This control can be efficiently realized by means of the Smith predictor [57].

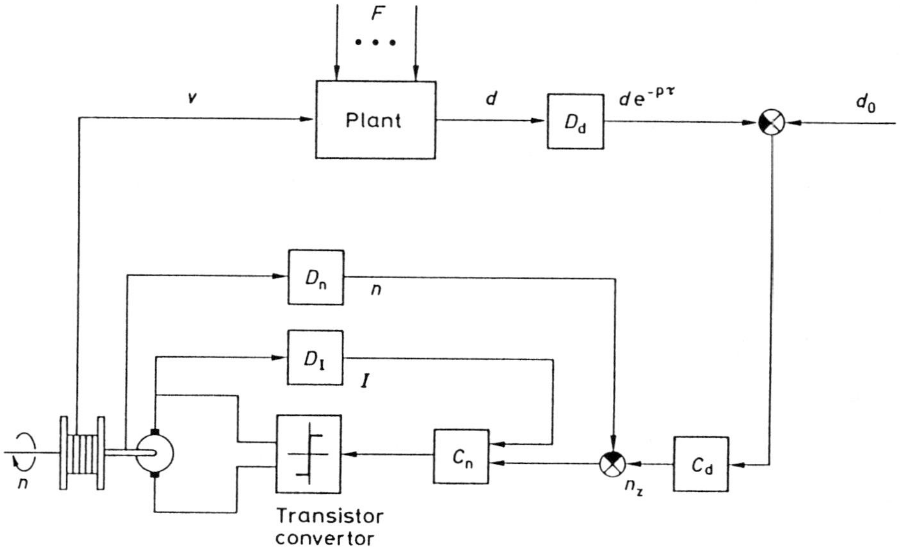

Fig. 23

This system designed and developed at the Moscow Textile Institute has undergone a complete cycle of tests and is operating with a pilot plant. It has enabled a reduction of control error from 8 to 3 through 4% for fiber diameter and from 1 to 0.3% for drafting speed, and a significant improvement in the insensitivity to disturbances.

Electric Arc Furnace

The control system of electric are furnace must support the desired current or voltage in the arc gap by varying the gap. The gap value is the output of the hydrolic booster, and it depends, in its turn, on the control valve position. As in the case discussed in the last section, the system is cascaded: the mismatch between the reference and actual values of the controlled parameter defines the reference input of the internal loop controlling the valve position. The sliding mode control algorithm used in this loop has made the system dynamic performance to be invariant to variations in plant parameters and to have simultaneously high-frequency oscillations necessary for the prevention of obliteration of the hydrolic booster control valve. Its block-diagram is shown in Fig. 24, and the system motion equations are

$$\dot{x}_1 = x_2,$$
$$\dot{x}_2 = f_1(x_1, x_2) + F_1(x_1, x_3) - F_2(x_1, x_4),$$
$$\dot{x}_3 = f_2(x_1, x_2, x_3) + b_1(x_1, x_3)u_1,$$
$$\dot{x}_4 = f_3(x_1, x_2, x_4) + b_2(x_1, x_4)u_2,$$

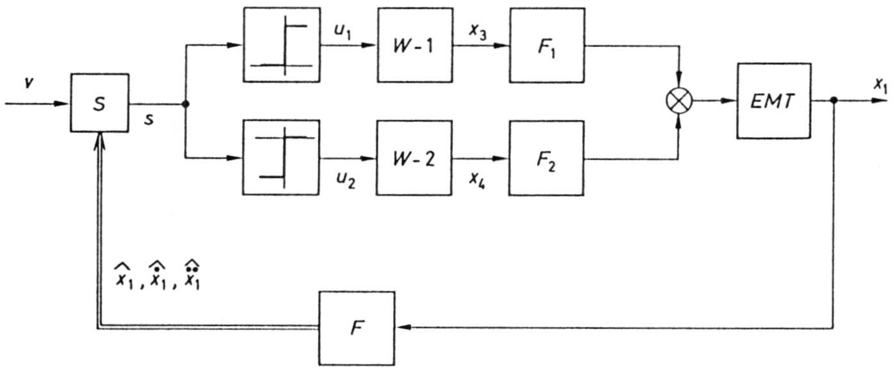

W-1, W-2 Windings
EMT Electromechanical transformer
F Filter
S Generator of switching function

Fig. 24

where x_1 is the control valve displacement; x_3 and x_4 are the currents in the electromagnet windings $W-1$ and $W-2$, respectively; $f_1, f_2, f_3, F_1, F_2, b_1, b_2$ are non-linear functions; $F_1 \geqq 0$, $F_2 \geqq 0$, $b_1 > 0$, $b_2 > 0$. The forces F_1 and F_2 cannot change their signs; therefore, in the physical installation the windings are situated so that the forces are opposite. The discontinuity surface of the control voltages u_1 and u_2 on the windings is

$$s = \ddot{x}_1 + c_2\dot{x}_1 - c_1(v - x_1) = 0$$

where v is the reference input of the internal loop defined by the mismatch with respect to the basic controlled parameter of the arc, the coefficients c_1 and c_2 are chosen so as to provide the desired properties of sliding mode motion on the surface $s = 0$ ($s = 0$ is also the motion equation). The controls $u_1 = MSgs$ and $u_2 = -MSgs$ ($M = \text{const}$, $M > 0$) induce sliding motion. The differentiating filter

$$\mu^3\dddot{\hat{x}}_1 + 3\mu^2\ddot{\hat{x}}_1 + 3\mu\dot{\hat{x}}_1 + \hat{x}_1 = x_1$$

with small parameter μ was used for estimating the derivatives of the displacement x_1.

In a system with physical differentiating filter instead of the ideal sliding mode, self-oscillatory modes occur that also must be induced by the control as it was noted above.

This algorithm has been implemented by the controller built at the Novosibirsk Electrotechnical Institute and installed on the electric are furnace DSP-50 of the "Energomashspetsstal" factory in Kramatorsk. The experience with these controllers gained since 1981 has demonstrated that owing to the reduction of current variations the input power is increased and melting time is reduced. As a result, the reduction in heat time is from 7 to 6.4 hours, in power consumption from 738 to 700 KW, in consumption of graphitized electrodes from 11 to 10.5 kg/t.

Set of Pneumatic Sliding Mode Controllers

In the petroleum, petrochemical and some other industries, there exists a wide class of processes that feature instability in critical (economically advantageous) modes, have non-stationarities and disturbances, numerous uncontrollable factors, interrelated controlled parameters and delays. In particular, the study of the dynamic of catalytic processes (alkylation by sulphuric acid is alkylbenzene production or copolymerization in production of butil rubber) and qualitative analysis of temperature stability of the non-linear models of chemical reactors have shown that the state plane of the plant has two or more singular points and always has at least one saddle point which is indicative of the substantial non-linearity in plant equations. The study of dynamics of catalytic process control systems demonstrates that the domain of initial conditions allowing linear controllers to maintain stability of non-linear processes is rather bounded. In the critical modes, the linear process control becomes impracticable. If process parameters vary over a wide range, the linear systems cannot provide high-quality control. For the processes under consideration the efficiency of control algorithms with deliberately induced sliding modes has been theoretically substantiated.

The institute "Neftekhimavtomat" in collaboration with the Institute of Control Sciences has designed a new "Universal set of pneumatic variable-structure devices" intended for the lower level of automation hierarchy of a wide class of processes in various industries where pneumatics has become accepted [1].

Notably, unlike the set SUPS [153] where sliding motion is used only in data handling units, in the new set sliding motions may be induced for the first time in the main loops of systems controlling a wide variety of processes. This is due to the fact that for the majority of petroleum and petrochemical processes the time constants are of the order of minutes, while the USEPPA[1] elements allows one to generate self-oscillations at much higher frequencies, and the diaphragm actuators are known to be basically able to operate in these modes.

The "Neftekhimavtomat" has brought the new set to the production status, and a batch of 42 complete sets has been installed at six Soviet plants for control of chemical processes.

The experience gained during several years of their operation in industrial environment, has demonstrated that these systems are more stable, reliable and maintainable, and in industrial environment have 2 to 6 times better control performance indices (dynamic accuracy, r.m.s. deviation, speed, etc.) as their linear counterparts.

4 Other Applications

Automation in Fishery

The mastering of the mid-ocean and dramatic changes in the raw material sources have generated a need for automatic facilities in order to improve fishing productivity. This is handicapped by the lack of a reliable controller of efficient warp winch using the positions of shoal and trawl, vessel speed, warp length and tension. The difficulty lies in the fact that the vessel-warp-trawl system is an essentially non-linear controlled plant with variable parameters and that its

[1] The set of pneumatic elements including controllers and units performing elementary operations such as summing, amplification, integration etc. which is produced in the USSR.

state and shoal position are sensed by radiolocation and acoustic methods with high level of noise.

The behaviour of the vessel-warp-trawl system is described by the following equations

$$\dot{v} = a_{11}(\alpha - R) - a_{12}v - a_{13}\omega r \cos\varphi,$$
$$\dot{\omega} = a_{21}(M - Tr),$$
$$\dot{l} = -r\omega$$

$$\dot{h} = a_{41} - a_{42}T\frac{h}{l} - a_{43}(v + \omega r \cos\varphi),$$

$$\dot{x} = v + \omega r \cos\varphi,$$
$$T = P_0 + P + k_1 v + k_1 \omega,$$

where v, ω, l, h, x and T are, respectively, vessel speed, angular winch drum speed, warp length and depth, distance covered by the warp and warp tension; R is a parameter depending on the warp length; $r(l)$, P_0, P are the drum radius, constant and variable warp tension components; a_{11}, a_{12}, a_{13}, a_{21}, a_{41}, a_{42}, a_{43}, k_1, k_2 are constant coefficients; $\cos\varphi = \sqrt{l^2 - h^2}/l$. The electrical torque M of the winch drive and the turning angle α of the variable-pitch propeller are control actions. Vessel speed and warp length should be controlled so as to reduce to zero the depth mismatch between the warp and shoal.

A controller based on the sliding mode algorithms has been designed and constructed for automation of this process. The sea trials in June 1983 in the White Sea had borne out its effectiveness: better guidance accuracy; prevention of fish escaping from the trawl; double reduction of the time required for the trawl to reach the desired level as compared with the common control where the desired vessel speed and trawl length first must be calculated and then separately controlled; the possibility of automatically controlled bottom fishing; prevention from the hooking up of fishing tools by submarine rises and pinnacles. Since measuring instrumentation and actuating mechanisms are already used on board, system's integration will not lead to appreciable capital expenditures.

Resonator Control System

Maintaining various parameters at the desired level with high accuracy is one of the most important problems in the automation of physical experiments. In order to maximize the energy of particle accelerators, in particular, the electromagnetic field frequency must be adjusted and stabilized in accordance with the natural frequency of the resonator that features a sharply pronounced resonance characteristic. This problem is complicated by plant's non-stationarity and non-linearity, incomplete knowledge of the state vector, and the fact that the resonator frequency characteristic is unknown and time-varying.

The equations of the resonator and electric heater connected by a pipe-line are as follows

$$c\frac{dT}{dt}=\frac{P_0}{1+k_0(T_0-T)^2}-\beta G(T-T_1),$$

$$\frac{dT_2}{dt}=\frac{2}{c_1}P-\frac{2\alpha G}{c_1}(T_1-T'_2),$$

$$\frac{dT_1}{dt}=\frac{2\alpha G}{c_2}(T_2-T'_2)-\frac{dT_2}{dt},$$

where c, c_1, c_2 are the resonator, heater and pipe-line heat capacities, respectively; T_0 is the resonator resonance temperature; T, T_1, T'_2, T_2 are water temperatures, respectively, with-in the resonator, at its input, at heater's input and output; P is the heater power; P_0 is the power input into the resonator; G is the cooling water consumption; k_0, β, α are constants. Both the parameters and resonance temperature T_0 in the plant equations are unknown and time-varying.

In the linear accelerator of the Nuclear Research Institute of the USSR Academy of Sciences, high-frequency power input of the resonator $P_r=P_0/(1+k_0(T_0-T)^2)$ is maximized by adjusting the resonator temperature T that is controlled by varying the consumption of cooling water G.

This approach was realized through sliding mode algorithms. The internal loop of the cascaded controller tracks the reference temperature input, the external loop automatically searches the reference input for the internal loop such that the temperature T is equal to the resonance one or that the directly measured energy of output particules is maximal. In the internal loop, a d.c. motor operating in sliding mode is used for varying the cooling water consumption. Self-optimization in the external loop is carried out by means of the algorithm of Sect. 13.3 that does not need any measurement of the gradient of the maximized function. The controller was experimented with a resonator and has proved to maintain the temperature to the accuracy of 0.05 °C within the given range of 22 through 28 °C which gives the frequency deviation 0.8 KHz at most, i.e. provides a sufficiently accurate determination of the resonance frequency which is 990 MHz.

References

1. Abdulayev, A.A., Abdulayev, F.S., Dzafarov, E.M., Utkin, V.I. and Fayazov, M.M.: Set of new variable-structure pneumatic controllers. Abstracts of 13th All-Union Conference on Pneumoautomation held in Donetsk. Moscow: Nauka, 1978 (in Runnian).
2. Aizerman, M.A. and Gantmakher, F.R.: On certain switching specifics in nonlinear automatic control systems with a piecewise-smooth characteristics of nonlinear element. Avtomatika i Telemekhanika, 8, No. 11, pp. 1017–1028, 1957 (in Russian).
3. Aizerman, M.A. and Pyatnitskii E.S.: Foundations of the theory of discontinuous systems I, II. Automation and Remote Control, 35, No. 7 (P. 1), pp. 1066–1079, and No. 8(P1), pp. 1242–1262, 1974.
4. Alimov, Yu.I.: On applying direct Lyapunov's method to differential equations with ambiguous righthand parts. Avtomatika i Telemekhanika, No. 7, pp. 817–830, 1961 (in Russian).
5. Andre, J. and Seibert, P.: Uber stückweise lineare Differential-gleichungen bei Regelungsproblem auftreten, I, II. Arch. der Math., 7, Nos 2 and 3, 1956.
6. Andreev, Yu.N.: Control of finite-dimensional linear plants. Moscow: Nauka, 1976 (in Russian).
7. Andronov, A.A., Vitt, A.A. and Khaikin, S.E.: Theory of oscillations. Moscow: Fizmatgiz, 1959 (in Russian).
8. Andronov, A.A. and Khaikin, S.E.: Theory of oscillations, Part 1. Moscow: ONTI NKTN, 1937 (in Russian).
9. Barbashin, E.A.: Introduction to the theory of stability. Moscow: Nauka, 1967 (in Russian).
10. Barbashin, E.A. and Geraschenko, E.I.: On speeding up sliding modes in automatic control systems. Differentzialniye Uravneniya, 1, No. 1, pp. 25–32, 1965 (in Russian).
11. Beiczy, A.K.: Robot arm dynamics and control. Jet Propulsion Lab. Tech. Memo., pp. 33–669, Feb. 12, 1974.
12. Bezvodinskaya, T.A. and Sabayev, E.F.: Study of the state space features in variable-structure automatic control systems. Automation and Remote Control, 34, No. 7(P. 1), pp. 1105–1108, 1973.

13. Bezvodinskaya, T.A. and Sabayev, E.F.: Stability conditions in the large for variable structure systems. Automation and Remote Control, 35, No. 10(P. 1), pp. 1596–1599, 1974.
14. Billingsley, P.: Convergence of probability measure, John Wiley and Sons, Inc., 1972.
15. Bingulac, S.P., Cuk, N.M. and Calovic, M.S.: Calculation of optimal feedback gain for output constrained regulators. IEEE Trans., Ac-20, pp. 164–166, 1975.
16. Blum, E., Trapeznikov, V.A., Zimonic, S. et al.: An asynchronous motor control device. USSR patent. Bulleten' Izobretenii, No. 36, p. 272, 1981 (in Russian).
17. Bogolyubov, N.N. and Mitropol'skii, Yu.A.: Asymptotic methods in nonlinear oscillations theory. Moscow: Fizmatgiz, 1963 (in Russian).
18. Bondarev, A.G., Bondarev, S.A., Kostyleva, N.E. and Utkin, V.I.: Sliding modes in systems with asymptotic state observers. Automation and Remote Control, 46, No. 6, (P. 1), pp. 679–684.
19. Bortzov, Yu.A., Sabinin, O.Yu., Yumger, I.B. et al.: Discontinuous control of multivariable electromechanical plants. Electric machine building and electric equipment. Rep Interbranch Proc. of Scientific and Technical Papers, pp. 40–48, Issue 35. Leningrad, 1982 (in Russian).
20. Braslavskii, A.D., Korovin, S.K., Kostyleva, N.E., Utkin, V.I. and Shubladze, A.M.: Sliding modes in problems of static optimization, programming and identification. Proc. 5th IFAC Congress, Paris, 1972. Public. IFAC by Inst. Soc. of America, Pittsburg, USA, 1972.
21. Braslavskii, A.D. and Shubladze, A.M.: Solution to problem of "fast" identificaton using multidimensional sliding modes. Automation and Remote Control, 41, No. 2 (P. 1), pp. 195–199, 1980.
22. Breger, A.M., Butkovskii, A.G., Kubyshkin, V.A. and Utkin, V.I.: Sliding modes for control of distributed-parameters entities subjected to a mobile multicycle signal. Automation and Remote Control, 41, No. 3(P.1), pp. 346–355, 1980.
23. Bucy, R.S.: Global theory of the Riccati equation. J. Comp. Systems Sci., 1, pp. 349–361, 1967
24. Butkovskii, A.G.: Theory of mobile control. Automation and Remote Control, 40, No. 6(P.1), pp. 804–813, 1979
25. Butkovskii, A.G., Darinskii Yu.V. and Pustyl'nikov, L.M.: Control of distributed sytems by displacement of the source. Automation and Remote Control, 35, No. 5(P.1), pp. 701–719, 1974
26. Butkovskii, A.G., Darinskii, Yu.V. and Pustylnikov, L.M.: Nonstationary control with systems having distributed parameters. Automation and Remote Control, 37, No. 2(P.1), pp. 133–142, 1976
27. Butuzov, V.F., Vasilyeva, A.B. and Fedoruk, N.V.: Asymptotic methods in theory of ordinary differential equations. Progress in Mathematics, New York: Plenum Press, Vol. 8, pp. 1–82, 1970
28. Buyakas, V.I.: Optimal control by systems with variable structure. Automation and Remote Control, 27, No. 4, pp. 579–589, 1967
29. Chan, Y.T.: Perfect model following with a real model. Proc. Joint Autom. Control Conf., pp. 287–293, 1973
30. Choi, S.S. and Sirisena, H.R.: Computation of output feedback gain for linear multivariable systems. IEEE Trans., AC-19, pp. 257–258, 1974
31. Chilikin, M.G., Klyuev, V.I. and Sandler, A.S.: Theory of automatic electric drive. Moscow: Energiya, 1979 (in Russian)
32. Chubarov, Ye.P.: Monitoring and control with movable local action. Moscow: Nauka, 1977 (in Russian)
33. Davison, E.J.: The output control of linear time-invariant systems with unmeasurable arbitrary disturbances. IEEE Trans., AC-17, No. 5, pp. 621–630, 1972
34. Davison, E.J. and Wang, S.H.: Properties and calculations of transmission zeroes of mutivariable systems. *Automatica*, 10, No. 6, pp. 643–658, 1974
35. Devand, F.M. and Caron, J.Y.: Asymptotic stability of model reference systems with bang–bang control. IEEE Trans., AC-20, pp. 694–696, 1975
36. Dolgolenko, Yu.V.: Sliding modes in relay indirect control systems. Proc. of 2nd All-Union Conference on Control Theory, 1. Moscow: AN SSSR Publ., pp. 421–438, 1955 (in Russian)
37. Dote, Y., Manabe, T. and Murakami, S.: Microprocessor-based force control of manipulator using variable structure with sliding mode. Proc. 3rd IFAC Symposium on Control in Power Electronics and Electrical Drives, Pergamon Press, 1983
38. Dote, Y., Takebe, M. and Ito, T.: Variable structure control with sliding mode for d.c. drive speed regulation. IEEE Power Electronic Spec. 13th Ann. Conf., Cambridge, Mass., June 14–17, 1982. New York, 1982
39. Drakunov, S.V.: On adaptive quasioptimal filter with discontinuous parameters. Automation and Remote Control, 44, No. 9 (P.2), pp. 1167–1175, 1983

40. Drazenovic, B.: The invariance conditions in variale structure systems. Automatica, 5, No. 3, Pergamon Press, pp. 287–295, 1969
41. Dudin, Ye.B. and Ulanov, G.M.: Combined systems of automatic control with variable structure. Proc. of the International Conference on Multivariable and Discrete Control Systems, Section B, Prague, 1965
42. Dudnikov, Ye.G.: Introduction to the theory of indirect control with constant closing speed. Avtomatika i Telemekhanika, No. 6, 1939 (in Russian)
43. Ezerov, V.B. and Utkin, V.I.: A variable-structure filter as an analog of forcing element. In: Systems with variable structures and their application to automation of flight. Moscow: Nauka, 1968 (in Russian)
44. Emelyanov, S.V., Matich, B. and Kostyleva, N.E.: A universal unified control system with variable structure. Pribory i Sistemy Upravleniya, P.1, No. 12, 1973; P.2, No. 1, 1974 (in Russian)
45. Eremin, I.I.: "Penalty" method in convex programming. DAN SSSR, 173, No. 4, pp. 748–751, 1967 (in Russian)
46. Feldbaum, A.A.: Foundations of automatic control systems. Moscow: Fizmatgiz, 1963 (in Russian)
47. Fiacco, A. and McCormick, G.: Nonlinear programming: sequential unconstrained minimization technique. New York: John Wiley and Sons, Inc., 1968
48. Filippov, A.F.: Differential equations with discontinuous righthand part. Mathematicheskii Sbornik, 51, No. 1, pp. 99–128, 1960 (in Russian)
49. Filippov, A.F.: Application of the theory of differential equations with discontinuous right-hand sides to non-linear problems of automatic control. Proc. of 1-st IFAC Congress, II, Butterworths, London, pp. 923–927, 1961
50. Filippov, A.F.: On some problems of optimal control. Vestnik MGU No. 2, pp. 25–32, 1962 (in Russian)
51. Fitzner, L.N.: Automatic optimization of complex systems with spatial distribution. Automation and Remote Control, 25, No. 5, pp. 575–582, 1964
52. Gabasov, R. and Kirillova, F.M.: Singular optimal controls. Moscow: Nauka, 1973 (in Russian)
53. Gamkrelidze, E.V.: On sliding optimal modes. DAN SSSR, 143, No. 6, 1962 (in Russian)
54. Golembo, B.Z., Yemel'yanov, S.V., Utkin, V.I. and Shubladze, A.M.: Application of piecewise-continuous dynamic systems to filtering problems. Automation and Remote Control, 37, No. 3(P.1), pp. 369–377, 1976
55. Golembo, B.Z., Emelyanov, S.V., Utkin, V.I. and Shubladze, A.M.: Application of variable-structure system methods to problems of adaptive filtering and differentiation. DAN SSSR, 226, No. 5, pp. 1034–1037, 1976 (in Russian)
56. Grishin, S.I. and Utkin, V.I.: On continuation of sliding modes in discontinuous systems. Differentzialniye Uravneniya, 16, No. 2, pp. 227–235, 1980 (in Russian)
57. Guretskii Kh.: Analysis and synthesis of control systems with a delay. Moscow: Mashinostroyeniye Publ., 1974 (in Russian)
58. Gurman, V.I.: Singular problems of optimal control. Moscow: Nauka, 1977 (in Russian)
59. Harashima, F., Hashimoto, H. and Maruyama: Practical robust control for robot arm using variable structure systems. Proc. of 1986 IEEE International Conference on Robotics and Automation. Computer Society Press, Vol. 1, pp. 532–539, 1986
60. Hadley, G.: Nonlinear and dynamic programming Moscow: MIR, 1967 (in Russian)
61. Horisberger, H.P. and Belanger, P.R.: A solution of optimal constant output feedback problem by conjugate gradient. IEEE Trans., AC-19, pp. 434–435, 1974
62. Hsu, J.C. and Meyer, A.U.: Modern control principles and applications. McGraw-Hill, 1968
63. Izosimov, D.B.: Multivariable nonlinear induction motor state identifier using sliding modes. In: Problems of Multivariable Systems Control. Moscow: Nauka, pp. 133–139, 1983 (in Russian)
64. Izosimov, D.B. and Lifshitz, Ya.M.: Sliding mode control in a thyristor d.c. drive. In: Electric and Hydraulic Drives of N/C machine Tools and Industrial Robots, ENIMS, pp. 50–52, 1982 (in Russian)
65. Izosimov, D.B., Matic, B., Utkin, V.I. and Sabanovic, A.: Using sliding modes in control of electric motors. Dokl. AN SSSR, 241, No. 4, pp. 769–772, 1978 (in Russian)
66. Izosimov, D.B. and Utkin, V.I.: On equivalency of systems with large gains to systems with discontinuous controls. Avtomatika i Telemekhanika, 41, No. 11, pp. 189–191, 1981 (in Russian)

67. Jameson, A. and O'Malley, R.E., Jr.: Cheap control of the time-invariant regulator. Appl. Math. Opt., 1, No. 4, pp. 337–354, 1975
68. Jamshidi, M.: Three-stage near-optimum design of nonlinear control pocesses. Proc. IEEE, 121, No. 8, pp. 886–892, 1974
69. Johnson, C.D.: Further study of the linear regulator with disturbances – the case of vector disturbance satisfying a linear differential equation. IEEE Trans., AC-15, pp. 222–228, 1970
70. Kagan, V.G. and Usachyov, A.P.: Wide-range speed control of d.c. drive with transistor inverter. Electrical Engineering Industry. "Electric Motor Drives" series, Issue 3 (96), pp. 13–16, 1981 (in Russian)
71. Kalman, R.E.: Contribution to the theory of optimal control. Bul. Soc. Mat. Mexicano, 5, pp. 102–119, 1960
72. Kalman, R.E.: On the general control systems theory. Proc. 1st IFAC Congress, Butherworth, London, 1961 (in Russian)
73. Kaynak, O., Harashima, F. and Hashimoto, H.: Variable structure systems theory applied to sub-time optimal position control with an invariant trajectory. Trans. IEE of Japan, 104, No. 3/4, pp. 47–51, March–April, 1984
74. Kelley, H.J.: Aircraft manouver optimization by reduced order approximations. Control and Dynamic Systems, pp. 131–178. New York: Academic Press, 1973
75. Khasminskii, R.Z.: A limit theorem for the solution of differential equations with a random righthand part. Teoriya Veroyatnostei i eyo Primeneniya, 11, No. 3, pp. 444–462, 1966 (in Russian)
76. Khasminskii, R.Z.: On the principle of averaging for stochastic differential equations. Teoriya Veroyatnostei i eyo Primeneniya, 13, No. 3, pp. 260–279, 1968 (in Russian)
77. Kokotovic, P.V., O'Malley, R.B., Jr. and Sannuti, P.: Singular perturbation and order reduction in control theory. Automatica, Pergamon Press, Vol. 12, pp. 123–132, 1976
78. Kornilov, Yu. G.: On effect of controller insensitivity on dynamics of indirect control. Avtomatika i Telemekhanika, 11, No. 1, pp. 39–57, (in Russian)
79. Korovin, S.K. and Utkin V.I.: Use of the slip mode in problems of static optimization. Automation and Remote Control, 33, No. 4 (P.1), pp. 570–579, 1972
80. Korovin, S.K. and Utkin, V.I.: Method of piecewise-smooth penalty functions. Automation and Remote Control, 37, No. 4 (P.2), pp. 39–48, 1976
81. Korovin, S.K. and Utkin, V.I.: Sliding mode based solution of static optimization and mathematical programming problems. Applied aspects. Preprints of IFAC-IFORS Symposium, Varna, Bulgaria, October 1974, pp. 1–8, 1974
82. Korovin, S.K. and Utkin, V.I.: Using sliding modes in static optimization and nonlinear programming. Automatica, Pergamon Press, Vol. 10, pp. 525–532, 1974
83. Kostyleva, N.E.: Recovery of differential equations of a plant by methods of variable-structure systems. Automation and Remote Control, 32, No. 10(P.1), pp. 1588–1597, 1971
84. Kouvaritakis, B. and McFarlane, A.G.J.: Geometric approach to analysis and synthesis of systems zeroes (Part 1): Square systems. Int. J. Contr., 23, No. 2, pp. 149–166, 1976
85. Krasovskii, A.A.: The dynamics of continuous self-tuning systems. Moscow: Fizmatgiz, 1963 (in Russian)
86. Krotov, V.F.: Methods for solving variational problems on the basis of the sufficient conditions for an absolute minimum. Automation and Remote Control, 23, No. 12, pp. 1473–1484, 1963
87. Kubyshkin, V.A.: Mathematical modelling and motion algorithms for a source with movable multicyclic action. Abstracts of papers presented on the 7th All-Union Conference on Mathematical Modelling Theory and Methods. Kuibyshev, USSR, September 1978. Moscow: Nauka, 1978 (in Russian)
88. Kulebakin, V.S.: To the theory of automatic vibrational regulators for electric motors. Teoreticheskaya i Eksperimental'naya Elektronika, No. 4, 1932 (in Russian)
89. Kulebakin, V.S.: On basic problems and methods of improving performance of automatic control systems. Proceedings of the 2nd All Union Conference on Automatic Control Theory, 2, Moscow: AN SSSR Publ., 1955 (in Russian)
90. Kukhtenko, A.I.: Problem of invariancy in automatic control. Kiev: Technical Literature State Publ., 1963 (in Russian)
91. Kwakernaak, H.: Asymptotic root-loci of multivariable linear optimal regulator. IEEE Trans., AC-21, pp. 378–392, 1976
92. Kwakernaak, H. and Sivan, R.: Linear optimal control system. Interscience, 1972

93. Laion, R.M.: Fast identification of linear and nonlinear systems. AIAA J., 5, No. 10, pp. 130–138, 1967
94. Landaw, I.D. and Courtiol, B.: Design of multivariable adaptive model following control systems. Automatica, Pergamon Press, Vol. 10, pp. 483–494, 1974
95. Letov, A.M.: Analytic design of controllers. Avtomatika i Telemekhanika, 21, No. 4, pp. 436–441, 1960 (in Russian)
96. Levine, W.S. and Athans, M.: On determination of optimal constant output feedback gains for linear multivariable systems. IEEE Trans., AC-15, pp. 44–48, 1970
97 Luenberger, D.G.: Observers for multivariable systems. IEEE Trans., AC-11, pp. 190–197, 1966
98. Lyapunov, A.M.: General problem of motion stability. Moscow: Fizmatgiz, 1959 (in Russian)
99. O'Malley, R.E., Jr.: Introduction to singular perturbations. New York: Academic Press, 1974
100. O'Malley, R.E., Jr. and Jameson, A.: Singular perturbations and singular arcs. IEEE Trans., AC-20, pp. 218–226, 1975
101. Markiewicz, B.R.: Analysis of the computed torque drive method and comparison with conventional position servo for a computercontrolled manipulator. Jet. Propulsion Lab. Tech. Memo., pp. 33-61, March 15, 1973
102. Meerov, M.V.: Automatic control systems stable with infinitely large gains. Avtomatika i Telemekhanika, 8, No. 4, pp. 225–242, 1947 (in Russian)
103. Meerov, M.V.: Design of high precision automatic control structures. Moscow: Fizmatgiz, 1959 (in Russian)
104. Meerov, M.V.: Multivariable control systems. Moscow: Nauka, 1967 (in Russian)
105. Mendel, J.M.: A concise derivation of optimal constant limited state feedback gains. IEEE Trans., AC-19, pp. 447–448, 1974
106. Murphy, J.M.D.: Thyristor control of a.c. motors. Pergamon Press, 1973
107. Neymark, Yu.I.: Note on A. Filippov's paper. Proc. 1-st IFAC Congress, II, Butterworth, London, 1961.
108. Neymark, Yu. I.: On sliding modes in relay control systems. Avtomatika i Telemekhanika, 18, No. 1, pp. 27–33, 1957 (in Russian)
109. Nikolski, G.N.: On automatic stability of a ship on a given course. Proc. of Central Communications Lab., No. 1, pp. 34–75, 1934 (in Russian)
110. Orlov, Yu. V.: Application of Lyapunov method in distributed systems. Automation and Remote Control, 44, No. 4 (P.1), pp. 426–430, 1983
111. Orlov, Yu.V. and Utkin, V.I. Control principles of infinite-dimensional dynamic plants using sliding modes. Dokl. AN SSSR, 283, No. 5, pp. 1116–1119, 1985 (in Russian)
112. Paul, R.C.: Modeling, trajectory calculation and servoing of a computer controlled arm. Stanford Artificial Intelligence Lab. Memo. AJM-177, Nov. 1972
113. Petrov, B.N.: On realizability of invariancy conditions. Proc. 1st Conference on Invariancy Theory, Kiev, 1958. Kiev, 1959 (in Russian)
114. Petrov, B.N. Rutkovskii, V.Yu., Krutova, I.N. and Zemlyakov, S.D.: Design principles of adaptive control systems. Moscow: Mashinostroyeniye Publ., 1972 (in Russian)
115. Pfaff, G., Wick, A.: Direkte Stromregelung bei Drehstromanrieben mit Pulswechselrichten. Regelungstechn. Prak., 25, No. 11, 1983
116. Plunkett, A.B. and Kliman, G.B.: Electrical vehicle AC drive development. SAE Techn. Pap. Ser., No. 800061, pp. 1–13, 1980
117. Popovski, A.M.: Linearization of sliding operation mode for a constant speed controller. Avtomatika i Telemekhanika, 11, No. 3, pp. 161–163, 1950 (in Russian)
118. Proceedings of the 2nd All-Union Conference on Theory of Automatic Control. Moscow: AN SSSR Publ., 1955 (in Russian)
119. Rastrigin, L.A. and Sytenko, L.V.: Multichannel static optimizers. Moscow: Energiya, 1973 (in Russian)
120. Rashevskii, P.K.: Geometrical theory of partial derivative equations. Moscow: Gostekhizdat, 1947 (in Russian)
121. Rey, R.L., Fel'dman, A.V. and Chabanov, A.I.: Wide-range main induction motor for N/C metalcutting machine tools. Elektrotekhnika, No. 9, pp. 40–42, 1981 (in Russian)
122. Rosembrock, H.H.: On a method of investigation of stability. Proc. II IFAC Congress, Basel, Switzerland, Vol. 1, pp. 590–594, 1963
123. Rosenbrock, H.H.: State space and multivariable theory. New York: Wiley, 1970
124. Sabanovic, A, and Izosimov, D.: Application of sliding modes to induction motor control. IEEE Trans., IA-17, No. 1, pp. 41–49, 1981.

125. Saridis, G.N. and Stephanon, H.E.: Hierarchically intelligent control of a bionic arm. Proc. 1975 IEEE Conf. Decision and Control, pp. 99–104, Dec. 5–7, Houston, 1975
126. Shaked, V. and Konvaritakis, B.: Asymptotic behaviour of rootloci of linear multivarible systems. Int. J. Contr., 23, No. 3, pp. 297–340, 1976
127. Shaked, V.: Design technique for high speed feedback gain stability, Int. J. Contr., 24, No. 1, pp. 137–144, 1976
128. Siljak, D.D.: Large-scale dynamic systems. Stability and structure. North Holland, 1978
129. Sirazetdinov, T.K.: Optimization of systems with distributed parameters. Moscow: Nauka, 1977 (in Russian)
130. Theory of variable-structure systems. Ed. S.V. Emelyanov. Moscow: Nauka, 1970 (in Russian)
131. Tikhonov, A.N.: Systems of differential equations with a small parameter multiplying derivatives. Matematicheskii Sbornik, 73, No. 31, pp. 575–586, 1952 (in Russian)
132. Tikhonov, A.N.: On one method of regularization of optimal control problems. DAN SSSR, 162, No. 4, pp. 763–765, 1965 (in Russian)
133. Tikhonov, A.N.: On regularization of ill-posed problems. DAN SSSR, 153, No. 1, 1963 (in Russian)
134. Tikhonov, A.N.: On some ill-posed problems of optimal planning. Zhurnal Vyssh. Matej. i Matemat. Fiziki, No. 1, pp. 81–89, 1966 (in Russian)
135. Tikhonov, A.N. and Samarskii, A.A.: Equations of mathematical physics. Moscow: Nauka, 1966 (in Russian)
136. Tzypkin, Ya.Z.: Theory of relay control systems. Moscow: Gostekhizdat, 1955 (in Russian)
137. Tzypkin, Ya.Z.: Adaptation and learning in automatic systems. Moscow: Nauka, 1968 (in Russian)
138. Ulanov, G.M.: Control by disturbances. Moscow: Energoizdat, 1960 (in Russian)
139. Utkin, V.I.: On compensation of the forced term of motion in variable-structure control systems. Izv. AN SSSR, "Tekhnicheskaya Kibernetika" series, No. 4, pp. 169–173, 1965 (in Russian)
140. Utkin, V.I.: Quasiinvariant control of forced motion in linear variable-parameter control plants. Tekhnicheskaya Kibernetika, No. 5, pp. 153–159, 1966 (in Russian)
141. Utkin, V.I.: Sliding modes in variable-structure systems. Teoretski Prilog, Automatika, Yugoslavija, No. 3, 1970
142. Utkin, V.I.: Equations of slipping regime in discontinuous systems, I, II. Automation and Remote Control, 32, No. 12 (P.1), pp. 1897–1907, and 33, No. 2 (P.2), pp. 211–219, 1971, 1972
143. Utkin, V.I.: Sliding modes and their applications in variable structure systems. Moscow: MIR, 1978 (translated from Russian)
144. Utkin, V.I.: Variable structure systems with sliding modes. IEEE Trans., AC-22, No. 2, pp. 212–222, 1977.
145. Utkin, V.I.: Application of equivalent control method to systems with large feedback gain. IEEE Trans., AC-23, pp. 484–486, 1978
146. Utkin, V.I.: Identification principles using sliding modes. Dokl. AN SSSR, 257, No. 3, pp. 558-561, 1981 (in Russian)
147. Utkin, V.I.: Variable structure systems: present and future. Automation and Remote Control, 44, No. 9, pp. 1105–1120, 1983
148. Utkin, V.I., Arkhangel'skii, N.A., Izosimov, D.B. et al.: A device to determine coordinaes of induction motor in a controlled electric motor drive. USSR patent No. 1039011. Bulleten Izobreteniy, No. 32, p. 209, 1983
149. Utkin, V.I. and Vostrikov, A.S.: Control systems with decoupling motions. A link between science and applications of automatic control. Preprints of the VII Triennial World IFAC Congress, Helsinki, Finland June 12–16, 1978. Pergamon Press, Vol. 2, pp. 967–973, 1978
150. Utkin, V.I., Vostrikov, A.S., Bondarev, S.A. et al.: Application of sliding modes in automation of technological plants. I.K.A., 53, No. 1, pp. 74–84, 1985 (in Russian)
151. Utkin, V.I. and Drakunov, S.V.: Stochastic regularization of discontinuous control systems. Dokl. AN SSSR, 272, No. 2, pp. 1069–1072, 1983 (in Russian)
152. Utkin, V.I., Izosimov, D.B., Andreev, G.I. et al.: Sliding mode controlled AC servo drives. In: Electric and hydraulic drives of N/C machine tools and industrial robots, ENIMS, pp. 63–70, 1982 (in Russian)
153. Utkin, V.I. and Kostyleva, N.E.: Principles of algorithm design for local automation systems with variable structure. I.K.A., 1(35), pp. 27–35, 1981 (in Russian)

154. Utkin, V.I. and Young, K.-K.D.: Methods for constructing discontinuity planes in multidimensional variable-structure systems. Automation and Remote Control, 39, No. 10, (P.1), pp. 1466–1470, 1978
155. Van Ness, J.E., Zimmer, H. and Culti, M.: Reduction of dynamic models of power systems. 1973 PICA Proceedings, pp. 105–112, 1973
156. Vasilyeva, A.B. and Butuzov, V.F.: Asymptotic decomposition of solutions to singularly perturbed equations. Moscow: Nauka, 1973 (in Russian)
157 Vasilyeva, A.B. and Butuzov, V.F.: Singularly perturbed equations in critical cases. Moscow: MGU Publ., 1978 (in Russian)
158. Vasilyeva, A.B. and Dmitriev M.G.: Singular perturbations and some optimal control problems. A link between science and applications of automatic control. Preprints of VII Triennial World IFAC Congress. Helsinki, Finland, June 12–16, 1978; Pergamon Press, Vol. 2, pp. 963–966, 1978
159. Venets, V.I.: A saddle point of Lagrange function and unsmooth penalty functions in convex programming. Automation and Remote Control, 35, No. 8(P.2), pp. 1306–1314, 1974.
160. Ventzel, A.D. and Freydlin, M.I.: Fluctuations in dynamic systems subjected to small random disturbance. Moscow: Nauka, 1979 (in Russian)
161. Vladimirov, V.S.: Equations of mathematical physics. Moscow: Nauka, 1976 (in Russian)
162. Warga, J.: Relaxed variational problems. J. Math. and Applic., 4, No. 1, 1962
163. Wilde, D. and Beightler, C.: Foundations of optimization. New York: Prentice Hall, Englewood Cliffs, 1967
164. Winsor, C.A. and Roy, R.J.: Design of model reference adaptive control by Lyapunov's second method. IEEE Trans., AC-13, p. 204, 1968
165. Wonham, W.M.: On pole assignment in multi-input controllable linear systems. IEEE Trans., AC-12, pp. 660–665, 1967
166. Young, K.-K.D.: Controller design for a manipulator using theory of variable structure systems. IEEE Trans., SHC-8, pp. 101–109, 1978
167. Young, K.-K.D.: Design of multivariable variable-structure model following control systems. Proc. 1977 IEEE Conf. Decision and Control. New Orleans, Dec. 7–9, 1977
168. Young, K.-K.D.: Asymptotic stability of model reference systems with variable structure control. IEEE Trans., AC-22, pp. 279–282, 1977
169. Young, K.-K.D., Kokotovic, P.V. and Utkin, V.I.: A singular perturbations analysis of high-gain feedback systems. IEEE Trans., AC-22, pp. 931–939, 1977
170. Zangwill, W.: Nonlinear programming via penalty functions. Manag. Sci. 13, pp. 344–358, 1967
171. Zohdy, M.A. and Apelvich, J.D.: A class of output feedback controller with min–max design. IEEE Trans., AC-21, pp. 614–615, 1976

Subject Index